24th International Conference on Adaptive Structures and Technologies (ICAST 2013)

AA005538

Aruba
6 – 9 October 2013

ISBN: 978-1-5108-5119-1

Printed from e-media with permission by:

Curran Associates, Inc.
57 Morehouse Lane
Red Hook, NY 12571

proceedings.com

Some format issues inherent in the e-media version may also appear in this print version.

Printed by Curran Associates, Inc. (2018)

Michigan Engineering
1221 Beal Avenue
Ann Arbor, MI 48109-2102
USA

Phone: +1 (734) 647-7000

engin-info@umich.edu

Additional copies of this publication are available from:

Curran Associates, Inc.
57 Morehouse Lane
Red Hook, NY 12571 USA
Phone: 845-758-0400
Fax: 845-758-2633
Email: curran@proceedings.com
Web: www.proceedings.com

ICAST 2013

October 6-9, ARUBA

24TH INTERNATIONAL CONFERENCE ON ADAPTIVE STRUCTURES AND TECHNOLOGIES

Multi-Objective Genetic Algorithm Optimization of the Fish Bone Active Camber Morphing Airfoil

B. K. S. Woods*, and M. I. Friswell

Swansea University, College of Engineering, Singleton Park, Swansea, SA2 8PP, United Kingdom

Abstract

This work presents the development of an optimization code for the geometry of the Fish Bone Active Camber morphing airfoil concept, which has been under development at Swansea University. Employing highly anisotropic structural compliance from a biologically inspired architecture, the smooth, continuous camber change generated by this concept has been shown in previous work to provide significant reductions in drag over traditional trailing edge flaps across a wide range of operating lift coefficients. Therefore, if the concept could be shown to have comparable weight, actuation requirements, and service life, then it would be a promising alternative for the wide range of applications where trailing edge flaps are currently used or are planned to be used. These include fixed wing aircraft, active rotorcraft blades, wind turbines, tidal stream turbines, and other aero and hydrodynamic applications. The objective of this current research is to extend a previously developed fluid-structure interaction analysis into a useful tool for establishing, to a first order approximation, the weight, drag, and actuation energy requirements of an optimized Fish Bone Active Camber morphing airfoil at a specific operating condition. A multi-objective optimization scheme is therefore established using a Genetic Algorithm which optimizes the internal core structure and elastic skin of the morphing airfoil for the three objectives of minimizing the added mass, the actuation energy requirement, and the operating drag coefficient required to achieve a design lift coefficient at a specified chord, angle of attack, and Mach number. Some initial results are shown which give Pareto fronts for the three objectives.

1. INTRODUCTION

The amount of camber, or curvature, present in an airfoil has a significant impact on the forces it will generate under fluid flow. Fluid dynamic systems often take advantage of this fact by employing structures with variable camber. Examples of this are manifold and present in nearly every category of aerodynamic and hydrodynamic applications. Fixed wing aircraft, for example, have used camber variation to control roll, pitch, and yaw motions for over 100 years.[1] Camber variation is also used during takeoff and landing to generate very high lift coefficients.[2] Examples of helicopters employing camber change for primary control can be found as early as 1946 in the machines of Charlie Kaman.[3] Furthermore, active camber change is commonly used in hydrodynamic systems such as ship rudders,[4] submarines[5] and hydrofoil boats[6] to provide directional control.

To date, the primary means of realizing variable camber has been through the use of discrete trailing edge flaps. In this approach, the trailing edge of the airfoil structure is essentially severed then re-attached with a rotary degree of freedom. An actuator is installed to generate torque on the flap so

* Corresponding author, email: b.k.s.woods@swansea.ac.uk

that it rotates up or down, thereby changing the effective camber of the airfoil and decreasing or increasing the amount of lift generated.

While this approach is conceptually very simple and certainly well proven, it is not without drawbacks. First and foremost, the presence of a sharp, discrete change in camber leads to a significant increase in drag over the baseline airfoil, particularly at high lift coefficients. This sharp camber discontinuity can also lead to early trailing edge flow separation, limiting the maximum lift coefficient. Furthermore, the flap, actuation system, and required structural reinforcements increase the weight, complexity and cost of the system.

Researchers have long pursued continuous changes to airfoil camber as an alternative due to the potential for significantly reduced drag. Additionally, arguments have been made that weight and perhaps complexity can be reduced, although this remains to be seen in practice. Many rotorcraft concepts have proposed smart material driven active camber changes, with the ambitious goals of vibration reduction, noise mitigation, and even potentially primary control.[7-9] Fixed wing examples of continuous camber change can be found throughout history,[10-12] and indeed even concepts for wind turbine blades have been explored.[13]

2. FISH BONE ACTIVE CAMBER CONCEPT

The Fish Bone Active Camber (FishBAC) mechanism provides an alternative design architecture to the traditional discrete flap. Introduced by Woods and Friswell[14], this design employs a biologically inspired compliant structure to create large, continuous changes in airfoil camber and section aerodynamic properties. The structure consists of a thin chordwise bending beam spine with stringers branching off to connect it to a pre-tensioned Elastomeric Matrix Composite (EMC) skin surface. Both core and skin are designed to exhibit near-zero Poisson's ratio in the spanwise direction. Pre-tensioning the skin in the chordwise direction significantly increases the out-of-plane stiffness and eliminates lower surface skin buckling when morphing. Smooth, continuous bending deflections are driven by a high stiffness, antagonistic tendon system. Actuators mounted in the D-spar drive a tendon spooling pulley through a non-backdrivable mechanism (such as a low lead angle worm and worm gear). A schematic overview of the FishBAC concept is shown in Figure 1. Since the tendon system is non-backdrivable, no actuation energy is required to hold the deflected position of the structure, reducing control action and power requirements. Furthermore, the automatic locking action of the non-backdrivable mechanism allows the stiffness of the tendons to contribute significantly to the chordwise bending stiffness of the trailing edge under aerodynamic load, without increasing the amount of energy required to deflect the structure.

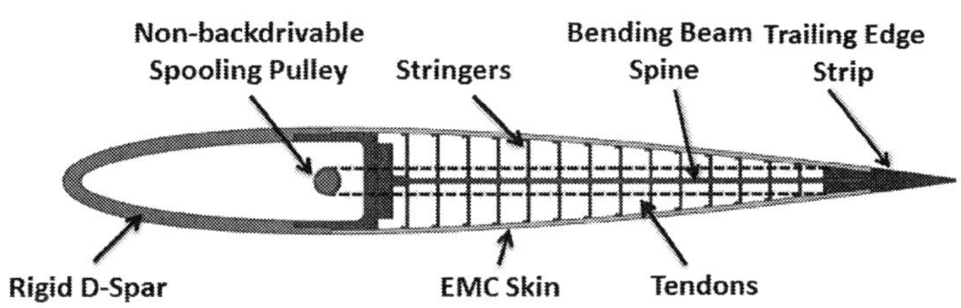

Figure 1. Fish Bone Active Camber concept

Wind tunnel testing of the prototype seen in Figure 2 found the FishBAC provided improved aerodynamic efficiency compared to traditional trailing edge flaps, with increases in lift-to-drag ratio of 25% being realized at equivalent lift conditions.[15] An increase in lift coefficient of $\Delta C_l = 0.72$ between unmorphed and morphed was measured at a freestream velocity of $V_\infty = 20$ m/s and an angle of attack of $\alpha = 0°$.

The large achievable deflections and continuous compliant architecture make this concept universally applicable to fixed wing applications ranging in scale from small UAVs to commercial airliners, and to rotary wing applications including wind turbines, helicopters, tilt-rotors, and tidal stream turbines.

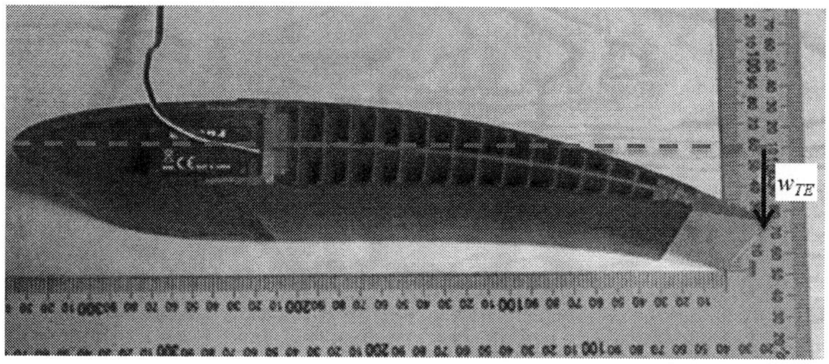

Figure 2. FishBAC wind tunnel test model showing deflected shape

3. FLUID-STRUCTURAL INTERACTION ANALYSIS

At the core of the optimization code presented here is a strongly coupled, partitioned fluid-structure interaction (FSI) analysis which predicts the deformed equilibrium shape, aerodynamic performance, and actuation requirements of the FishBAC structure under internal actuation loading and external aerodynamic loading. This code has been presented in detail in previous work,[16] and so is only outlined here to provide sufficient understanding of its operation with respect to the optimization run with it. Figure 3 shows schematically the components of this FSI code, and the manner in which the structural and aerodynamic partitions are coupled and iterated until convergence.

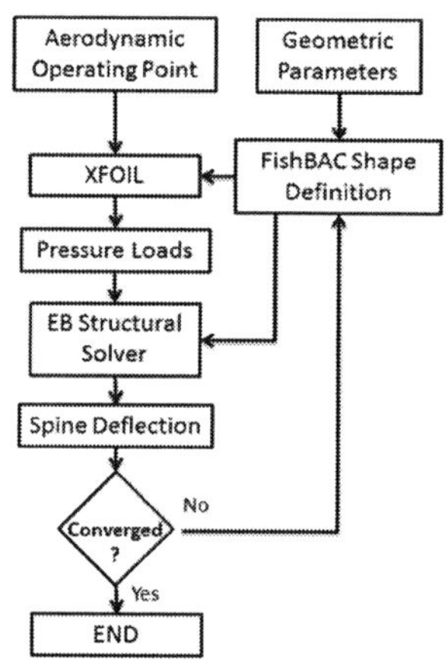

Figure 3. Schematic overview of the fluid-structure interaction analysis

3.1 Geometry Definition

The definition of the FishBAC geometry used within the FSI analysis is relevant here since its parameters form the design variables which define the individuals in the Genetic Algorithm optimization. In this work, the bending beam spine, stringers, and skin are considered as separate elements which can be varied to create significantly different FishBAC configurations with widely varying mechanical and aerodynamic properties. Figure 4 shows the physical meaning of the variables used to define the geometry. The thicknesses of the skin, stringers and bending beam spine form three of the design variables, the start and stop point of the morphing along the chord are two more, and the final design variable is the number of stringers, for a total of six design variables required to define the FishBAC geometry.

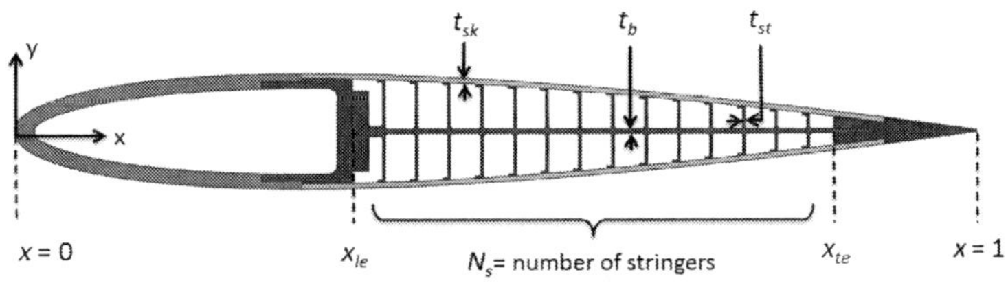

Figure 4. Geometry definition for current study

3.2 Structural Model

This section introduces the low fidelity, Euler Bernoulli (EB) beam theory structural model developed for the FishBAC. This model allows for very quick analysis of design configurations and for parameter driven modifications to the geometry and loading. Since the derivation and validation of this model is the subject of previous work,[16,17] the basic approach will just be outlined here and the inputs and outputs discussed.

The low fidelity structural model is analytical and derived from Euler-Bernoulli beam theory. The low chordwise bending stiffness and high length-to-thickness ratio of the bending spine, low in-plane stiffness of the skin, and the continuous loading along the span make Euler-Bernoulli theory a good initial approximation for analysis. Furthermore, the high stiffness stringers branching off from the spine help to enforce the 'plane sections remain plane and normal to the neutral axis' assumption which underlies the derivation of Euler-Bernoulli beam theory.[18]

The relationship between net aerodynamic pressure, p, flexural rigidity, EI, and vertical displacement, w, is given by the Euler-Bernoulli beam equation:[18]

$$\frac{d^2}{dx^2}\left(EI(x)\frac{d^2w}{dx^2}\right) = p(x) \tag{1}$$

The bending deflections of the morphing structure are found by integrating the aerodynamic pressure distribution in Equation (1) to find shear distribution, integrating that to find moment distribution, integrating again to find the distribution of slope, and then integrating one final time to find the deflection of the neutral axis.

The bending moment produced by the tendons is included into the integration of the Euler-Bernoulli equations. Since the strain, and therefore force, in the tendons depends on the amount of deformation in the spine the solution must be iterated until convergence. The known thickness distribution for the NACA 0012 airfoil used is then superimposed onto the spine, producing the net airfoil shape.

The flexural rigidity (EI) distribution of the FishBAC structure is formulated as a linear superposition of its components in this analysis. As described in further detail elsewhere,[17] the flexural rigidity of the various FishBAC components are calculated individually then summed along the chord.

This structural model has been shown to provide good levels of accuracy over a range of loading conditions, with prediction error less than 5% compared to experiments.[17] The very low computational cost of this method compared to a high fidelity finite element analysis makes it ideal for use in the optimization code developed here.

3.3 Aerodynamic Model

The aerodynamic pressure distribution acting on the FishBAC is found using the XFOIL panel method code. This code includes a viscous boundary layer component to predict skin friction drag and flow separation, offering a more complete drag prediction than inviscid codes. The code requires as inputs the aerodynamic conditions, [Mach number, M, angle of attack, α, and Reynolds number, Re] and the non-dimensionalized airfoil skin coordinates. Using a viscous formulation of linear vorticity potential flow theory, XFOIL then calculates the distribution of pressure coefficient over the airfoil. The lift, drag and moment coefficients for the current airfoil shape are found by XFOIL by integrating this pressure coefficient distribution.

For the structural component of this FSI problem, we consider the distribution of the dimensional pressure acting from the start of the morphing section to the trailing edge ($x_{le} < x < 1$). Furthermore, since deformations in the thickness direction are not considered in this analysis (since the thickness distribution is maintained by the stringers), the pressure on the lower and upper skins can be combined to create a net pressure which acts on the spine as a chordwise varying distributed load, p, as shown in Figure 5. The effect of the skin surface pressures on local deformations of the skin between stringers is considered separately, as discussed below. For the FSI structural analysis, the net pressure distribution is applied directly to the bending spine.

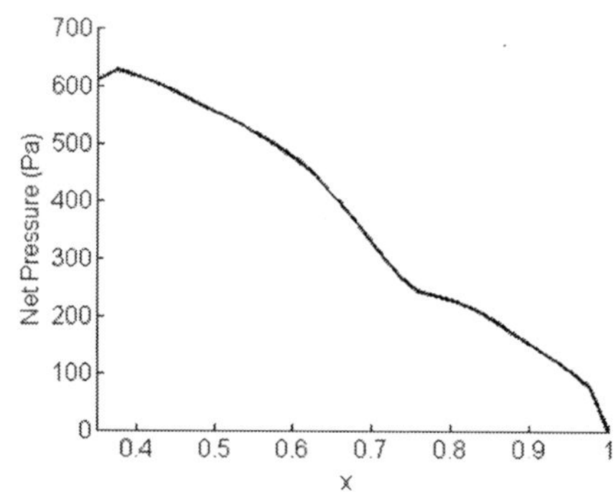

Figure 5. Representative net pressure distribution

4. FISHBAC MULTI-OBJECTIVE GEOMETRY OPTIMIZATION

The FishBAC Fluid-Structure-Interaction code is used in this work inside of a Genetic Algorithm (GA) optimization scheme to determine the combinations of FishBAC geometric parameters which give optimized performance against three different objective functions at a given operating condition and under specific constraints. The details of this analysis will be presented in this section. First the selection of a GA optimizer will be explained, followed by details of the operating condition chosen for the present analysis. Then the derivation and significance of the three objective functions chosen will be presented, the upper and lower bounds placed on the optimized variables explained, and the constraint function implemented to ensure that the out-of-plane deflections on the skin under aerodynamic loading are appropriately small will also be derived.

4.1 Optimization Algorithm

Firstly, it is important to explain the choice of a Genetic Algorithm (GA) for the current optimization work, as many optimization schemes are available in the literature with varying degrees of applicability to the current problem. The FSI code being optimized has highly non-linear behavior due to the presence of non-linear aerodynamic effects (separation, stall, etc) and non-linear structural interactions between the antagonistic drive tendons and the base compliant structure, which also includes geometry non-linearities. Furthermore, there is no *a priori* knowledge of the fitness landscape, with local minima likely present. These factors complicate the use of gradient based methods. Purely stochastic methods are not expected to efficiently search the design space in this case, and the non-negligible amount of computation time required for each converged solution of the FSI (roughly 20-30 seconds on an Intel i7 processor) necessitates a reasonable amount of efficiency in searching out optima. In order to make this code useful for real life design problems, it is desired to have convergence times measured in hours instead of days on reasonably powerful computers. Brute-force (i.e. exhaustive) searchers are consequently prohibitively expensive. The characteristics of this problem therefore make it amenable to the use of GAs, although the size of the design space and the desire for tightly mapped Pareto frontiers will require large population sizes.

This work uses a multi-objective genetic algorithm optimizer included in the Global Optimization toolbox for Matlab R2012b. Known as "gamultiobj", this function uses a controlled elitist genetic algorithm which is a variant of NSGA-II.[19] As the underlying FSI code is written in Matlab, its integration into the GA is straightforward. The control parameters for the GA were experimented with initially to find combinations that provided good population diversity and dense population along the Pareto frontiers. Sufficient mutation was ensured by setting the crossover fraction to 0.4 and a Pareto fraction of 0.7 was used to encourage the retention of individuals on the Pareto frontiers, a minimum of 300 individuals were used in all cases.

4.2 Operating Condition

In order to allow for a meaningful insight into the mechanics of the FishBAC concept through optimization, it is necessary to keep certain parameters constant over the course of the study. It is meaningless, for example, to have one member of your population operating at $M = 0.3$ with a lift coefficient of $c_l = 0.5$ and another at $M = 0.1$ with a lift coefficient of $c_l = 0.9$ inside of the optimization. The parameters which were set before optimization are shown in Table 1 and can be thought of as the parameters that define the operating condition in which we would like to introduce a FishBAC morphing airfoil as a means of camber control. They are therefore application specific, but an effort has been made to keep the operating condition fairly generic. This is achieved by using a chord length and Mach number that could easily be present on the control surfaces of a UAV-scale fixed wing aircraft, midspan on a helicopter blade, or at the tip of a wind turbine blade. The chosen operating lift coefficient is similarly intended to be representative; of a roll maneuver on a fixed wing aircraft, forward flight of a helicopter, or a low wind speed operating condition of an active wind turbine.

Table 1. Prescribed parameters for the current study

Parameter	Value
baseline airfoil	NACA 0012
chord (c)	305 mm
span (b)	150 mm
Mach number (M)	0.3
angle of attack (α)	5 deg
operating lift coefficient (c_l)	1.0
skin elastic modulus (E_{sk})	3.18 MPa
skin density (ρ_{sk})	1020 kg/m^3
spine elastic modulus (E_b)	2.14 GPa
spine density (ρ_b)	1300 kg/m^3
stringer elastic modulus (E_{st})	2.14 GPa
stringer density (ρ_{st})	1300 kg/m^3

Note also that the material properties for both the skin and the compliant core (bending beam spine and stringers) have been prescribed in this study. While material choice is a meaningful variable to consider for an optimization scheme such as this, it can be problematic to implement due to constraints which are difficult to quantify: things such as manufacturability, stress/strain limits, and fatigue, since realistic numbers are hard to obtain or sensitive to detailed aspects of the design. Therefore for this initial, low fidelity, study two materials were chosen which have been used in other work by these authors. Previous work with the carbon fiber reinforced silicone Elastomeric Matric Composite (EMC) and the 3D printed Acrylonitrile Butadiene Styrene (ABS) plastic specified here has shown them to be useful materials for the types of engineering prototypes under consideration.[15] Transitioning to detailed developmental models and flight test articles would most likely require a reconsideration of the material choices, and a corresponding re-optimization of the structural geometry. It is expected though that the underlying mechanics of the morphing concept and the optimization code presented here would remain similar, although higher fidelity structural and aerodynamic models may be desired at that stage.

4.3 Objective Functions

The three objective functions which are optimized concurrently in this analysis are section drag coefficient, FishBAC morphing section mass, and energy required to overcome the aerodynamic and structural stiffness. The derivation of these three quantities will be shown, and their importance to the active camber design problem discussed.

The first objective function, J_1, is derived from the desire to maximize the lift-to-drag ratio of the airfoil shape at the design condition. Since the chosen optimization algorithm works by minimizing the objective functions, the lift-drag ratio can be maximized by minimizing its inverse, as shown in Equation (2). Furthermore, since the lift coefficient in this study is prescribed as $c_l = 1$, J_1 reduces to the section drag coefficient, c_d.

The section drag coefficient is a simple objective value to obtain as it is immediately available from the output of the XFOIL aerodynamic code used within the FSI analysis. Since improved lift-to-drag ratio through reduced drag (compared to trailing edge flaps) is the main motivation for considering morphing camber concepts in general and the FishBAC in particular, understanding the interplay between this and the other objectives is important.

$$J_1 = \frac{1}{c_l/c_d} = {}^{c_d}/_{c_l} \xrightarrow{c_l=1} J_1 = c_d \tag{2}$$

The mass of the morphing structure is also a very important metric for system level comparison of flapped and morphing solutions. In order to be successful, the weight added by a morphing solution (relative to a flapped solution) must not be so large as to outweigh the benefits of the improved aerodynamic performance. To address this, a mass estimation for the major components of the FishBAC structure is used as the second objective function, J_2, minimized by the optimizer. At the start of analysis for each individual configuration generated by the genetic algorithm, the mass of the spine, m_b, stringers, m_{st}, and skin, m_{sk}, are calculated from derived cross sectional areas and the prescribed span and material densities outlined in Table 1. Additionally, provisions for skin bonding are included in the mass estimate. Specifically, the small skin bonding flanges at the top and bottom of the stringers (as visible in Figure 1) and the additional length of skin required past the extents of the morphing section for bonding to the host structure are included. These components make up the total mass m_{tot}, as defined in Equation (3).

$$J_2 = m_b + m_{st} + m_{sk} = m_{tot} \tag{3}$$

It is important to note that this mass metric does not include the mass of the non-morphing portion of the airfoil. This was done to avoid making the results of this analysis overly specific. Estimating the mass of the non-morphing portion of the airfoil is a complex problem that is highly sensitive to the specifics of a given design, including aspects such as aircraft type, size, weight, flight speed, design load factors, and material choice, amongst others. While some empirical sizing formulas are available,[20] these are hard to find with the required level of detail (particularly with respect to the chordwise distribution of mass) and are highly dependent on all of the design specifics, as previously stated. Furthermore, such empirical weight estimates would be for non-morphing airfoils, and therefore of dubious use given the impact the morphing section is likely to have on the distribution of stiffness and strength, and of the need for additional structure at interfaces between morphing and non-morphing regions. Therefore this analysis considers the added mass of the FishBAC itself, since it can be approximated well to the first order and since it should ideally be minimized.

The third objective for the FishBAC optimization is actuation energy. The total actuation energy, E_{tot}, calculated in this analysis is a combination of the energy spent overcoming the aerodynamic pressure loading during morphing and the elastic strain energy in the compliant structure. This term is found by integrating the tendon moment applied to the trailing edge over the range of spooling pulley rotation angles (δ_F) required to achieve the design lift coefficient, $c_{l,des}$, as shown in Equation (4).

$$J_3 = \int_0^{\delta_{c_{l,des}}} T \, \partial\delta_F = E_{tot} \tag{4}$$

The total actuation energy is shown schematically in Figure 6, where the increasing tendon moment with rotation can be seen, up until the point where the design lift coefficient is reached. Note that the integration is performed with δ_F in radians, and tendon moment is in Nm.

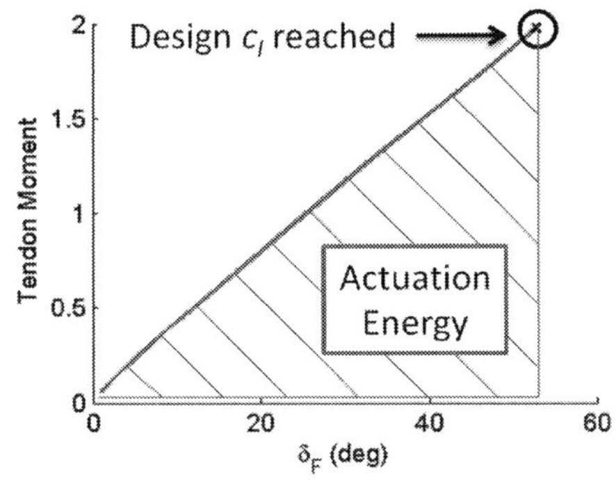

Figure 6. Graphical representation of actuation energy

Together, these three objective functions provide the parameters needed to build an understanding of the impact that the morphing camber airfoil will have on the overall vehicle performance. They quantify the main benefit of the approach, drag reduction, while estimating to a first order the major costs; weight and energy. Of course assessing overall vehicle performance requires careful system level analysis, including factors such as actuator sizing, fuel burn savings from reduced drag, and mission profile analysis to identify sensitivity of the benefits to different types of missions with differing amounts of cruise, loiter, maneuver, etc. Furthermore, a conventional flap system needs to be analyzed in parallel to establish the relative improvement that morphing may allow. This is clearly a significant effort which is outside of the scope of the current work, but which will rely heavily on the optimization code developed here. Therefore, while the results presented below are interesting in their own right, providing insight into the design space available to the FishBAC concept, they are presented with an eye towards future use of the code in system level analyses.

4.4 Variable Bounds and Constraints

The upper and lower bounds of the optimization variables were set primarily by practical physical limitations of the current scale of airfoil being considered, but these bounds were intentionally set fairly wide so as to not limit the optimizer or exclude non-obvious solutions.

For the example results shown below, it was decided to fix the location of the start of the morph to $x_{le} = 0.5$. The position of the end of the morph, x_{te} was then allowed to vary between 0.55 and 0.95 (that is 55% and 95% of chord). This ensured a minimum of 5% chord available for the morphing section. The end position x_{te} is limited to a maximum of 0.95 because of a lack of thickness at the trailing edge. The antagonistic tendon drive system creates bending moments through tension forces acting at a moment arm from the neutral axis, and at the trailing edge there is very little airfoil thickness, and therefore very little moment arm. Even ignoring practical considerations such as the need for enough space to anchor the tendons, the forces required to generate a given moment with such a small moment arm quickly become prohibitive. There is therefore a rigid portion of trailing edge with a length of at least 5% chord in all cases. The bounds for the other variables are presented in Table 2, and are generally chosen to be reasonable for the given scale of morphing airfoil considered.

Table 2. Design variable bounds

Design Variable	Symbol	Lower Bound	Upper Bound	Units
Skin thickness	t_{sk}	1	4	mm
Spine thickness	t_b	0.5	5	mm
Stringer thickness	t_{st}	0.5	3	mm
# of stringers	N_s	4	16	–
Start of morph	x_{le}	0.5	0.5	–
End of morph	x_{te}	0.55	0.95	–

4.5 Out-of-Plane Deflection Constraint

In this analysis a constraint was included to ensure that the out-of-plane deflections arising from the aerodynamic pressure loading on the skin surface were reasonable. This is required to keep the optimizer from creating overly thin skins in an attempt to reduce mass and actuation energy. The procedure for calculating out-of-plane displacement will be presented, followed by discussion of the threshold chosen for the current study.

The FishBAC skin is treated as a tensioned member under distributed pressure loading with linear material elasticity. The tension in the membrane due to the combined effects of initial pre-strain (which is 9% in this study) and bending strains due to the camber morphing is calculated for the worst case on both the upper and lower skins. The upper skin experiences an increase in membrane tension with morphing while the lower skin decreases with increasingly positive camber. Additionally the unsupported length of the skin between each adjacent pair of stringers increases on the upper surface but decreases on the lower surface. The inside of the airfoil is assumed to be at atmospheric pressure, which affects the net pressure acting on the lower skin and upper skins.

In order to reduce computation time, the full distribution of out-of-plane displacement along the chord for the upper and lower skins is not determined, instead the worst case situation for both sides is determined and only those two deflections are solved for. For the upper skin surface the worst case is the maximum pressure felt along the morphing section applied to the skin section with the lowest amount of tension. While in actuality the highest pressure on the upper skin rarely occurs at the point of minimum tension, this approach creates a conservative estimate of upper skin deflection with only a single calculation per individual. In the case of the lower skin the highest magnitude pressure usually does occur at the point of lowest membrane tension (on the skin section nearest the leading edge) since the bending induced compressive strains which reduce skin tension are largest at this point. In both cases the calculation proceeds as follows. First the maximum magnitudes of net pressure (surface pressure minus atmospheric pressure, p_{atm}) acting on the upper and lower skins surfaces are found. In the case of the upper skin the net pressure is negative and deforms the skin outwards, whereas on the lower skin the net pressure is positive and deforms the skin inwards. For this analysis the direction of deflection is not constrained, only its magnitude, hence the absolute value symbol in Equation (5).

$$p_{max} = \max\left(|p(x) - p_{atm}|\right) \tag{5}$$

The distribution of bending induced strain along the chord for both the upper and lower skins are then calculated using Euler Bernoulli theory:

$$\varepsilon_b = \frac{M_b\, y_s}{EI} \tag{6}$$

with the bending moment, M_b, and flexural rigidity, EI, and distance of the skin from the neutral axis, y_s, already known for each individual from the results of the FSI analysis presented in Section 3.2. In addition to the bending strain, there is an initial pre-strain, ε_p, present when the skin is bonded to the core to increase out-of-plane stiffness as discussed in Section 2. For this study, this pre-strain is fixed in to 9%:

$$\varepsilon_p = 0.09 \tag{7}$$

The total skin strain is then found by adding the minimum bending induced strain to the pre-strain:

$$\varepsilon_{tot} = \varepsilon_p + \varepsilon_b \tag{8}$$

The length of the unsupported skin between stringers, l_{su}, is then found from the design variables and bending strain according to:

$$l_{su} = \frac{(x_{te} - x_{le})(1 + \varepsilon_b)c - t_{st} N_s}{N_s}$$

The axial stiffness, k_a, and the tension in the skin due to the total strain, T_ε, are then found from:

$$k_a = \frac{E_s t_s b}{l_{su}} \tag{9}$$

$$T_\varepsilon = k_a (l_{su} \varepsilon_{tot}) \tag{10}$$

The distribution of out-of-plane displacement for a single skin panel under pressure loading can then be found from the analytical equation for the out-of-plane displacement distribution, *w(x)*, of a tensioned membrane:

$$w(x) = \frac{C_1}{\alpha} e^{-\sqrt{\alpha}x} + \frac{C_2}{\alpha} e^{\sqrt{\alpha}x} + C_3 x + C_4 - p_{max} \frac{x^2}{2T_\varepsilon} \tag{11}$$

where the coefficients $C_1 - C_4$ are:

$$C_1 = p_{max} \frac{e^{\sqrt{\alpha}\, l_{su}} - 1}{T_\varepsilon (e^{\sqrt{\alpha}\, l_{su}} - e^{-\sqrt{\alpha}\, l_{su}})} \tag{12}$$

$$C_2 = p_{max} \frac{1 - e^{-\sqrt{\alpha}\, l_{su}}}{T_\varepsilon (e^{\sqrt{\alpha}\, l_{su}} - e^{-\sqrt{\alpha}\, l_{su}})} \tag{13}$$

$$C_3 = p_{max} \frac{l_{su}}{2T_\varepsilon} \tag{14}$$

$$C_4 = -p_{max} \frac{1}{T_\varepsilon \alpha} \tag{15}$$

with

$$\alpha = \frac{T_\varepsilon}{E_{sk} I_{sk}}$$

and the second moment of area of the skin sheet, I_{sk} found from:

$$I_{sk} = \frac{1}{12} b t_{sk}^3 \tag{16}$$

Once the distribution of out-of-displacement over the worst case skin panel for both the upper and lower skins are calculated, the overall maximum magnitude of displacement on both skins is found and checked to make sure that it is less than the prescribed threshold. If the threshold is exceeded, then arbitrarily large values of the objective functions are assigned to the individual, ensuring its removal from the optimizer at the end of the generation.

For this analysis, the maximum displacement threshold was set to 1mm, which at 0.33% of the chosen chord length, is fairly small. This value is nominal and meant to be representative; future work will look more closely into the impact of surface deformations on performance using high fidelity CFD and wind tunnel tests to allow for more precise constraints to be set.

5. RESULTS AND DISCUSSION

Initial results from running the optimizer will be shown in this section. As this is currently a work in progress, a full suite of results for different design conditions is not yet available. Due to time restrictions, the number of generations was limited to 20, implying that the results are likely not fully converged. For this reason the entire data set of 300 individuals is shown, instead of the extracted Pareto front, since the available Matlab functions for extracting Pareto fronts for three dimensional data give erroneous results if the solution is not fully converged. However, the location of the Pareto fronts are still visible, and the results show interesting trends.

It is useful to consider the 3D Pareto frontier through a series of two dimensional projections with only two of the objectives being compared at once, as is done in Figure 7 through Figure 9. This helps to reduce the difficulty of pulling off values from a 3D plot. In Figure 7 the values of the population for the mass objective function are shown plotted against the drag objective function. Here a Pareto frontier can be seen which has an interesting shape. Generally, decreasing the drag coefficient can be achieved at the expense of adding mass, as is expected. However, there is one region in particular around $c_d = 0.012$ with a particularly steep slope, implying a significant amount of mass must be added to decrease drag even slightly. This aspect of the optimal objective space is perhaps due to non-linearities in the aerodynamic behavior, and warrants further investigation.

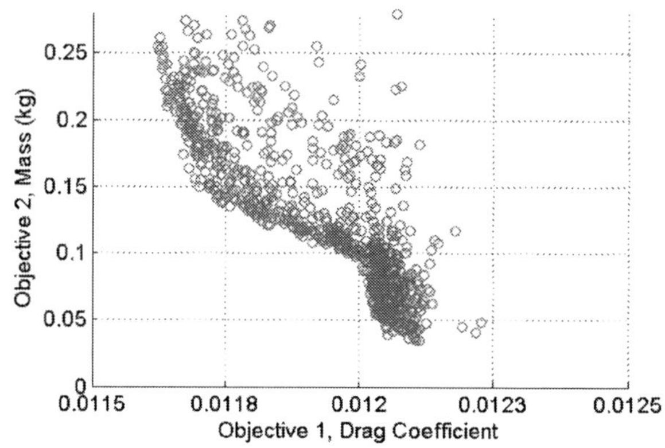

Figure 7. Pareto frontier for drag coefficient and mass

Next, consider the second objective relative to the third. Figure 8 plots the actuation energy required to hit the design lift coefficient against the mass of the FishBAC compliant structure. In this case there is a more clearly defined corner in the lower left hand region of the Pareto front, indicating less of a tradeoff between the two objectives and more of a best-of-both situation.

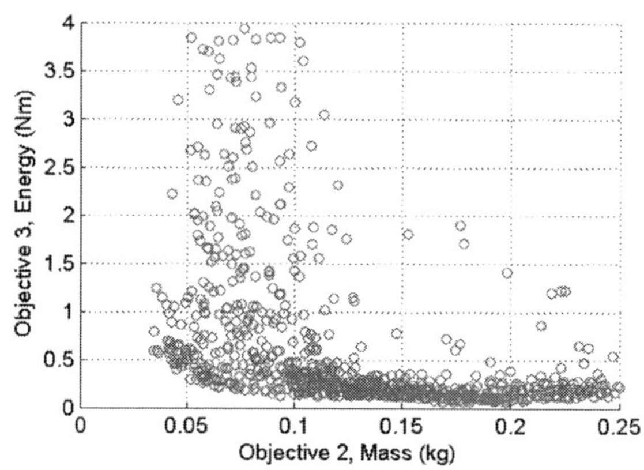

Figure 8. Pareto frontier for mass and actuation energy

Finally, in Figure 9 the results for actuation energy are compared to the drag coefficient values. The behavior seen in this case is quite different from the other two cases. While there is a well-defined Pareto front for the drag coefficient which approaches a sharp corner in the lower left hand side, the results for actuation energy do not approach this corner in the same smooth manner. Instead there is a lack of solutions which have high energy and low drag. Several things could explain this situation. The first would be an intrinsic coupling between low drag and low energy. This would imply that the solutions which have the best drag ratios are low stiffness geometries which occupy larger portions of the chord. These solutions would therefore also have low actuation energy due to the low bending stiffness inherent in the same parameters. A second possibility is that the optimizer is hitting the bounds of the variables, such that it is not able to produce geometries which map to the "empty" portion of the objective space of low drag and high energy. However, inspection of the points nearest this region shows that they are in fact quite far from the bounds. Generally the low drag solutions have small values for the parameters (with the exception of x_{te} which approaches the trailing edge), and so it might be expected that simply increasing these values would lead to stiffer solutions with similar drag properties, but this does not appear to happen. This further implies that the deformed geometries that obtain the lowest drag coefficients are unique to the low energy cases. Further investigation of this is clearly needed, and future extended runs of the optimizer will investigate it.

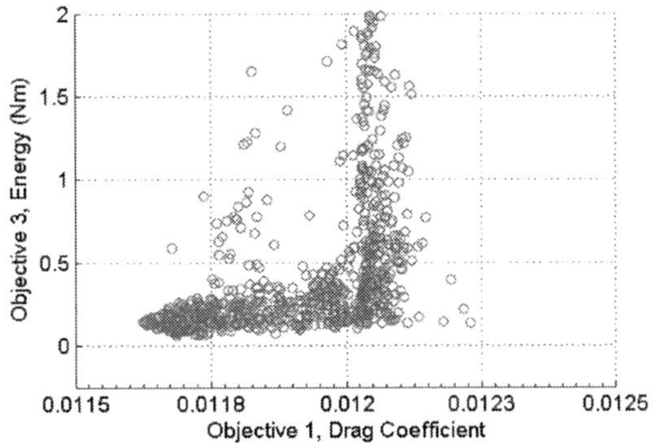

Figure 9. Pareto frontier for drag and actuation energy

6. CONCLUSIONS

This work has presented a newly developed analysis tool which wraps a multi-objective genetic algorithm based optimization around a fluid-structure interaction analysis of the Fish Bone Active Camber morphing airfoil concept. An overview of the structural and aerodynamic modeling methods has been given, along with detailed discussion of the optimization method and parameters. This code allows for examination of the interplay between three critical aspects of the design of a compliant morphing solution through use of a multi-objective optimization scheme; drag, mass, and actuation energy. Preliminary results were shown for the optimization of a morphing FishBAC section starting at 50% chord, and the resulting Pareto fronts were discussed.

Acknowledgments

The research leading to these results has received funding from the European Research Council under the European Union's Seventh Framework Programme (FP/2007-2013) / ERC Grant Agreement n. [247045].

References

[1] Abzug, M.J., and Larrabee, E.E., *Airplane Stability and Control, Second Edition; A History of the Technologies That Made Aviation Possible*, Cambridge University Press, Cambridge, UK, 2002, pg. 3.

[2] Raymer, D.P., *Aircraft Design: A Conceptual Approach*, AIAA, Washington, D.C., 1992.

[3] Kaman, C.H., "Aircraft of Rotary Wing Type," U.S. Patent No. 2,455,866, filed August 19, 1946, issued December 7, 1948.

[4] Overgaard. O., and Livingston, J., "Recent Development of the Ship Rudder with Particular Reference to the 'Flettner Rudder'," *Trans. Soc. Naval Arch. And Marine Eng.*, Vol. 34, 1926, pg. 205.

[5] Burcher, R., and Rydill, L., *Concepts in Submarine Design*, Cambridge University Press, Cambridge, England, 1995.

[6] Shinners, S., *Modern Control System Theory and Design*, John Wiley & Sons, Inc, New York, NY, 1998, pp. 22-23.

[7] Chopra, I, "Review of State of Art of Smart Structures and Integrated Systems," *AIAA Journal*, 2002, 40:2145-2187.

[8] Giurgiutiu, V., "Recent Advances in Smart-Material Rotor Control Actuation," Proceedings of the AIAA/ASME/ASCE/AHS/ASC 41st Structures, Structural Dynamics and Materials Conference, AIAA-2000-1709, Atlanta, GA, April 3-6 2000.

[9] Grohmann, B., Maucher, C., Prunhuber, T., Janker, P., Dieterich, O., Enenkl, B., Bauer, M., Ahci, E., Altmikus, A., Baier, H., "Multidisciplinary Design and Optimization of Active Trailing Edge for Smart Helicopter Rotor Blade," *Mechanics of Advanced Materials and Structures*, 2008, Vol. 15, No. 3, 307-324.

[10] Antoni, U., "Construction of Flexible Aeroplane Wings Having a Variable Profile," US Patent No. 1,886,362, filed August 24, 1931, issued November 8, 1932.

[11] Barbarino, S., Bilgen, O., Ajaj, R.M., Friswell, M.I., Inman, D.J., "A Review of Morphing Aircraft," *Journal of Intelligent Material Systems and Structures*, June 2011, Vol. 22, No. 9 pp. 823-877.

[12] Bartley-Cho, J.D., Wang, D.P., Martin, C.A., Kudva, J.N., and West M.N., "Development of high-rate, adaptive trailing edge control surface for the smart wing phase 2 wind tunnel model". *Journal of Intelligent Material Systems and Structures* 2004, 15: 279–291.

[13] Daynes, S., and Weaver, P.M., "Morphing Blade Fluid-Structure Interaction", Proceedings of the 53rd AIAA/ASME/ASCE/AHS/ASC Structures, Structural Dynamics, and Materials Conference, 2012.

[14] Woods, B.K.S., and Friswell, M. I., "Preliminary Investigation of a Fishbone Active Camber Concept," *Proceedings of the ASME 2012 Conf. on Smart Materials, Adaptive Structures and Intelligent Systems*, Sept 19-21, 2012, Stone Mountain, GA.

[15] Woods, B.K.S., Bilgen, O., and Friswell, M.,I., "Wind Tunnel Testing of the Fishbone Active Camber Morphing Concept," Proceedings of the 23rd International Conference on Adaptive Structures and Technologies, October 11-13, 2012, Nanjing, China.

[16] Woods, B.K.S., and Friswell, M.I., "Fluid-Structure Interaction Analysis of the Fish Bone Active Camber Mechanism," Proceedings of the 54th AIAA Structures, Structural Dynamics, and Materials Conference, Boston, MA, April 8-11, 2013.

[17] Woods, B.K.S., and Friswell, M.I., "Structural Analysis of the Fish Bone Active Camber Concept," Presented at the 22nd AIDAA Italian Association of Aeronautics and Astronautics Conference, Naples, Italy, September 9-12, 2013.

[18] Meirovitch, L., *Fundamentals of Vibration*, McGraw-Hill, New York, NY, 2001.

[19] Deb, K., *Multi-Objective Optimization using Evolutionary Algorithms*, John Wiley & Sons, New York, NY, 2001.

[20] Raymer, D., *Aircraft Design: A Conceptual Approach*, 4th Edition, American Institute of Aeronautics and Astronautics, Reston, VA, 2006.

Experimental and finite element analysis on an SMA-based actuator for aerodynamic load control on aircraft wings

A. Lara-Quintanilla[1*], H. E. N. Bersee[1]

[1] Structure Integrity and Composites, Faculty of Aerospace Engineering
Delft University of Technology, Delft, The Netherlands
*Corresponding author: a.laraquintanilla@tudelft.nl

Abstract

Fixed wing aircraft are designed to work optimally within flight envelope (altitude, aircraft weight, Mach number...). However, most of these parameters suffer alterations during the flight. For that reason, new wing design concepts are being developed in the last years. Most of these concepts consist of providing the wing with a set of sensors, controllers, actuators and devices that can modify the wing geometry or the characteristics of the airflow. This adaptive response of the wing to the flight conditions is intended to get the optimum configuration of the wing to fly in such conditions.
In the present work, a variable camber trailing edge application is presented. It consists of an actuator, activated by Shape Memory Alloys (SMA) wires. The novel actuator was subjected to a proof-of-concept in order to ensure its performance. Fuzzy Logic Controllers were used to control the actuator, achieving a good performance for sinusoidal response in a range of frequencies from 0.1Hz to 0.6Hz.
The main weakness of SMAs for actuation purposes is their bandwidth, what is limited by their heating and cooling rate. Therefore, the Müller-Achenbach-Seelecke SMA model was implemented in a Finite Element Analysis software (COMSOL Multiphysics®) to have a better understanding of the performance of the SMA wires when they are embedded in the flexible surface. The ultimate aim is to increase both the accuracy and the working rate of the SMA-based actuator.

1. INTRODUCTION

Shape Memory Alloys (SMAs) have attracted the attention of scientists and engineers over the last two decades. These alloys have certain features that make them unique smart materials. Apart from some remarkable characteristics like high power density, attainable strokes (up to 8%) [1, 2], lightness and noiseless operation among others, they show Shape Memory Effect (SME) when they are deformed at low temperatures and Superelasticity (SE) when they are deformed at high temperatures, see Figure 1.

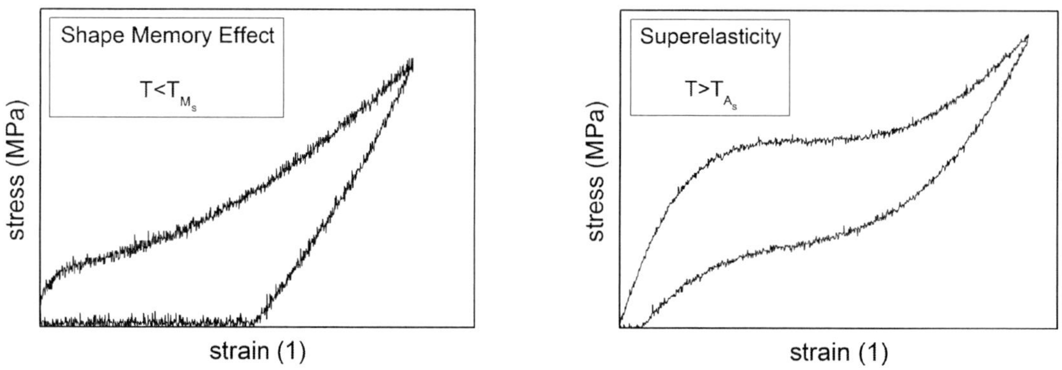

Figure 1: Strain-stress relationship showing Shape Memory Effect (left) and Superelasticity (right)

The underlying system of SME and SE consists of a sequence of transformations between different phases, martensite and austenite. At low temperatures, two types of variants of martensite can coexist: Twinned martensite, which is a mixture of self-accommodated martensite variants, *i.e.*, there is not a dominant martensite variant; in detwinned martensite, however, there is a dominant variant. This phase is stable at temperatures below T_{M_s} and is the only stable one below T_{M_f}. The austenite phase, also called the parent phase, is stable at temperatures above T_{A_s}, and the only stable one above T_{A_f}. The mentioned temperatures are called transition temperatures and vary linearly with the stress applied on the SMA. Further information about this transition temperatures and SMAs' operation is given by Lagoudas in [3].

One of the main interests about SMAs in the aerospace field is the development of adaptive wings. These wings adapt their shape to reach optimum geometries at different surrounding flight conditions, thus yielding important improvements on the performance of the aircraft. Adaptive wings can be designed and manufactured using a wide range of devices that modifies the shape of the wing on many ways. We focus our attention on the out-of-plane deformations of the trailing edge since it is one of the best ways to modify the aerodynamic characteristics of the wing while keeping attached flow; more specifically, we focus on the development of a bending surface actuated by SMA wires, what is called "SMA based actuator" (or simply "actuator") throughout this article.

Several researchers have used different ideas to bend flexible cantilever beams by means of eccentrically attached SMA wires. However, there are some applications – like the one discussed here – in which embedding the SMA wires in the host material is a more appropriate option. SMA wires can be directly embedded, bonding the wires to the host material. In a different way, they can be indirectly embedded, placing the wires into sleeves or empty channels through the host material. Researchers working with directly embedded SMA wires have found difficulties during the manufacturing process [4], and debonding problems during actuation [4, 5]. Indirectly embedding SMA wires facilitates the manufacturing process because the SMA wires are placed into the host material after producing the beam. Furthermore, this configuration permits the use of active cooling, what is one of the main features of the actuator.

This actuator consists of a flat and flexible surface made out of anionic polyamide-6 (APA-6), a form of Nylon. That surface has several sets of SMA wires eccentrically embedded along empty channels that are created during manufacture through it. The wires are capable of producing the necessary force to bend it, upwards and downwards. A complete description of the actuator, as well as its proof-of-concept, can be found on [6]. On that article, Lara-Quintanilla et al. explained the design, manufacture and control of the SMA based actuator, observing a satisfactory performance of the actuator upon tracking different kind of signals. The actuator showed an accurate response tracking sinusoidal signals of amplitude up to 12mm peak-to-peak at frequencies of 0.6 Hz and attained frequencies of 1Hz for smaller deflections. The article emphasises the importance of the active cooling as a part of the strategy control of the actuator. Active cooling is a crucial feature since it allows the actuator to perform faster, i.e., to increase its bandwidth. In order to implement an effective active cooling system, further research must be carried out on its performance.

In this paper, we study the performance of active cooling on a SMA based actuator. A constitutive model for SMA has been implemented in a Finite Element Analysis software. This model and experimental data have been used to determine the limitations of a single SMA wire in terms of actuation rate under different working conditions of applied stress, heating and cooling rate.

2. SMA MODEL

SMAs exhibit a strong thermo-mechanical coupling. Moreover, the relationship between strain, the stress and the temperature is nonlinear and hysteretic, what complicates their modelling. There are a number of articles describing different models for SMA, as well as their strengths and weakness.

The one-dimensional model used here was initially proposed by Achenbach [7] and expanded afterwards by Seelecke and Müller [8]. This model possesses several features that were found attractive. First of all, it captures the behaviour of the wire reasonably well [9, 10], even its single crystal version, which is used in this paper. Another strength is the simplicity to get the parameters of the SMA that are necessary for the model. As explained by Crews et al. [10], only two isothermal tests at two different temperatures above the austenitic temperature are necessary to obtain the required parameters. Finally, this model has shown a great computational efficiency [11]. This model has already been used for different purposes. For example, Yang and Seelecke [12] analysed a bio-inspired bone-joint system of a bat-like wing, and Crews and Buckner [9] optimised the design of a SMA-based catheter by means of this SMA model. The equations used in this model and its implementation into COMSOL can be found in [9, 10, 13].

3. EXPERIMENTAL

First, two isothermal experiments were run to calculate the mechanical parameters from experimental data with the method described in [10]. The parameters and the validation of the model are presented in [13]. Afterwards, isobaric experiments were carried out. As explained in the introduction, active cooling is a crucial feature of the actuator. The aim of these isobaric experiments was to investigate two related phenomena. Firstly, the convective heat transfer between the SMA and the surrounding different forced airflows was to be found. Different cooling rates were applied through the actuator while recording the temperature of the wire. Secondly, the contraction rate under different loads, heating and cooling conditions wanted to be found. Consequently, a Linear Variable Differential Transformer (LVDT) was used to measure the displacement of the hanging weight.

Similar thermal conditions to those in real performing of the actuator wanted to be reproduced. For that reason, a looped SMA wire was placed throughout one of the channels of the actual flexible beam. The two ends of the looped wire were attached above and out of the beam. The loop that formed in the middle of the wire, below and out of the beam, was used to hang different weights. A detail of this joint is shown at the left hand of Figure 3. The airflow inlet was at the top of the beam and the outlet at the bottom. The electrical connections were done at the two ends of the wire. The setup is sketched in Figure 2 for a better understanding. Figure 3 shows a detail of the thermographic camera's measuring point.

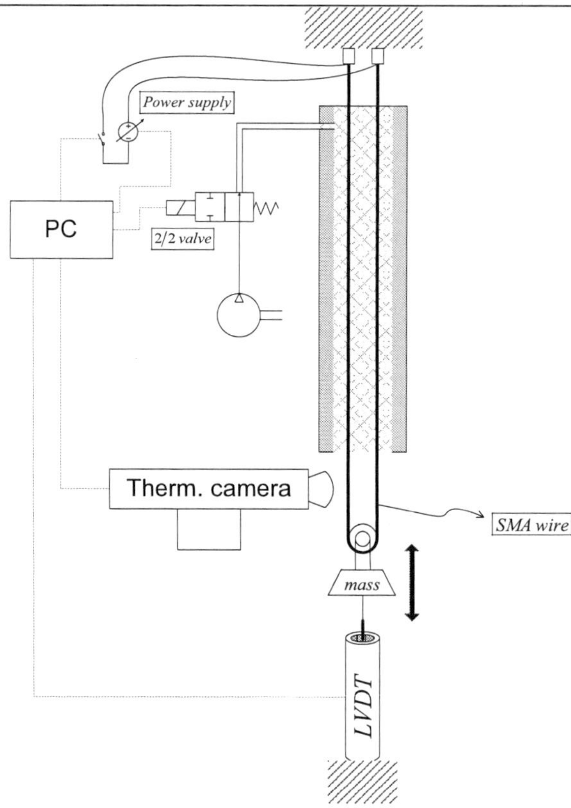

Figure 2: Schematics of the isobaric setup

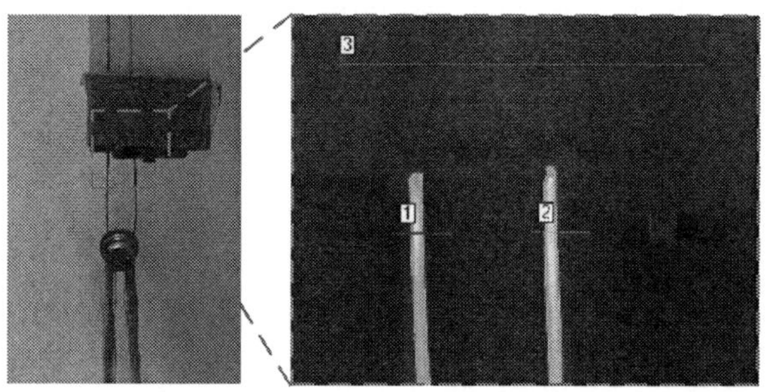

Figure 3: Detail of the thermographic camera's measuring points. (1) and (2) measure the temperature on the wire and (3) the temperature of the beam

The heating and the cooling of the SMA wire were controlled via a DC power Supply and an electronically controlled valve respectively. These devices were controlled in turn by a PC running LabVIEW® [National Instruments Corporation, Austin, TX, USA]. A summary of the equipment used in these experiments is listed on Table 1.

Table 1: Equipment used on isobaric tests

DC Power Supply	SM 120-50 DC Power Supply	Delta Elektronika BV
Valve	2/2-way proportional valve Type 2826	Bürket Fluid Control Systems
Thermographic camera	FLIR SC7000 Series	FLIR
LVDT	SE 373/50	SE Labs
Flow meter	Molded Flow meter Series MR3000	Key Instruments
Data acquisition	NI 9215 – Analog Input Module	National Instruments
Power supply and valve control	NI 9263 – Analog Output Module	National Instruments

The experiment to measure the temperature decay on the SMA wire consisted of two steps. First, the wire was heated by means of an electric current of 1.5A. The current was kept constant until the wire had a uniform and fairly constant temperature around 393K, what guaranteed that the SMA reached complete transformation into austenite. Subsequently, the current was cut off at the same time that the valve was opened. A timing diagram is shown in Figure 4.

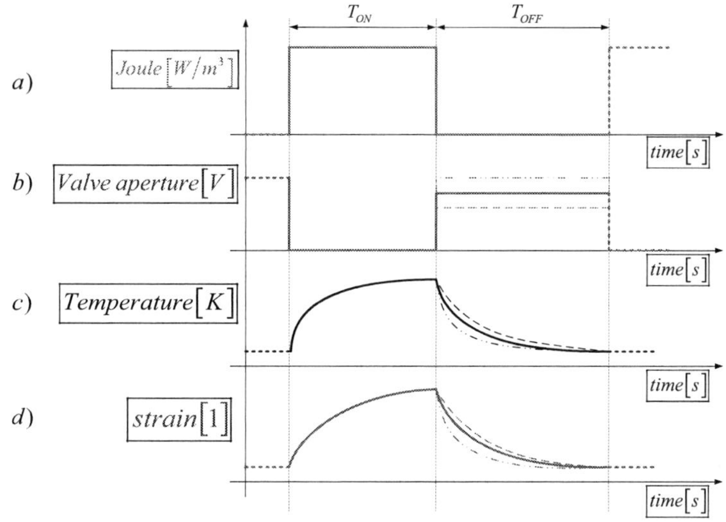

Figure 4: From top to bottom: Sequence of actions performed during the isobaric tests, a) Joule heating and b) Different valve aperture; Sequence of measurements, c) Temperature and d) Strain. Solid and dashed lines show different measurements for different values of valve aperture.

In order to find the convective heat transfer coefficient, the temperature decay had to be analysed. In a convective phenomenon, it is governed by

Equation 1: Time evolution of the temperature

$$T_{SMA}\left(t\right) = T_{\infty} + \left(T_0 - T_{\infty}\right) e^{\frac{-h_{conv} A_{SMA}}{v_{SMA} \rho_{SMA} c_p} t}$$

where T_{SMA} is the temperature on the SMA wire, T_{∞} is the room temperature, what is the same that the airflow's temperature, T_0 is the temperature at $t = 0$. A_{SMA} is the area of the wire, v_{SMA} its volume, ρ_{SMA} is its density and c_p its specific heat. Equation 1 can be fitted to an exponential equation of the type

Equation 2: Exponential fitting equation

$$y = y_0 + ae^{bt}$$

All the parameters in Equation 1 are known except h_{conv}, which is deducted from the coefficient b in Equation 2.

4. RESULTS AND DISCUSSION

4.1 Convective heat transfer coefficient

The convective heat transfer coefficient has been calculated for both, natural and forced convection. MATLAB has been used to detect the temperature decays on the thermographic camera and to fit its data to Equation 2. The results of the fitting are presented in Table 2.

Table 2: Measured airflow and Fitting results

Valve $[V]$	*Airflow* $[l/min]$	y_0	a	R_0	R^2	h_{conv} $[W/(m^2K)]$
0	0.0	45.22	66.08	-0.36	0.99	116
0.5	1.4	48.13	61.18	-0.40	0.99	128
1	5.9	45.43	65.39	-0.90	0.99	290
1.5	9.1	44.17	70.13	-0.95	0.99	308
2	11.4	45.50	70.33	-1.95	0.99	629
2.5	14.3	45.81	65.56	-2.32	0.99	748
3	16.3	48.64	62.31	-2.43	0.99	783
3.5	18.6	46.89	67.60	-2.64	0.99	851
4	20.8	47.06	62.45	-2.87	0.99	926
4.5	22.6	47.46	66.79	-3.05	0.99	985
5	24.1	47.46	56.38	-2.95	0.99	952

Three measuring points can be seen in Figure 3. The heat transfer coefficient h_{conv} has been calculated from the maximum temperature measured along Line 1. Line 2 has been used to compare visually the temperature on the other part of the wire to ensure that the measurements were reliable. Line 3 has been set to ensure that the temperature of the beam was not rising up significantly and also to guarantee that the beam was at room temperature at the beginning of each experiment.

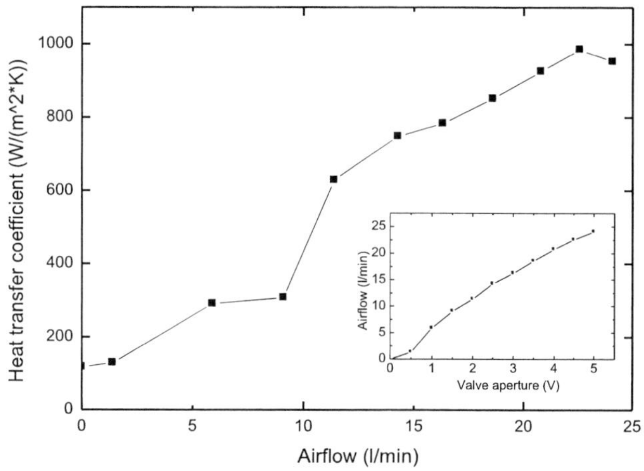

Figure 5: Relationship between the heat transfer coefficient and the airflow. The embedded graph shows the corresponding airflow for each value of aperture of the valve

The relationship between the valve aperture and the airflow (embedded in Figure 5) is almost linear (a proportional valve was used) as well as the relationship between the airflow and the convective heat transfer coefficient. The importance of the active cooling is clearly seen in Figure 5,which shows that the heat dissipation is 8.5 times higher at the maximum airflow ($985W/\left(m^2K\right)$) than that with natural convection ($116W/\left(m^2K\right)$).

4.2 Attainable working rate

The model was subsequently used to predict the maximum attainable working rate of a single SMA under different cooling and load conditions. For this experiment, the inputs to the model are a constant load applied to the free end of the wire (stress) and the electrical energy supplied to it (Joule heating). With those inputs, the model is able to yield the time evolution of the strain and the temperature. The strain does not change significantly at temperatures close to room temperature or much higher than T_{A_S}. For that reason, the recovery time has been calculated as the time that it takes for the wire to recover from the 10% to the 90% of its maximum strain (point A and B in Figure 6).

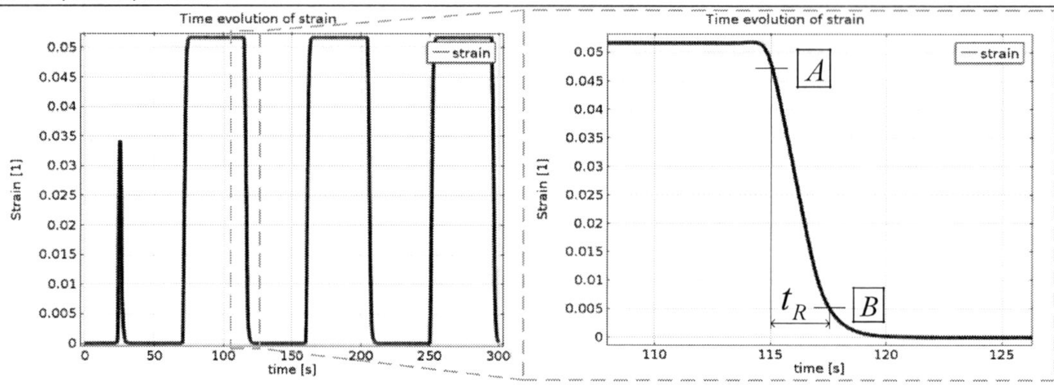

Figure 6: Time evolution of the strain simulated with COMSOL

The recovery time has been calculated in a similar way for the different airflows listed on Table 2 and different loads (1kg, 2kg and 3kg), both for experimental tests and model simulations. The resulting recovery times are shown in Figure 7.

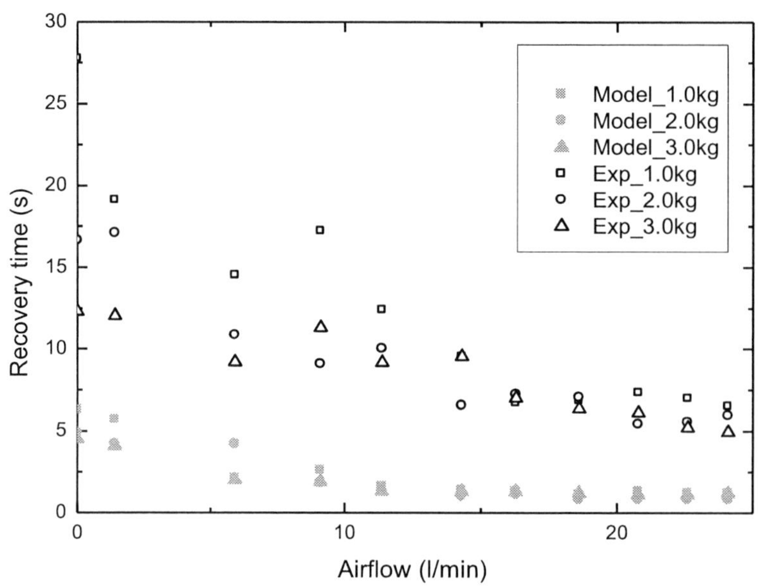

Figure 7: Recovery times for model simulations and experimental tests

As can be seen in Figure 7, there is an overall mismatch between the model and the experiments. However, there are also some similarities. In both cases, and more noticeable on the experimental data, recovery times for natural convection and small airflows are much longer than those at high airflows, as expected. In addition, it can be observed that from a certain value of airflow, increasing the airflow does not have a significant consequence on the recovery time. This maximum effective airflow is $16.3\ l/min$ for the experimental data and $11.4\ l/min$ on the simulations. Because the difference on strain comes from the phase transformations, this limit is determined by the time that it takes for the austenitic phase to transform into martensite. Furthermore, it can be seen that the heavier the weight is, the less noticeable the contribution of the airflow on the recovery time is. It comes from the fact that the load applies a tensile stress on the wire that cooperates with the cooling process, contributing to its elongation.

The reason for the disagreement between experimental and simulated data can be explained by comparing the behaviour of the model and the SMA wire for an isobaric test. As shown in Figure 8, the transformation temperatures and therefore the temperature's hysteresis loop are different in both cases. It can be seen that the range of temperatures in which the transformation from austenite to martensite happens is much wider for the experimental test than for the model simulation. This explains that all the recovery times from the simulations are shorter than from the experiments. Moreover, the difference between the maximum strain observed on both dataset comes from the use of a single crystal model, whereas the tested SMA wire is polycrystalline. The single crystal model considers a single grain oriented on the direction in which the contraction happens. However, the polycrystalline SMA wire has multiple grains oriented on random directions. This makes the overall strain to be noticeably smaller than the strain if all the grains were aligned with the strain direction.

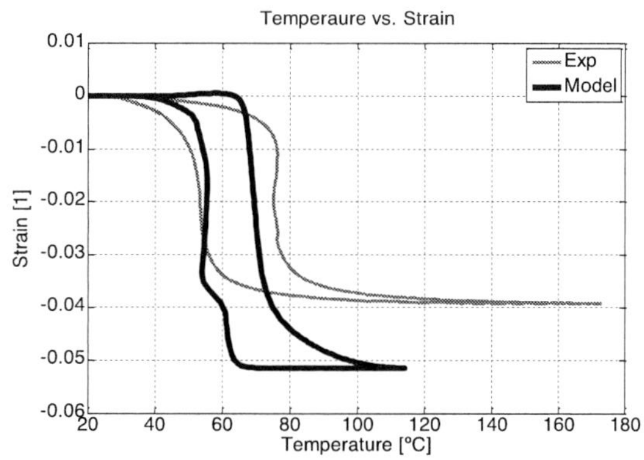

Figure 8: Model and experimental results for an isobaric experiment at T = 263K with a mass of 1.25 kg

Regarding to the heating rate, it has been determined as the time that it takes for the wire to contract from the 90% to the 10% of its maximum strain upon Joule heating. It has been tested in different ways using COMSOL and experimental data. The heating in COMSOL is power controlled whereas it is current controlled in the lab setup. Therefore, data yield by both setups cannot be directly compared. The method used here to study different heating rates is similar to the method used to study different recovery times (see Figure 4), but applying different values of Joule heating and keeping the cooling at a high value to guarantee room temperature on the SMA wire at the beginning of each experiment.

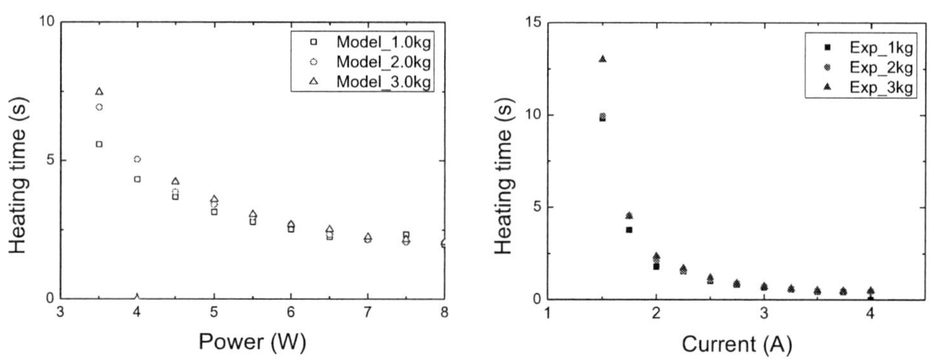

Figure 9: On the left, heating times tested at different power inputs obtained from COMSOL simulations (from 3.5W to 8W in steps of 0.5W). On the right, heating times tested at different current inputs obtained from the experimental setup (from 1.5A to 4A in steps of 0.5A)

Figure 9 shows the resulting heating times for both simulations and experimental data. Although the values obtained cannot be compared, some conclusions can be drawn. As expected, the heating time decreases exponentially with the electric power given to the SMA wire. In both cases, there is a value of power from which the decrease of the heating time is insignificant. Moreover, the weight affects the heating time in a similar manner that affected the recovery time. Heavier weights imply longer heating times because the tensile force that the weight exerts opposes to the strain.

Finally, linking together the times that it takes for the wire to elongate and contract, the maximum attainable working rate can be calculated.

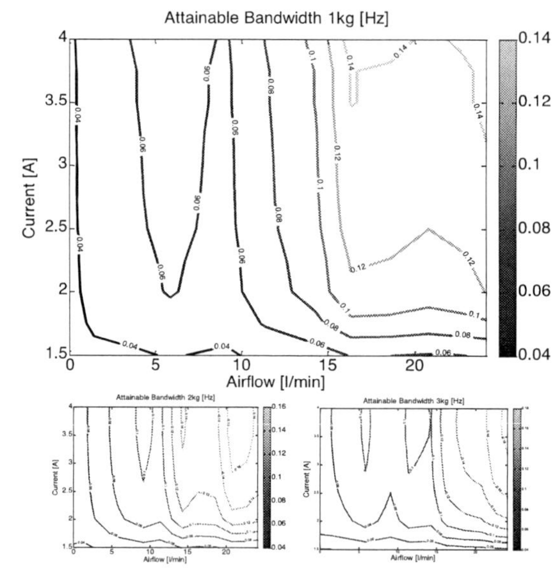

Figure 10: Attainable actuator working rates (in Hertz) for different applied loads: a weight of 1kg (top), 2kg (bottom left) and 3kg (bottom right)

As can be seen in Figure 10, the maximum working rate for a load of 1kg is 0.14Hz, for 2kg it is 0.16Hz and for 3kg it is 0.18Hz. The results observed here suggest that, if an increase of the SMA's bandwidth is wanted, the wire must be placed in such a manner that it can easily release heat to the surroundings. For low working rates, the limiting factor is the current input to the system. However, airflow is the limiting factor for higher working rates. In the three cases shown in Figure 10 the airflow gains importance for values of current above 2 amperes.

The maximum actuation rates shown in Figure 10 have been calculated from the range going from the 10% to the 90% of the maximum strain under the previously described experimental conditions. However, the calculated actuation rate depends on the strain range of actuation. Shorter strain ranges implies faster actuation rates. The experimental method proposed here can be used to find actuation rates (heating and cooling rates) at different working strain ranges.

5. CONCLUSIONS

A prototype was first designed, manufactured and tested in order to study the feasibility of this beam-like actuator with embedded Shape Memory Alloy (SMA) wires, heated by Joule heating and cooled by means of a forced airflow system. The need of a cooling system to increase the actuator's bandwidth was shown. It motivated further research on the cooling system. The heat losses by convective heat transfer to the surrounding forced airflow were study and it was found that it increases the convective heat transfer coefficient from $116W/(m^2K)$ (natural convection) to $985W/(m^2K)$ (maximum aperture of the valve that controls the airflow). The importance of the cooling system has been proven with an increase of 8.5 times on the convective heat transfer.

Subsequently, the heating and recovery times were studied; these are the times that it takes for the wire to contract from the elongated form and strain again respectively. The combination of the heating and the recovery times yielded the overall time that it takes for the wire to complete one cycle. This time was used to calculate the maximum actuation frequency for each combination of input power and airflow along the wire. It was found that it depends on the input electric energy for low actuation rates. However, for values of current above 2 amperes, the airflow becomes the delimiting factor. This method can be used to find actuation rates (heating and cooling rates) at different working strain ranges. In addition, an SMA one-dimensional model was implemented in COMSOL. The mechanical behaviour of the SMA wire was validated, but the model did not perform accurately when simulating thermo-mechanical SMA's behaviour and working rates. The reason is that the model considers a single crystal SMA wire whereas the tested SMA wire is polycrystalline. Some inaccuracies could happen upon getting the model parameters from experimental data. However, further efforts should be made to obtain an accurate model, because it could be used to predict the maximum attainable bandwidth of SMA based actuators on different configurations.

References

1. Callister, W.D. and D.G. Rethwisch, *Materials Science and Engineering: An Introduction*. 2009: John Wiley & Sons Canada, Limited.
2. Pons, J.L., *Emerging Actuator Technologies: A Micromechatronic Approach*. 2005: Wiley.
3. Lagoudas, D.C., *Shape Memory Alloys: Modeling and Engineering Applications*. 2008: Springer.
4. Zhou, G. and P. Lloyd, *Design, manufacture and evaluation of bending behaviour of composite beams embedded with SMA wires*. Composites Science and Technology, 2009. **69**(13): p. 2034-2041.
5. Doran, C.J. *The effect of prestrain on the actuation performance of embedded shape memory alloy wires*. in *European Conference on Smart Structures and Materials*. 1994. DTIC Document.
6. Lara-Quintanilla, A., A.W. Hulskamp, and H.E.N. Bersee, *A high-rate shape memory alloy actuator for aerodynamic load control on wind turbines*. Journal of Intelligent Material Systems and Structures, 2013. **24**(15): p. 1834-1845.
7. Achenbach, M., *A model for an alloy with shape memory*. International Journal of Plasticity, 1989. **5**(4): p. 371-395.
8. Seelecke, S. and I. Müller, *Shape memory alloy actuators in smart structures: Modeling and simulation*. Applied Mechanics Reviews, 2004. **57**(1): p. 23-46.
9. Crews, J.H. and G.D. Buckner, *Design optimization of a shape memory alloy–actuated robotic catheter*. Journal of Intelligent Material Systems and Structures, 2012. **23**(5): p. 545-562.
10. Crews, J.H., et al., *Data-driven techniques to estimate parameters in the homogenized energy model for shape memory alloys*. Journal of Intelligent Material Systems and Structures, 2012. **23**(17): p. 1897-1920.
11. Heintze, O., *A Computationally Efficient Free Energy Model for Shape Memory Alloys - Experiments and Theory*. 2004, North Carolina State University.
12. Yang, S. and S. Seelecke, *FE analysis of SMA-based bio-inspired bone–joint system*. Smart Materials and Structures, 2009. **18**(10): p. 104020.
13. Lara-Quintanilla, A. and H.E.N. Bersee, *Work in progress*.

On the Characterization of Effective Properties of Piezoelectric Macro-Fibre Composites

Marcelo A. Trindade

Department of Mechanical Engineering, São Carlos School of Engineering, University of São Paulo, Av. Trabalhador São-Carlense, 400, São Carlos-SP, 13566-590, Brazil

Abstract

This work presents recent results on the characterization of effective properties of d_{31} and d_{15} piezoelectric MFCs using a finite element homogenization technique, accounting in particular for the effect of electrode and protective layers on the effective elastic, piezoelectric and dielectric coefficients. Then, an analysis of the effect of the fibre volume fraction and electric boundary conditions on the main effective properties of the d_{31} MFC is presented. For the d_{15} MFC, an analysis of the electric field dependence of its properties is performed and compared to experimental results. It is shown that electric boundary conditions (electrodes) play an important role in the predicted response of the MFC and that nonlinear response may be reasonably predicted using piezoelectric fibre electric field dependence. It is also shown that the mechanical behaviour of electrode and protective layers must be accounted for in order to well predict the MFC effective properties.

1. INTRODUCTION

Macro-Fibre Composites (MFC) have become very popular since they combine the conformability of epoxy-matrix composites and the electromechanical energy density of piezoceramic materials [1]. MFCs that are commercially available use either the 33 or 31 response modes, meaning that the piezoelectric constant that characterizes their actuation/sensing behaviour is respectively the d_{33} or d_{31}, and therefore are named as d_{33} or d_{31} MFCs. Schematic representations of the d_{33} and d_{31} MFCs are shown in Figure 1. The major differences between the two commercially available MFCs are the electrode design and the poling direction of the fibres. In the d_{31} MFC, the poling direction is perpendicular to the fibres (along their thickness) so that transverse extension of the fibres can be obtained through the d_{31} constant. On the other hand, in the d_{33} MFC, the poling direction follows a complex non uniform distribution according to the disposition of the interdigitated positive and negative electrodes (IDE) aiming to yield an equivalent macroscopic poling direction parallel to the fibre length, so that longitudinal extension of the fibres can be obtained through the d_{33} constant. The main problem of the d_{33} MFC is that the equivalent macroscopic longitudinal poling direction, and thus its performance, increases with the distance between interdigitated electrode fingers which leads to the requirement of a high actuation voltage (up to 1200 V) to generate the necessary electric field. The d_{31} MFC, on the other hand, requires much lower actuation voltages since the electric field is uniform and in the direction of the actuator thickness, which can be made quite small, but its piezoelectric constant d_{31} is much smaller than d_{33}.

In order to profit from a higher piezoelectric coefficient, d_{15}, an alternative MFC design based on piezoceramic fibres working in the thickness-shear mode was proposed recently [2]. The d_{15} thicknes-shear mode can be obtained by the application/measurement of an electric field that is perpendicular to the remanent polarization direction. The transverse shear mode in commercially available piezoceramic patches is normally obtained by polarization along the length or width direction and a dominant electric field, either applied or induced, in the thickness direction. In the last two decades, thickness-shear piezoceramic materials have been considered for smart structures applications , including active, passive and active-passive vibration control [3]. The d_{15} MFC proposed by [2] considered macro-fibres oriented perpendicular to the direction of motion and poled widthwise, such that the effective poling direction is parallel to the direction of motion (Figure 2a). More recently, a prototype of a d_{15} MFC poled lengthwise was manufactured by Smart Material Co. for its numerical characterization and experimental performance evaluation [4]. This prototype, shown schematically in Figure 2b, profits from the existing d_{31} MFC (P2 type) manufacturing procedure.

Figure 1. Schematic representation of longitudinal "d_{33}" (left) and transverse "d_{31}" (right) MFCs.

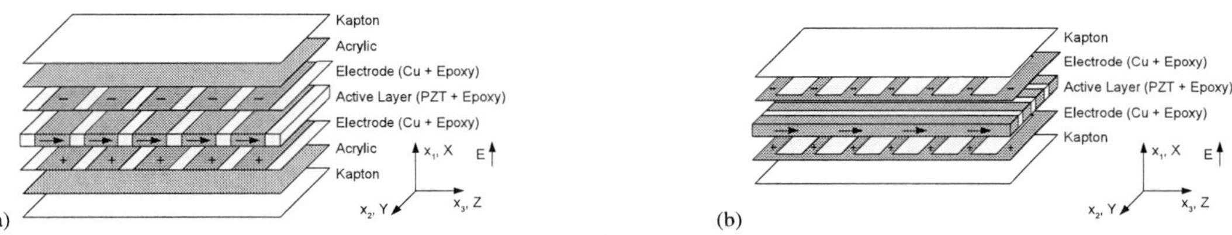

(a) (b)

Figure 2. Schematic design of widthwise (left) and lengthwise (right) thickness-shear d_{15} MFCs.

Since MFC transducers are heterogeneous and made of several different materials (piezoceramic fibres, epoxy matrix, electrode and protective layers), a methodology is required to relate fibres material properties to the effective MFC ones and, thus, estimate the effective piezoelectric coefficients d_{ij} and e_{ij}. Substantial research effort has been directed to the characterization of their effective material properties, as in [5, 6] for d_{31} and d_{33} MFCs and [4, 7, 8, 9] for d_{15} MFCs. Some of these studies [4, 8] have shown that the effective properties depend not only on the piezoceramic fibres, epoxy materials and fibre volume fraction of the active layer but also on the geometrical and material properties of the protective and electrode layers. Indeed, analytical and numerical homogenization techniques have been proposed for the active layer only [5, 7] but they fail to capture the effect of the electrode and protective layers which seems to be particularly important for 'force actuators' applications.

This work presents recent results on the characterization of effective properties of d_{31} and d_{15} piezoelectric MFCs using a finite element homogenization technique, accounting in particular for the effect of electrode and protective layers on the effective elastic, piezoelectric and dielectric coefficients. Then, an analysis of the effect of the fibre volume fraction and electric boundary conditions on the main effective properties of the d_{31} MFC is presented. For the d_{15} MFC, an analysis of the electric field dependence of its properties is performed and compared to experimental results.

2. d_{31} MFC EFFECTIVE MATERIAL PROPERTIES EVALUATION AND ANALYSIS

The d_{31} MFC considered is made of an active layer, composed of piezoceramic macro-fibres embedded in an Epoxy matrix. The active layer is then covered by electrode and protective layers, as shown in Figure 1. The piezoelectric material is considered to be poled in x_3 direction with electrodes fully covering top (x_3^+) and bottom (x_3^-) surfaces such that a preferential or dominant x_3 direction is imposed for the electric field and displacement. Therefore, it can be assumed that $E_1 = E_2 = 0$ and the following reduced constitutive equations can be written

$$\begin{Bmatrix} T_1 \\ T_2 \\ T_3 \\ T_4 \\ T_5 \\ T_6 \\ \hline D_3 \end{Bmatrix} = \left[\begin{array}{cccccc|c} c_{11}^E & c_{12}^E & c_{13}^E & 0 & 0 & 0 & -e_{31} \\ c_{12}^E & c_{22}^E & c_{23}^E & 0 & 0 & 0 & -e_{32} \\ c_{13}^E & c_{23}^E & c_{33}^E & 0 & 0 & 0 & -e_{33} \\ 0 & 0 & 0 & c_{44}^E & 0 & 0 & 0 \\ 0 & 0 & 0 & 0 & c_{55}^E & 0 & 0 \\ 0 & 0 & 0 & 0 & 0 & c_{66}^E & 0 \\ \hline e_{31} & e_{32} & e_{33} & 0 & 0 & 0 & \epsilon_{33}^S \end{array}\right] \begin{Bmatrix} S_1 \\ S_2 \\ S_3 \\ S_4 \\ S_5 \\ S_6 \\ \hline E_3 \end{Bmatrix}, \tag{1}$$

where T_p and S_q, with $p,q = 1,\ldots,6$, denote the six components of mechanical stress and strain vectors in Voigt notation. D_3 and E_3 denote the transverse electric displacement and field. For the e-form, c_{pq}^E, e_{3j} $(j = 1,2,3)$ and ϵ_{33}^S denote the elastic stiffness (at constant electric fields), piezoelectric and electric permittivity (at constant mechanical strains) coefficients.

It can be noticed that the electromechanical coupling of major importance (in the longitudinal direction) is between the electric field and displacement in x_3 direction, E_3 and D_3, and x_1 normal stress and strain, T_1 and S_1. Therefore, it is essential to evaluate the piezoelectric e_{31}/d_{31}, elastic c_{11}^E/s_{11}^E and dielectric $\epsilon_{33}^S/\epsilon_{33}^T$ coefficients in order to qualify these materials as sensors and actuators.

In order to evaluate the effective coefficients that are important to a d_{31} MFC, as discussed in the previous section, a finite element homogenization technique similar to that proposed in [8] is used. It consists in evaluating the average stresses, strains, electric fields and electric displacements in a Representative Volume Element (RVE) of the MFC when subjected to properly chosen mechanical and electrical boundary and periodicity conditions. Then, using the assumed constitutive equations of an equivalent piezoelectric material, as described in the previous section, it should be possible to evaluate effective elastic, piezoelectric and dielectric coefficients of the MFC. Since only a RVE of the MFC is used to perform the calculations, it is important to account for symmetric or periodic mechanical and electrical boundary conditions in directions x_1 and x_2.

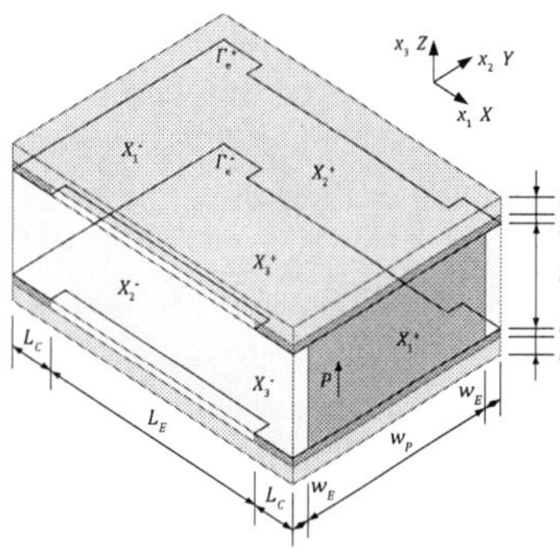

Figure 3. Representative volume element (RVE) of the d_{31} MFC.

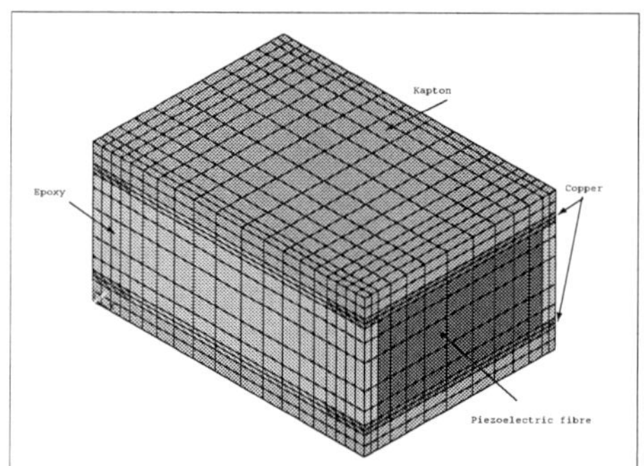

Figure 4. Finite element mesh for the full 5-layered d_{31} MFC RVE with FVF=0.86.

The RVE considered in this work for the d_{31} MFC is presented in Figure 3. It is composed of a so-called 'Active Layer', with one longitudinal section of the piezoceramic macro-fibre (through-thickness poled and electroded) between two Epoxy inserts. The upper and lower surfaces of the Active Layer are then covered by one thin electrode layer each, with two Copper and one Epoxy layers representing a section of the fingered electrode layer. The electrode layers are then covered by a Kapton protective layer. The boundaries of the RVE are denoted X_1^-, X_1^+, X_2^-, X_2^+, X_3^- and X_3^+ according to Figure 3. For a more realistic analysis, additional electric potential boundary conditions are imposed at the electrodes on surfaces Γ_e^- and Γ_e^+.

Average strains and electric fields can therefore be imposed to the RVE using displacements and voltage boundary conditions. Then, the resulting average strains, stresses, electric fields and electric displacements in the RVE are approximated in Ansys$^{(R)}$ by a sum over averaged element values multiplied by the respective element volume divided by total volume of the RVE.

The elastic c_{11}^E and piezoelectric e_{31} coefficients are approximated by imposing unitary normal strains S_1 to the RVE and evaluating, respectively, the average normal stress T_1 and electric displacement D_3. A short circuit electric boundary condition is ensured by setting the electric voltages at surfaces Γ_e^- and Γ_e^+ to zero. The dielectric coefficient of the d-form, ϵ_{33}^T, is approximated by imposing unitary through-thickness electric field to a mechanically free RVE and evaluating the resulting electric displacement D_3. In all cases, symmetric (periodic) mechanical and electric boundary conditions are also considered.

Other coefficients may be evaluated by post-processing the previous ones. The d-form piezoelectric coefficients d_{31}, d_{32} and d_{33} are approximated by inverting the three first lines of (1), such that $\mathbf{d} = [\mathbf{c}^E]^{-1}\mathbf{e}$. The effective electromechanical coupling coefficient k_{31}^2 may be evaluated as $k_{31}^2 = d_{31}^2/s_{11}^E\epsilon_{33}^T$.

2.1 Comparison with previous results for MFC properties

The finite element mesh used in Ansys$^{(R)}$ to solve the local problems and evaluate effective properties of a 5-layered d_{31} MFC RVE, using a FVF of 0.86 for the PZT+Epoxy Active Layer, is shown in Figure 8. The 3D 20-Node coupled-field solid finite element SOLID226 was used to mesh all volumes. It has the three displacement translations along the Cartesian coordinate system and the electric potential as nodal degrees of freedom.

3211 finite elements were used considering 19 divisions in x_1 direction (11 divisions for the Epoxy layer and 4 divisions for each electrode layer), 13 divisions in x_2 direction (9 divisions for the PZT fibre and 2 divisions for the Epoxy layer) and 13 divisions in x_3 direction (5 divisions for the PZT layer, 2 divisions for each electrode layer and 2 divisions for each protective layer). The dimensions considered for this RVE (Figure 3) are $h_P = 180$ μm, $h_K = 40$ μm, and $h_C = 18$ μm, $L_E = 420$ μm, $L_C = 80$ μm, $w_E = 28.35$ μm and $w_P = 348.3$ mm [5]. The material properties for the piezoceramic material Sonox P502, taken from [5], are: $s_{11}^E = s_{22}^E = 18.5$ pm^2/N, $s_{33}^E = 20.7$ pm^2/N, $s_{12}^E = -6.29$ pm^2/N, $s_{13}^E = -6.23$ pm^2/N, $s_{44}^E = s_{55}^E = 33.2$ pm^2/N, $s_{66}^E = 52.3$ pm^2/N, $d_{31} = d_{32} = -185$ pC/N, $d_{33} = 440$ pC/N, $d_{15} = d_{24} = 560$ pC/N, $\epsilon_{11}^T = \epsilon_{22}^T = 1950\epsilon_0$, $\epsilon_{33}^T = 1850\epsilon_0$. The material properties for the other constituents, taken from [4], are: Epoxy – $Y = 2.5$ GPa, $\nu = 0.42$ and $\epsilon = 4.25\epsilon_0$; Kapton – $Y = 2.5$ GPa, $\nu = 0.34$ and $\epsilon = 3.4\epsilon_0$; and Copper electrode – $Y = 110$ GPa, $\nu = 0.34$ and $\epsilon = 5\epsilon_0$.

In order to validate the finite element homogenization, the evaluated effective coefficients of the d_{31} MFC were compared to previous analytical and experimental results found in the literature. Using a mixing rules technique, Deraemaeker et al. [5] evaluated effective plane-stress coefficients for the active layer of the same d_{31} MFC considered in this work. Thus, the finite element homogenization procedure presented previously was adapted to a finite element model of the active layer only by removing the electrode (Copper+Epoxy) and protective (Kapton) layers. To better compare with the results presented in [5], the Epoxy material properties provided therein were used ($Y = 2.9$ GPa, $\nu = 0.3$ and $\epsilon = 4.25\epsilon_0$) and the finite element homogenization was performed removing the boundary conditions imposing $S_3 = 0$. Also, to allow comparison with the full 5-layered MFC, the electric potentials are still applied to the surfaces Γ_e^+ and Γ_e^-, shown in Figure 3. The first and second columns of Table 1 display, respectively, the results obtained by the finite element homogenization proposed in the present work and the mixing rules technique of [5]. It is noticeable that the present results agree very well with the mixing rules ones. It is worthwhile to notice from Table 1 that the presence of the Epoxy filling diminishes the effective piezoelectric coefficients d_{31} and e_{31} of the active layer, compared to those of the PZT fibre alone, and also leads to an orthotropic active layer (and no longer transversally isotropic as the PZT fibre). In particular, the active layer is more effective in the longitudinal direction such that d_{31} and e_{31} are higher than d_{32} and e_{32}.

Then, the results obtained for the full 5-layered MFC with the proposed approach were compared to data provided by the manufacturer of a commercial d_{31} MFC (Smart Material Corp.) [10], with the same nominal geometrical and material properties for its constituents considered in the present work. Notice that not all coefficients are provided by the manufacturer. Nevertheless, it can be concluded that the proposed approach represents very well the actual (manufactured) d_{31} MFC. The only coefficient presenting a more important error (13%) compared to the data from manufacturer is the transversal elastic modulus Y_2^E, but this may be explained by the fact that the actual width of the Copper electrode finger (L_c) is not known and was only visually approximated from a microphotography presented by the original prototype developer at NASA Langley Research Center [1]. A wider Copper electrode finger would lead to a higher transversal elastic modulus Y_2^E. It could also affect the transversal piezoelectric coefficients e_{32} and d_{32}. The results obtained for the most important effective coefficients, Y_1^E, d_{31} and ϵ_{11}^T, are all below 4% error (3.5%, 1.3% and 0.9% respectively) compared to the data provided by the Smart Material Corp. [10].

Notice that the electrode and protective layers diminish substantially the effective properties of the MFC, compared to the active layer alone (see first and third columns of Table 1). In particular, the piezoelectric coefficients d_{31} and $\bar{e}_{31}$ are reduced by 6% and 33%, respectively, while the effective electromechanical coupling coefficient k_{31}^2 is reduced by 39%.

Table 1. Effective short-circuit material properties of d_{31} MFC and its active layer (*: values obtained post-processing data provided in the reference).

	PZT-Epoxy Active Layer		Full 5-Layered MFC	
Coefficient	Finite element	Mixing rules [5]	Finite element	Manufacturer [10]
Y_1^E (GPa)	46.89	47.17	31.40	30.34
Y_2^E (GPa)	17.92	16.98	19.80	15.86
G_{12}^E (GPa)	5.87	6.03	5.34	5.52
ν_{12}^E	0.39	0.39	0.36	0.31
d_{31} (pC/N)	-183.1	-183	-172.2	-170
d_{32} (pC/N)	-167.0	-153	-133.8	–
e_{31} (C/m^2)	–	–	-6.04	–
$\bar{e}_{31}$ (C/m^2)	-10.35	-10.23*	-6.93	–
$\bar{e}_{32}$ (C/m^2)	-4.53	-4.05*	-4.23	–
$\epsilon_{33}^T/\epsilon_0$	1609.9	1600	1572.1	1558.6
k_{31}^2 (%)	11.0	11.2*	6.7	6.4*

2.2 Analysis of the effect of FVF and electric boundary conditions

In this section, the finite element homogenization technique was used to evaluate the effect of the fibre volume fraction on the main effective properties of the d_{31} MFC, namely the longitudinal Young's modulus Y_1, the piezoelectric coefficient d_{31} and the free dielectric coefficient ϵ_{33}^T. In addition, the results obtained are compared to those for the MFC Active Layer to quantify the effect of the protective and electrode layers. The results obtained for the MFC Active Layer are also compared to those obtained using a Mixture Rules technique proposed in [5].

According to [5], the effective properties of a d_{31} MFC Active Layer, composed of piezoelectric fibre (p) and Epoxy matrix (e), can be written as

$$Y_1^E = \rho Y_1^{Ep} + (1-\rho) Y_1^e \, , \; d_{31} = \rho (E_1^p/E_1) d_{31}^p \, , \; \epsilon_{33}^T = \rho \epsilon_{33}^{Tp} + (1-\rho)\epsilon_{33}^e . \tag{2}$$

where Y_1^{Ep} and Y_1^e are the longitudinal Young's moduli of the piezoelectric fibre (at constant electric field) and Epoxy matrix, respectively. ϵ_{33}^{Tp} and ϵ_{33}^e are the dielectric coefficient of the piezoelectric fibre (at constant stress) and Epoxy matrix, respectively. d_{31}^p is the piezoelectric coefficient of the piezoelectric fibre and ρ is the fibre volume fraction of the active layer.

It should be noted that in order to obtain (2), it is assumed that the electrodes cover the active layer entirely, which is not generally the case. Most d_{31} MFC use fingered electrodes such as the one shown in Figure 1. In this practical case, part of the Epoxy matrix is not covered by electrodes. This effect is highlighted here by performing an additional finite element homogenization of the active layer with continuous electrodes (CE) fully covering the upper and lower surfaces of the active layer. For that, the electric potential boundary conditions are applied to surfaces $X_3^+ (x_3 = h_P)$ and $X_3^- (x_3 = 0)$ instead of surfaces Γ_e^+ and Γ_e^-.

Hence, four sets of results are presented in the following analyses, namely those obtained for the active layer only using mixture rules (MR), finite element homogenization with continuous electrodes

(FE/CE) and finite element homogenization with fingered electrodes (FE), and for the full (5-layer) d_{31} MFC using finite element homogenization with fingered electrodes.

Figure 5a shows the longitudinal Young's modulus for the MFC Active Layer and for the full 5-layered d_{31} MFC. It is possible to notice that the three results for the active layer only agree satisfactorily, meaning that the electrode configuration does not affect very much the elastic behaviour of the active layer. On the other hand, the inclusion of the relatively softer protective and electrode layers leads to a significant reduction of the MFC longitudinal elastic modulus, of around 32% compared to that of the active layer. Notice that the effective longitudinal elastic modulus of the standard d_{31} MFC could be increased of around 9% by augmenting the FVF from the standard value of 86% up to 95%.

The relative dielectric coefficient $\epsilon_{33}^T/\epsilon_0$ for the MFC Active Layer and for the full 5-layered d_{31} MFC is shown in Figure 5b. In this case, it is noticeable that the electric boundary conditions do affect the results for the active layer. Indeed, the hypothesis of continuous electrodes leads to smaller dielectric coefficients, if compared to fingered electrodes (FE Active Layer), as found using both the Mixture Rules technique (MR) and finite element homogenization (FE/CE), in particular for smaller FVFs. One may also notice that the inclusion of the protective and electrode layers also leads to a reduction of the effective MFC dielectric coefficient, compared to that of the active layer (FE Active Layer). This reduction decreases with increasing FVF from 8.5%, for FVF=45%, down to 2%, for FVF=95%.

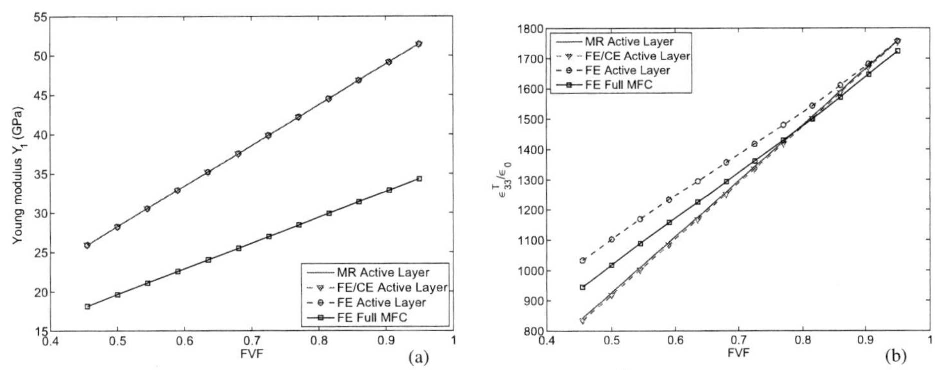

Figure 5. Young's modulus Y_1 (a) and dielectric coefficient ϵ_{33}^T (b) as function of the fibre volume fraction. Comparison between results found using mixture rules (active layer only) and finite element (active layer and full d_{31} MFC).

The effect of the fibre volume fraction on the piezoelectric coupling efficiency of the d_{31} MFC was also analysed. Figures 6a and 6b show its effect on the effective piezoelectric coefficient d_{31} and effective electromechanical coupling coefficient k_{31}^2, respectively.

From Figure 6a, it is possible to notice that the piezoelectric coefficient d_{31}, as the Young's modulus, is not very much affected by the electrode configuration and, thus, both finite element models (with continuous, FE/CE, and fingered, FE, electrodes) lead to similar results for the active layer. They also both agree well with the results found using the Mixture Rules technique. The inclusion of the protective and electrode layers leads to a reduction of the effective piezoelectric coefficient d_{31} that goes from 9.7%, for FVF=45%, down to 5.7%, for FVF=95%. Also, if compared with elastic modulus and dielectric coefficient, the MFC effective piezoelectric coefficient d_{31} is much less affected by the fibre volume fraction. Hence, augmenting the FVF from 86% up to 95% yields an increase of only 1% on d_{31}.

The effect of the fibre volume fraction on the effective electromechanical coupling coefficient k_{31}^2 was also analysed and is shown in Figure 6b. The k_{31}^2 coefficient was obtained by post-processing of the results found using Mixture Rules and finite element homogenization techniques. Due to the behaviour of the dielectric coefficient, the coupling coefficient k_{31}^2 is also affected by the electrode configuration and, thus, a continuous electrode must be considered in the finite element model (FE/CE) in order to obtain results similar to those obtained using Mixture Rules technique (MR) for the active layer. Indeed, by considering a fingered electrode configuration (FE Active Layer in Figure 6b), a significantly smaller coupling coefficient may be predicted for the active layer by the finite element homogenization, although the difference decreases substantially for increasing FVF. For FVF=86%, the coupling coefficient predicted with the fingered electrode model is 1.5% smaller than the one obtained with the continuous electrode model.

Figure 6 also shows the electromechanical coupling coefficient for the full d_{31} MFC where it is possible to notice that the inclusion of the protective and electrode layers reduce substantially, by around 39%, the effective coupling coefficient, compared to the active layer. It is also possible to observe that the k_{31}^2 coefficient is not substantially augmented by increasing the FVF from 86% ($k_{31}^2 = 6.7\%$) up to 95% ($k_{31}^2 = 6.8\%$).

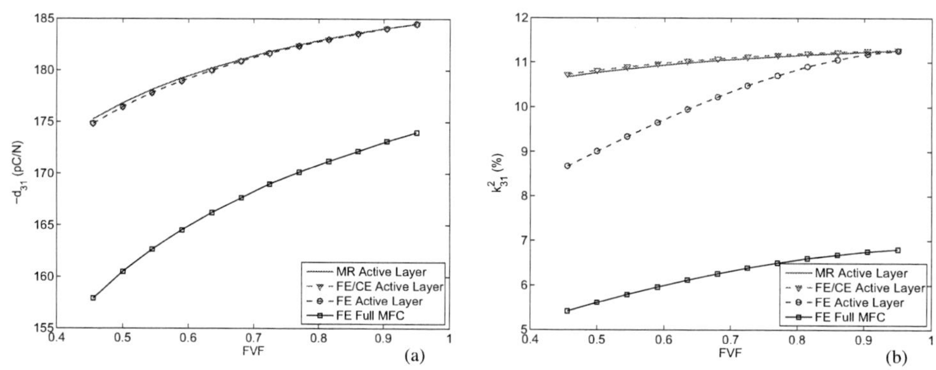

Figure 6. Piezoelectric coefficient d_{31} (a) and electromechanical coupling coefficient k_{31}^2 (b) as function of the fibre volume fraction. Comparison between results found using mixture rules (active layer only) and finite element (active layer and full d_{31} MFC).

3. d_{15} MFC EFFECTIVE MATERIAL PROPERTIES EVALUATION AND ANALYSIS

The d_{15} thickness-shear mode of piezoceramics requires that poling vector P and main electric field E to be perpendicular to each other. For practical reasons, the electric field is normally applied/measured in the thickness direction and, thus, the poling direction should be in the longitudinal direction, as shown in Figure 2. In the present work, the lengthwise d_{15} recently prototyped [4], shown schematically in Figure 2b is studied.

As shown in Figure 2, the piezoelectric material is considered to be poled in x_3 direction with electrodes fully covering top (x_1^+) and bottom (x_1^-) surfaces such that a preferential or dominant x_1 direction is imposed for the electric field and displacement. Therefore, it can be assumed that $E_2 = E_3 = 0$ and, hence, reduced constitutive equations can be written in either e-form or d-form types, respectively, as

$$\left\{\begin{array}{c} T_1 \\ T_2 \\ T_3 \\ T_4 \\ T_5 \\ T_6 \\ \hline D_1 \end{array}\right\} = \left[\begin{array}{cccccc|c} c_{11}^E & c_{12}^E & c_{13}^E & 0 & 0 & 0 & 0 \\ c_{12}^E & c_{22}^E & c_{23}^E & 0 & 0 & 0 & 0 \\ c_{13}^E & c_{23}^E & c_{33}^E & 0 & 0 & 0 & 0 \\ 0 & 0 & 0 & c_{44}^E & 0 & 0 & 0 \\ 0 & 0 & 0 & 0 & c_{55}^E & 0 & -e_{15} \\ 0 & 0 & 0 & 0 & 0 & c_{66}^E & 0 \\ \hline 0 & 0 & 0 & 0 & e_{15} & 0 & \epsilon_{11}^S \end{array}\right] \left\{\begin{array}{c} S_1 \\ S_2 \\ S_3 \\ S_4 \\ S_5 \\ S_6 \\ \hline E_1 \end{array}\right\}, \quad (3)$$

where T_p and S_q, with $p,q = 1,\dots,6$, denote the six components of mechanical stress and strain vectors in Voigt notation. D_1 and E_1 denote the electric displacement and field. For the e-form, c_{pq}^E, e_{15} and ϵ_{11}^S denote the elastic stiffness (at constant electric fields), piezoelectric and electric permittivity (at constant mechanical strains) coefficients. It can be noticed that the electromechanical coupling is then between the electric field and displacement in x_1 direction, E_1 and D_1, and $x_1 - x_3$ shear stress and strain, T_5 and

S_5. Therefore, it is essential to evaluate the piezoelectric d_{15}/e_{15}, elastic c^E_{55}/s^E_{55} and dielectric $\epsilon^T_{11}/\epsilon^S_{11}$ coefficients in order to qualify these materials as sensors and actuators.

A finite element homogenization technique is applied to the lengthwise d_{15} MFC recently proposed [4]. The RVE considered in this case is shown in Figure 7 and for which the boundaries considered for the finite element analysis are $X_1^-: x_1 = 0$, $X_1^+: x_1 = h_P + 2(h_N + h_C + h_K)$, $X_2^-: x_2 = 0$, $X_2^+: x_2 = w_P + 2w_E$, $X_3^-: x_3 = 0$, $X_3^+: x_3 = 2L_C + L_E$. In addition, electric potential boundary conditions are imposed at internal electrodes on surfaces $X_1^{e-}: x_1 = h_K + h_C + h_N$ and $X_1^{e+}: x_1 = h_K + h_C + h_N + h_P$.

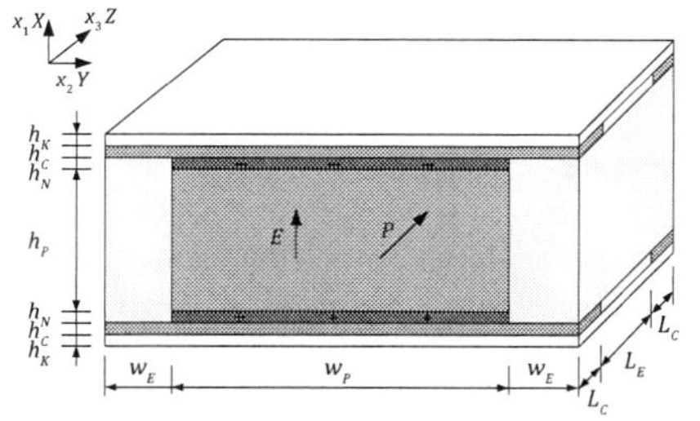

Figure 7. RVE considered for the lengthwise d_{15} MFC recently proposed in [4].

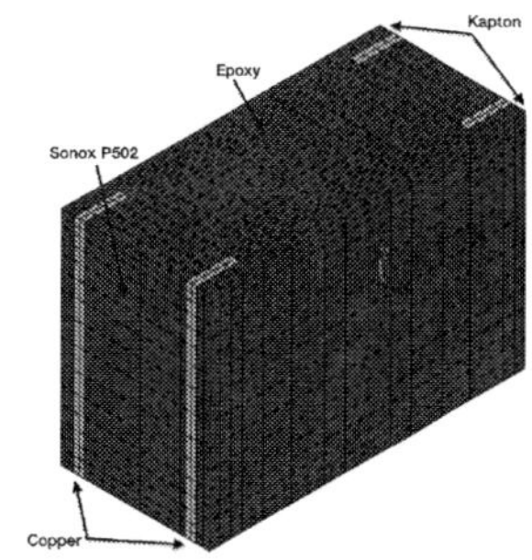

Figure 8. Isometric view of the FE mesh for the RVE of the Lengthwise d_{15} MFC in ANSYS.

Following the procedure presented for the d_{31} MFC, the elastic c^E_{55} and piezoelectric e_{15} coefficients are approximated by imposing xz shear strains to the RVE and evaluating, respectively, the resulting average shear stress T_5 and electric displacement D_1. The dielectric coefficient of the d-form, ϵ^T_{11}, is approximated by imposing unitary through-thickness electric field to a mechanically free RVE and evaluating the resulting electric displacement D_1. As before, symmetric (periodic) mechanical and electric boundary conditions are also considered.

Figure 8 shows the finite element mesh used in Ansys$^{(R)}$ for the 5-layered d_{15} MFC RVE using a fibre volume fraction (FVF) of 0.86 for the PZT+Epoxy Active Layer. The 3D 20-Node coupled-field solid finite element SOLID226 was used to mesh all volumes. It has the three displacement translations along the Cartesian coordinate system and the electric potential as nodal degrees of freedom. 7392 finite elements were used considering 21 divisions in x_1 direction (13 divisions for the PZT layer and 2 divisions for each electrode and protective layer), 16 divisions in x_2 direction (12 divisions for the PZT layer and 2 divisions for each Epoxy layer) and 22 divisions in x_3 direction (14 divisions for the Epoxy layer and 4 divisions for each electrode layer). According to Figure 7, the dimensions considered for this RVE are $h_P = 180$ μm, $h_K = 25$ μm, $h_N = 2$ μm and $h_C = 18$ μm, $L_E = 420$ μm, $L_C = 80$ μm, $w_P = 350$ μm, $w_E = 27.5$ μm [4]. The material properties are the same presented previously for the d_{31} MFC. Due to their very small thicknesses, the Nickel electrode layers over upper and lower surfaces of the piezoceramic fibre are neglected. Instead, electric boundary conditions were imposed on the corresponding surfaces of the fibre to emulate the equipotential condition that would be provided by the Nickel electrodes.

Table 2 shows the effective properties, transformed to Voigt notation, of the d_{15} MFC using a fibre volume fraction of 0.86 obtained by the finite element local problems. The effective electromechanical coupling coefficient of the RVE k_{15} is evaluated from the effective elastic, strain piezoelectric and free dielectric shear-mode response properties, $s^E_{55} = 1/G_{13}$, d_{15} and ϵ^T_{11}, using $k^2_{15} = d^2_{15}/s^E_{55}\epsilon^T_{11}$.

Table 2 also shows that, compared to the piezoceramic fibre alone, the strain piezoelectric d_{15} and free dielectric ϵ^T_{11} coefficients of the d_{15} MFC are not significantly reduced by the MFC packaging with 1% and 13% reduction, respectively. On the other hand, the piezoelectric coefficient e_{15} is reduced by 90%, which is also the reduction in the shear modulus G_{13}. Therefore, it is suggested here that smaller reductions in the last two coefficients could be obtained by stiffer protective layers or shear stiffeners in the MFC. These effective properties yield a much smaller electromechanical coupling coefficient k_{15} than the PZT (Sonox P502) alone, such that k^2_{15} decreases by 88%, from 55% to 6.3%.

Table 2. Effective short-circuit material properties of d_{15} MFC compared to the PZT fibre alone.

d_{15} MFC	Sonox P502
Y_1 = 11.46 GPa	Y_1 = 54.05 GPa
Y_2 = 22.22 GPa	Y_2 = 54.05 GPa
Y_3 = 31.34 GPa	Y_3 = 48.31 GPa
G_{23} = 6.86 GPa	G_{23} = 30.12 GPa
G_{13} = 3.10 GPa	G_{13} = 30.12 GPa
G_{12} = 2.54 GPa	G_{12} = 19.12 GPa
d_{15} = 554.02 pC/N	d_{15} = 560.00 pC/N
e_{15} = 1.72 C/m^2	e_{15} = 16.87 C/m^2
ϵ_{11}^T = 15.02 nF/m	ϵ_{11}^T = 17.27 nF/m
k_{15}^2 = 6.3 %	k_{15}^2 = 55 %

3.1 Evaluation of electric field dependence of d_{15} MFC effective material properties

In the present work, it is also desired to represent the effective piezoelectric and dielectric coefficients increase with applied electric field. Mueller and Zhang [11] presented experimental observation of the electric field dependence for various soft PZTs and proposed the following power law that seems to fit reasonably well their experimental data for all materials they studied, above the threshold electric field of 10 V/mm,

$$d_{15}(E_1) = d_{15}^{lin}[1 + (\gamma_d E_1)^\beta], \qquad \epsilon_{11}^T(E_1) = \epsilon_{11}^{T\,lin}[1 + (\gamma_\epsilon E_1)^\beta]. \tag{4}$$

An effective exponent β of 1.2 ± 0.1 seems to fit well the electric field dependence of all materials considered and, thus, the nonlinear coefficients γ_d and γ_ϵ, for piezoelectric d_{15} and dielectric ϵ_{11}^T coefficients, respectively, could be evaluated from a curve-fit of the experimental data. Since the electric field dependence of the Sonox P502 is not known, to the authors knowledge, the nonlinear parameters of PZT-5A, which is a similar piezoceramic material, were used $\gamma_d = \gamma_\epsilon = 2.8\ 10^{-3}$ mm/V and $\beta = 1.2$. The linear (or low-field) piezoelectric d_{15}^{lin} and dielectric $\epsilon_{11}^{T\,lin}$ coefficients for the Sonox P502 were the same used previously, that is $d_{15}^{lin} = 560$ pC/N and $\epsilon_{11}^{T\,lin} = 1950\epsilon_0$.

Then, the finite element homogenization technique presented previously is combined with the power laws (4), as suggested by Mueller and Zhang [11], to account for the electric field dependence of fibre material properties. This is done here by performing a finite element homogenization for each value of applied electric field in the range $\{0, 500\}$ V/mm, for which the Sonox P502 fibre material properties were updated following the power law presented previously (4).

Then, d_{15} MFC effective material properties were evaluated as functions of the applied electric field E_1. As expected, effect was only observed for xz shear related material coefficients d_{15}, e_{15} and ϵ_{11}^T. However, as shown in Figure 9, the effective piezoelectric coefficient e_{15} increases substantially with increased electric field. Figure 9a also allows to compare the MFC effective e_{15} with the one of the fibre alone. It is noticeable that they increase in a similar manner but in a totally different scale and there seems to be always a factor of ten between the fibre e_{15} and the MFC effective e_{15}.

A similar analysis was performed for the piezoelectric d_{15} and dielectric ϵ_{11}^T coefficients, both for MFC and fibre, and their dependence on the electric field. Figures 9b and 10 show that the MFC effective properties d_{15} and ϵ_{11}^T increase substantially with increased electric field, following the increase of the PZT fibre properties. However, differently from the previous e_{15} case, the values for MFC and PZT fibre are of the same order. Indeed, from Figure 9b, it can be observed that the MFC effective d_{15} coefficient is very close to the fibre one, meaning that this coefficient is much less affected by packaging and, thus, may fully profit from the increase of PZT fibre d_{15} coefficient. As for the free dielectric coefficient ϵ_{11}^T, it can be noticed that its MFC effective value is always somewhat reduced, compared to the fibre corresponding coefficient, due to packaging (Figure 10a).

In order to evaluate attainable energy conversion performance, the linear definition of the electromechanical coupling coefficient k_{15}^2 may be used in terms of the varying d_{15} and ϵ_{11}^T coefficients as function of the electric field. Notice that this is only an approximative analysis for the k_{15}^2 since it does not correspond to the efficiency in energy conversion through a full energy cycle. Figure 10b shows that the k_{15}^2

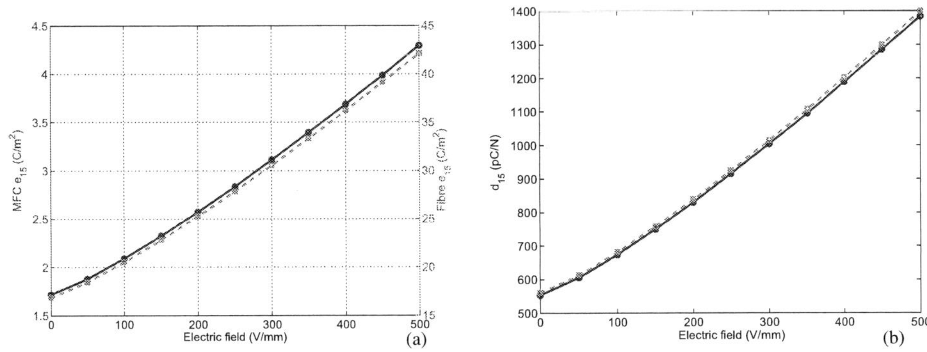

Figure 9. Piezoelectric coefficients e_{15} (left) and d_{15} (right), of MFC (solid) and PZT fibre (dashed), as function of electric field E_1.

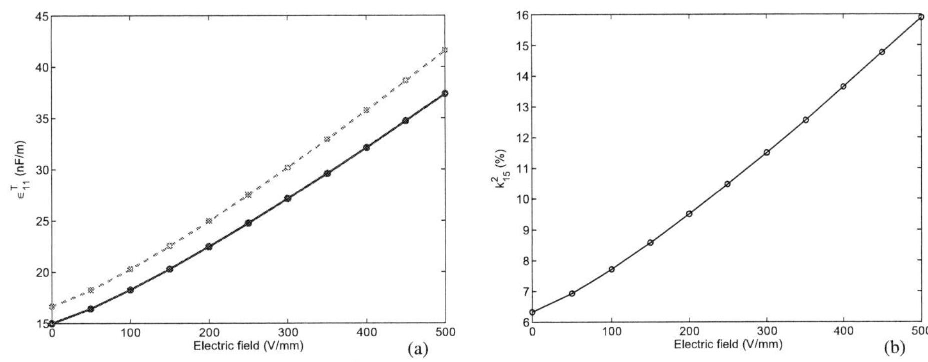

Figure 10. Free dielectric coefficient ϵ^T_{11}, of MFC (solid) and PZT fibre (dashed), and electromechanical coupling coefficient k^2_{15} as function of electric field E_1.

can increase from 6.3% up to 16%.

3.2 Comparison with experimentally observed nonlinear behaviour

The numerical procedure presented previously to evaluate the electric field dependence of d_{15} MFC effective material properties was tested in comparison with experimentally observed nonlinear behaviour of a d_{15} MFC manufactured by Smart Material Co. An experimental test was performed to evaluate the pure shear deformation capability of the d_{15} MFC when subjected to increasing values of applied voltage. This was done by applying a known electric voltage to a MFC transducer which was bonded to a rigid base on one side and to a small rigid block on the other side. The displacement of the small block was measured using a laser vibrometer and then plot against the applied electric voltage. More details on the experimental tests and results can be found in [4].

In order to compare the experimental results with those predicted by the finite element homogenization combined to the power laws (4) for the electric field dependence of fibre material properties, the following approximation was considered for the shearing mechanical displacement of the free surface $u \approx S_5 h_P \approx d_{15}(E_1)E_1 h_P$, where it is considered that all periodic cells of the MFC are free of shear stresses and present the same displacement. A schematic representation of the approximation to the experimental test is depicted in Figure 11. Notice that, although the voltage applied to the MFC transducer is quite low (between 5 and 90 V), the corresponding electric field is substantial since the transducer is very thin ($h_P = 180$ μm). Indeed, $E_1 = V_{ac}/180\,10^{-3} = 5.56V_{ac}$ (V/mm), which means that only 2 V are enough to put the material above the threshold electric field value of 10 V/mm (100 V/cm).

The predicted response for the shearing displacement induced by the applied electric voltage together with the experimental results presented in [4] are shown in Figure 12. It can be noticed that the predicted

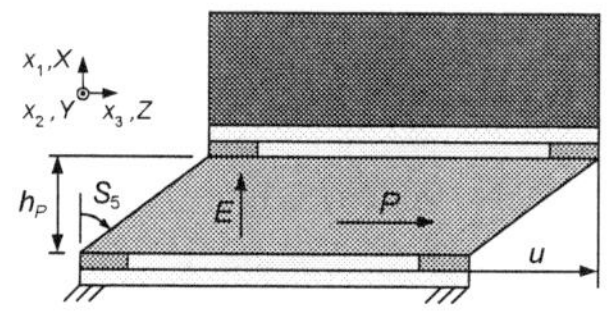

Figure 11. Sketch of the experimental test [4] to measure effective piezoelectric material coefficient through displacement induced by an applied voltage.

response using the nonlinear field-dependent laws (4) is closer to the experimental one, although the nonlinear behaviour of the PZT fibre (Sonox P502) was not known and, thus, is only approximated by that of the similar piezoceramic material PZT-5A. Nevertheless, it is suggested here that these results at least indicate that most of the nonlinear behaviour observed in experimental tests with the d_{15} MFC could be explained by a nonlinear behaviour of the PZT fibres. Notice also that the predicted response using constant low-field coefficients for the fibres (Figure 12) is very far from the experimental results.

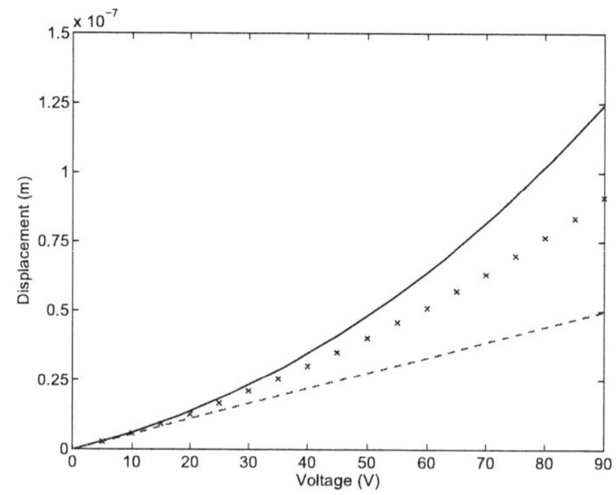

Figure 12. Shearing displacement versus applied voltage amplitude: experimental (x) predicted response evaluated using nonlinear field-dependent (solid) and constant low-field coefficients (dashed).

3.3 Effect of active layer thickness on MFC performance

From the analysis presented previously, it becomes evident that the shearing displacement provided by the d_{15} transducer could be increased by augmenting the active layer thickness. This can be observed in Figure 13a that shows the displacement versus the applied electric field for increasing values of the fibre thickness $\{180, 360, 540\}$ μm, while maintaining the other geometric properties and design of the transducer. This is, of course, primarily due to the direct increase of the effective shearing displacement induced for a given shear deformation. However, it is also due to the fact that the PZT-to-Kapton thickness ratio is increased and, thus, the effective material properties are changed. It is also suggested that for higher values of PZT-to-Kapton thickness ratio, the effective shear modulus of the transducer is increased and, thus, the transducer might be more adequate for actuation. To illustrate this effect, the e_{15} piezoelectric coefficient was evaluated in terms of the applied electric field for the three fibre thicknesses considered (Figure 13b). The results indicate that indeed there may be an increase in the actuation performance of the d_{15} MFC for thicker active layers. Notice, however, that the electric voltage required to induce such electric fields also increases with the fibre thickness and can reach up to 270 V (for 500 V/mm applied to a 540 μm thick active layer).

4. CONCLUDING REMARKS

This work has presented recent results on the characterization of effective properties of d_{31} and d_{15} piezoelectric MFCs using a finite element homogenization technique, accounting in particular for the

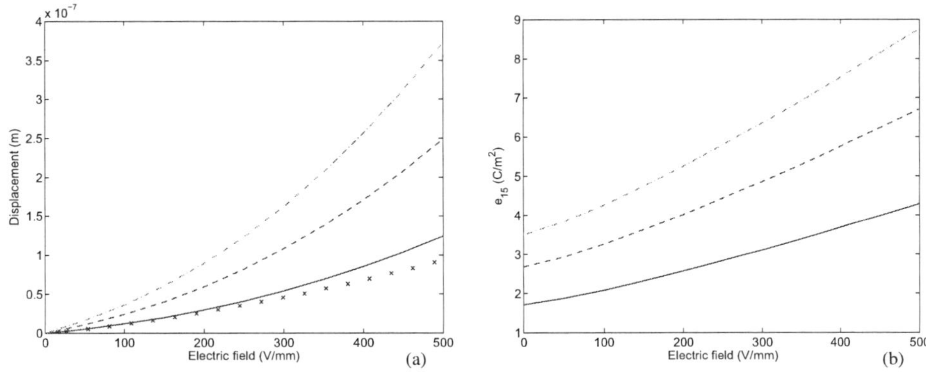

Figure 13. Shearing displacement versus applied electric field (left) and e_{15} coefficient (right) for various values of PZT thickness h_P in μm: 180 (solid), 360 (dashed), 540 (dash-dot).

effect of electrode and protective layers on the effective elastic, piezoelectric and dielectric coefficients. It was shown that electric boundary conditions (electrodes) play an important role in the predicted response of the MFC and that nonlinear response may be reasonably predicted using piezoelectric fibre electric field dependence. It was also shown that the mechanical behaviour of electrode and protective layers must be accounted for in order to well predict the MFC effective properties.

ACKNOWLEDGEMENTS

Support of MCT/CNPq/FAPEMIG National Institute of Science and Technology on Smart Structures in Engineering, grant 574001/2008-5, is acknowledged.

REFERENCES

[1] W. K. Wilkie, G. R. Bryant, and J. W. High. Low-cost piezocomposite actuator for structural control applications. In *Proceedings of SPIE Int. Symp. on Smart Structures and Materials*, Newport Beach, 2000.

[2] S. Raja and T. Ikeda. Concept and electro-elastic modeling of shear actuated fiber composite using micro-mechanics approach. *Journal of Intelligent Material Systems and Structures*, 19:1173–1183, 2008.

[3] A. Benjeddou. Shear-mode piezoceramic advanced materials and structures: A state of the art. *Mechanics of Advanced Materials and Structures*, 14(4):263–275, 2007.

[4] B. Kranz, A. Benjeddou, and W.-G. Drossel. Numerical and experimental characterizations of longitudinally polarized piezoelectric d_{15} shear macro-fiber composites. *Acta Mechanica*, 2013.

[5] A. Deraemaeker, H. Nasser, A. Benjeddou, and A. Preumont. Mixing rules for the piezoelectric properties of macro fiber composites. *Journal of Intelligent Material Systems and Structures*, 20(12):1475–1482, 2009.

[6] A. Deraemaeker and H. Nasser. Numerical evaluation of the equivalent properties of macro fiber composite (MFC) transducers using periodic homogenization. *International Journal of Solids and Structures*, 47(24):3272–3285, 2010.

[7] A. Benjeddou and M. Al-Ajmi. Analytical homogenizations of piezoceramic d_{15} shear macro-fibre composites. In M. Kuna and A. Ricoeur, editors, *Proceedings of the IUTAM Conference on Multiscale Modelling of Fatigue, Damage and Fracture in Smart Materials*, Freiburg (Germany), 2009.

[8] M. A. Trindade and A. Benjeddou. Finite element homogenization technique for the characterization of d_{15} shear piezoelectric macro-fibre composites. *Smart Materials and Structures*, 20(7):075012, 2011.

[9] M. A. Trindade and A. Benjeddou. Finite element characterization and parametric analysis of the nonlinear behaviour of an actual d_{15} shear MFC. *Acta Mechanica*, 2013.

[10] Smart Material Corp. Macro-fibre composites technical data. www.smart-material.com, Sarasota, 2009.

[11] V. Mueller and Q. M. Zhang. Shear response of lead zirconate titanate piezoceramics. *Journal of Applied Physics*, 83(7):3754–3761, 1998.

Energy Harvesting from Flow-Induced Waving Motion of a Piezoelectric Grass Array

Jared D. Hobeck[1*] & Daniel J. Inman[2]

[1,2] Department of Aerospace Engineering, University of Michigan, Ann Arbor, USA

Abstract

This paper discusses the design and development of a novel deploy-and-forget, lightweight, robust energy harvester for remote, unmanned, and uncontrolled environments. The biologically inspired harvester design concept consists of an array of over one hundred piezoelectric cantilevers, and has been aptly named *piezoelectric grass*. Presented in this paper are the results of a thorough experimental analysis on a piezoelectric grass harvester prototype. These results include trends in power output as a function of flow speed, number of array elements, size of array elements, and packing density. It will be shown that for a given array configuration an output of approximately 1.9 watts per square meter of area covered by the array can be achieved at a mean flow velocity of 14.3 m/s. Modeling techniques and theory of operation will also be discussed.

1. INTRODUCTION

Recent work by the authors has shown that significant amounts of useful power can be extracted from piezoelectric cantilevers in highly turbulent cross-flow [1–4]. Until now the focus of earlier studies has been on experimental and theoretical analysis of single cantilever harvesters. This paper will focus on the dynamics of large arrays of up to 116 cantilevers.

The proposed harvester prototype, called *piezoelectric grass,* is designed specifically to extract energy from low velocity, highly turbulent, or highly unsteady fluid flows. Such fluid flow environments include small rivers, streams, tidal flows, ducts, and ventilation systems. Although the initial interest of this work was for water applications, early experiments showed that piezoelectric grass had exceptional performance in air.

A primary function of the energy harvester is to eliminate the need for batteries and provide power used for autonomous self-sustaining sensor networks. Specific application areas for the piezoelectric grass harvester include powering surveillance and security systems, and sensor networks for structural, environmental, and industrial monitoring. An example application area of the piezoelectric grass concept is illustrated in figure 1.

Initial proof of concept experiments were able to show that the dynamic characteristics of large arrays of flexible structures placed in a flowing fluid are extremely attractive for flow-induced vibration energy harvesting applications. The most advantageous feature of these large array dynamics is that at a certain

* Corresponding author, email: jdhobeck@umich.edu

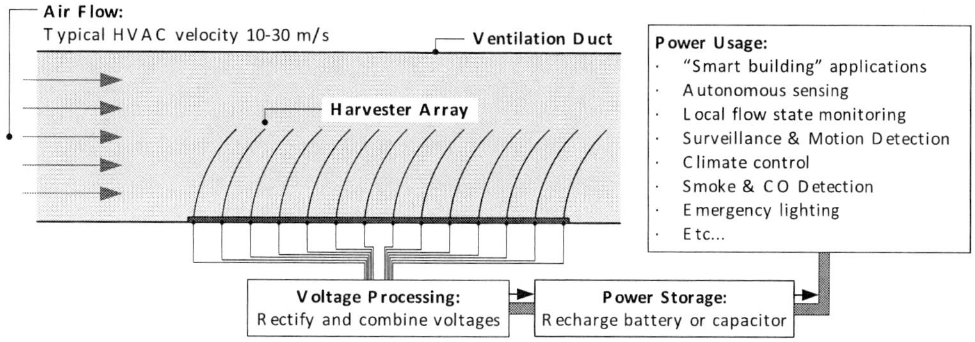

Figure 1. Conceptual schematic showing and listing potential application areas for *piezoelectric grass* harvester arrays.

flow velocity the entire array experiences a resonance condition where all array elements undergo persistent large amplitude vibrations. It has been shown that this resonance state is achieved and maintained over a large range of flow velocities even when the flow is spoiled with high-intensity turbulence [4]. This robust flow-induced vibration excitation mechanism was observed decades ago in waving wheat crops; however, it has never before been used for energy harvesting applications.

2. THEORY OF OPERATION

The earliest experimental investigation on the dynamics of large arrays of flexible structures in flow was presented by Inoue in 1955 [5]. This early investigation focused on the formation of large amplitude (sometimes damaging) vibrations in crops such as wheat, grass or corn caused by wind. These vibrations are what give a field of grass the appearance of ocean-like waves traveling through it on a windy day. Inoue called this waving phenomenon 'honami'.

More than two decades after Inoue documented his observations, Finnigan *et al* presented results from an extensive experimental study on the waving motions of wheat [6–9]. Finnigan showed that large coherent structures develop and travel across the top of the wheat with a periodicity of 5 to 8 times the height of the wheat. He proposed that vortices having a width several times the wheat height travel across the top of the wheat at a velocity of approximately twice the mean free-stream air velocity. When the vortex passing frequency equals or becomes close to the natural frequency of the wheat, a resonance condition occurs causing large waving motions.

The following relationship can be made from Finnigan's observations,

$$\omega_v = 2\pi\left(\frac{\overline{U}k}{L}\right) \quad for \quad k = (0.4 - 0.25) \tag{1}$$

where w_v is the vortex passing frequency in rad/s, and k represents an experimentally determined coefficient which provides upper and lower bounds of the vortex passing frequency at a given mean flow velocity $\overline{U}$. Because the waving motion in the array is a result of array elements being forced into a resonance condition, it is important that all cantilevers in the array have the same fundamental frequency as well as the same length. The energy harvester presented in this work can therefore be modeled as an array of similar cantilevers all having a length L and a fundamental frequency w_1. For a given application in an environment where the average air velocity is known, a harvester can easily be designed such that its fundamental frequency matches the vortex passing frequency e.g., $w_1 = w_v$. The first energy harvesting investigation that took advantage of this type of flow-induced vibration was presented by the authors in 2012 [4].

3. EXPERIMENTAL METHODS

Numerous wind tunnel experiments were performed on the piezoelectric grass harvester. The purpose of these experiments was to identify and quantify trends in power output versus flow velocity for many different array configurations. This section discusses experimental details such as: array materials and design, terminology and coordinate system, data processing, testing facilities, instrumentation, circuitry, and software.

3.1. Wind Tunnel Facilities

All experiments presented in this paper were performed in an open loop, single stage wind tunnel. The test section has a length of 1.22 m, and has a cross-section of 30.5 x 30.5 cm. With exception of the flow straightener and contraction section, the authors designed the entire wind tunnel and assembled all of its components. (See figure 2.) Fan speed or flow velocity can be controlled manually or automatically using a 3-phase, 3-hp, variable frequency drive (VFD) controller with build-in PID feedback control (Mitsuishi Electric, Model: FR-E720-110-NA).

The wind tunnel is able to produce flow velocities ranging from 0 to 35 m/s (approx. 80 mph). Accurate velocity measurements can be made from 35 m/s down to 0.2 m/s using a pitot static tube attached to a differential pressure transducer with a full range of +/-748 Pa (Omega, Model: PX653-03D5V). LabVIEW development software was used to design a virtual control panel which allows for a PC to interface with National Instruments hardware for data acquisition and control. A 14-bit, multifunction DAQ with 8 analog inputs and 2 analog outputs (National Instruments, Model: USB-6009) was used as the dedicated data acquisition and control hardware. The LabVIEW virtual control panel was used for displaying, controlling, and saving real-time data from the pressure transducer and VFD along with any other external inputs

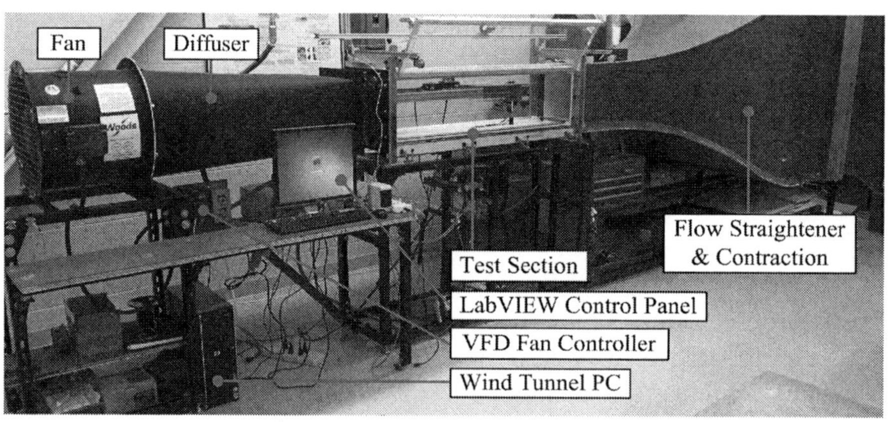

Figure 2. Wind tunnel testing facility

Table 1. Design parameters for passive and active element types

Type	Layer	Material	Length (cm)	Width (cm)	Thickness (mm)	Natural Frequency (Hz)
Passive	Substrate	Steel	14.610	2.540	0.254	9.3
	Active layer	-	-	-	-	
Active	Substrate	Steel	14.610	2.540	0.254	16.3 ($M_t = 0$ gm)
	Active layer	PZT	4.295	2.096	0.254	9.3 ($M_t = 2.9$ gm)

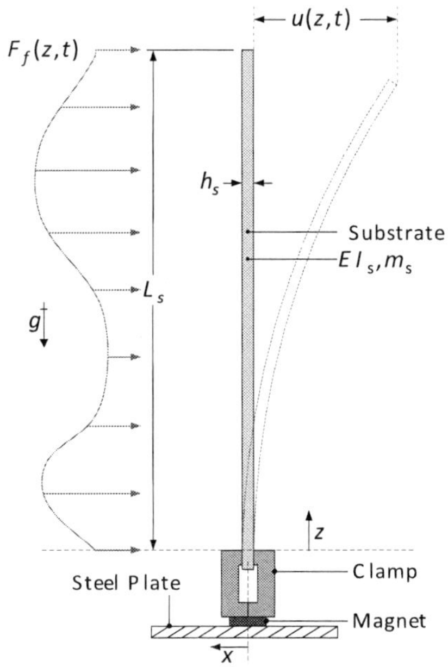

Figure 3. Schematic of passive array element

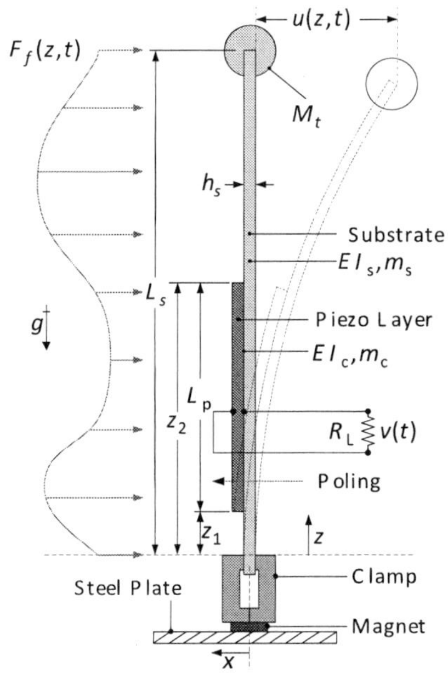

Figure 4. Schematic of active array element with added tip mass and piezoelectric material.

3.2. Array Design & Configuration

Considering factors such as assembly time, material expense, availability of data acquisition hardware, and quantity of data, the authors decided against building an entire array of piezoelectric cantilevers. Rather, only ten cantilevers consisted of a steel substrate with piezoelectric material bonded to them while all other cantilevers in the array were steel with uniform mass and stiffness. The cantilevers with piezoelectric material will be referred to as *active* array elements while all others will be called *passive*. Table 1 summarizes the design parameters for both element types.

Figure 3 and figure 4 show schematics of the passive and active array element assemblies respectively. Compared to the passive elements, active elements had increased mass and bending stiffness caused by the bonded piezoelectric material. This increased stiffness caused an increase in natural frequency. Because all elements must have approximately the same natural frequency for array resonance (*honami*) to occur, a tip mass was added to all active elements. Each tip mass was adjusted until the active

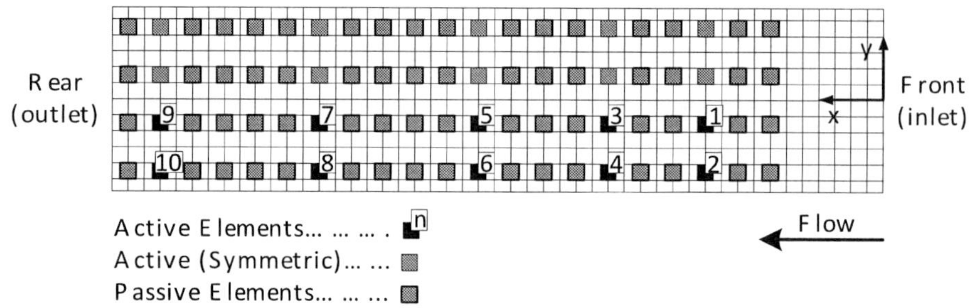

Figure 5. Schematic of a top view of the steel plate (drawn to scale) showing the coordinate system, the 2.54 cm square grid, and the locations of all array elements for a given array configuration.

elements had the same natural frequency as the passive elements. A tip mass of approximately 2.9 grams was found reduced the natural frequency of the active elements from 16.3 Hz to 9.3 Hz.

All array elements were fixed in individual clamps, and all clamps were attached to a high-grade neodymium magnet. The magnets provided enough attractive force to a steel plate to hold all elements in place during a test, yet allowed for convenient repositioning between tests. The steel plate was centered on and attached to the test section floor and had a 2.54 cm square grid drawn on it for a reference. The grid was used as the x-y plane where the positive x direction was the same as the flow direction. Figure 5 shows the coordinate system and defines common terminology used for all large array tests presented in this paper.

In order to estimate the total power output that could be generated from a given array, active elements were systematically positioned among passive elements. The following is a list of observations made from initial large array experiments. Included below each of these observations is the corresponding conclusion used to determine the placement of active elements for all array configurations in this study.

a) The first row produced a negligible amount of power compared to all other rows:
  ⇒ No active elements were positioned on the first row.

b) Power output changed rapidly as a function of x-position toward the front of the array:
  ⇒ More active elements were placed toward the front of the array.

c) The array configuration and power output was symmetric about the y-axis:
  ⇒ Output from all active elements was assumed to be symmetric about the y-axis.

d) Power output toward the back of the array showed little change as a function of x-position:
  ⇒ Active element spacing increased as a function of x-position.

Figure 5 shows a typical array configuration where one can see how active elements were placed among passive elements in a manner determined by the previous list of conclusions.

3.3. General Testing Procedure

All elements were positioned on the steel plate in the wind tunnel test section to a specified x and y spacing. Then, the velocity was increased from 0 to 15 m/s in 15 evenly spaced increments. At each velocity increment, the wind tunnel was allowed come to steady average velocity before data was gathered. All time-series data, including wind tunnel speed settings, velocity, and load voltages from the ten active elements was sampled at 2.0 kHz for 60 seconds and recorded in a LabVIEW measurement file (.lvm). Data recording was then paused while the velocity was adjusted to the next velocity increment.

Table 2. Piezoelectric grass wind tunnel test matrix.

Parameter	No. of Increments	Range
Velocity	15	(0 - 15) m/s
x-spacing	10	(3.01 – 15.24) cm
y-spacing	5	(3.49 – 6.35) cm

This procedure was automated within LabVIEW. After all 15 velocity increments were completed, the next array configuration was set by adjusting the x or y spacing of the array. This procedure was repeated for ten different x-spacing values ranging from 3.01 cm to 15.24 cm, and at each x-spacing there were five y-spacing values ranging from 3.49 cm to 6.35 cm. In total, experiments were performed on 50 different array configurations which yielded 750 data sets, and 9.1 GB of data.

Table **2** gives a test parameter summary for all experiments performed during this study.

3.4. Data Processing

All post-processing of the raw data was performed using MATLAB software. Raw data refers to unfiltered, time-series voltages. These voltages were first digitally filtered, then converted into their respective physical properties via predetermined calibration equations. For each 60 second test, average flow velocity, fan speed, and the active element RMS load voltages were calculated and stored.

Recall that symmetry was assumed about the y-axis; therefore, the 10 active elements yielded 20 average power output values for a given array configuration and flow velocity. A smooth 3D surface was fitted to these 20 power values across the array using a *shape-preserving* piecewise cubic interpolation scheme. (See figure 8 and figure 9.) Average power output for all elements could then be estimated as if all passive elements were replaced with active elements having the same design parameters as those listed in table 1. For example: given the x-y location of a passive element, its average power output could be estimated from the interpolated 3D surface if that element were replaced with an active element. Finally, the average power values for all elements are combined to yield the total estimated power output for a given array configuration and flow velocity. This total power output is then normalized by the total area covered by the array to generate plots such as those in figure 10 and figure 11.

4. RESULTS

Given the large amount of data recorded in this study, results from all experiments will not be shown in this paper. Rather, the results presented in this section will focus first on only two (out of 50 total) specific array configurations that have notable differences. Then the results from 20 array configurations each with 15 velocity increments (300 tests total) will be summarized and discussed.

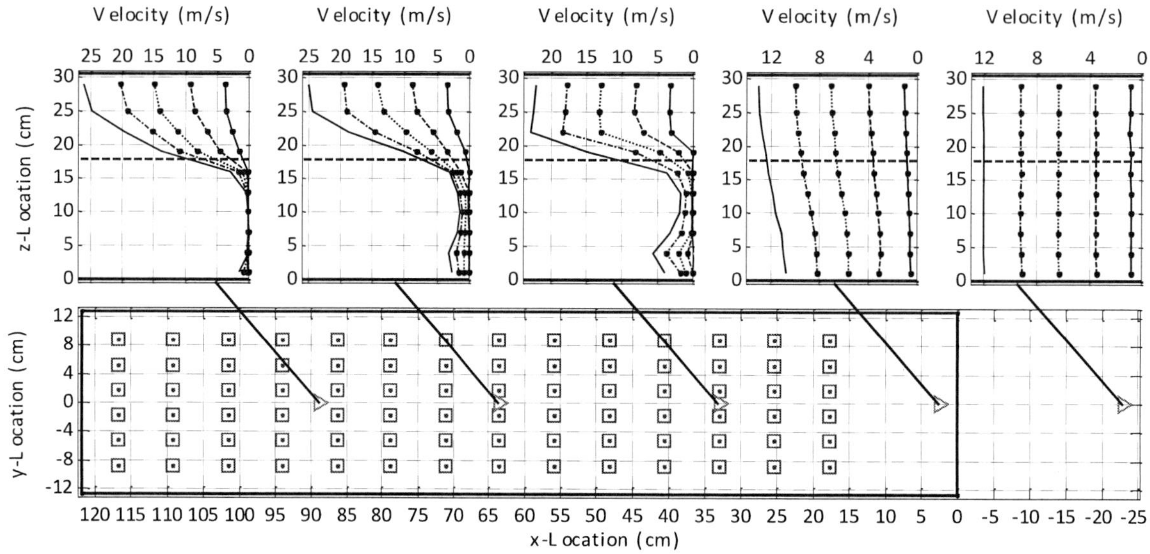

Figure 6. Velocity profiles as functions of array height shown at five x-locations in the center of a 14x6 element array having an x and y spacing of 7.62 cm and 3.49 cm respectively. The horizontal bold dotted line indicates the array element height while the horizontal bold solid lines indicate the floor and ceiling of the test section. Locations of the elements (blue squares) and the pitot tube tip (red triangles) are shown on the lower plot. The bold solid lines in the lower plot indicate the edge of the steel grid plate.

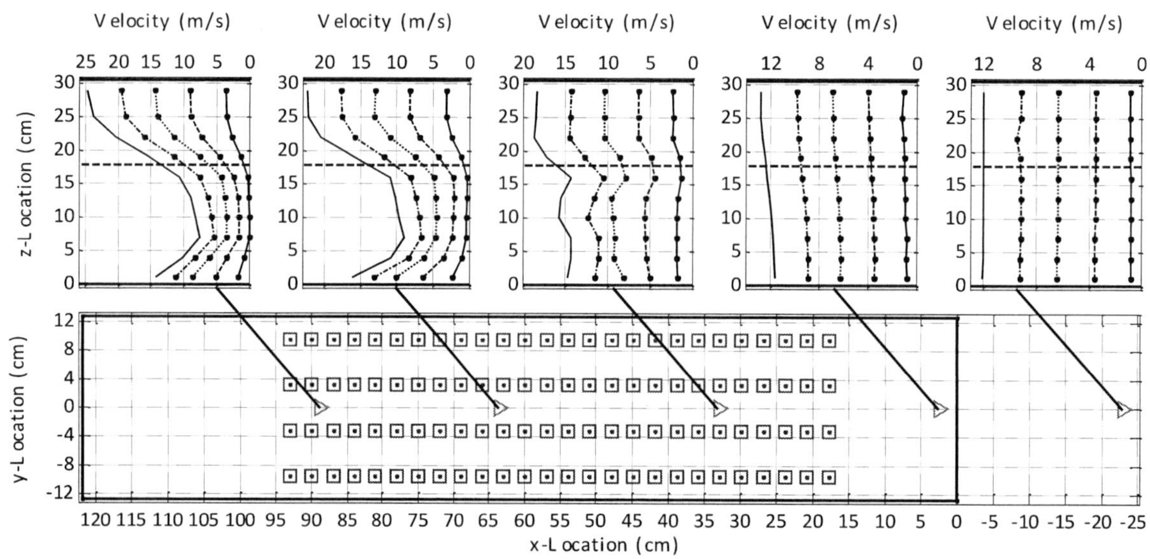

Figure 7. Velocity profiles as functions of array height shown at five x-locations in the center of a 26x4 element array having an x and y spacing of 3.01 cm and 6.35 cm respectively. The horizontal bold dotted line indicates the array element height while the horizontal bold solid lines indicate the floor and ceiling of the test section. Locations of the elements (blue squares) and the pitot tube tip (red triangles) are shown on the lower plot. The bold solid lines in the lower plot indicate the edge of the steel grid plate.

4.1. Test Cases: Single Array Configurations

Average velocity profiles and power output profiles for two different array configurations are presented here. Average velocity profiles were measured at 5 x-locations along the middle of the test section (at y = 0 cm). Each velocity profile was made by taking local average velocity measurements with the pitot-static tube at 10 evenly spaced z-locations. Velocity profiles were made at each x-location for 5 evenly spaced free-stream flow velocities.

4.1.1. Velocity profiles

Figure 6 and figure 7 show velocity profiles and their measurement locations throughout the test section. The horizontal bold dotted line in each of the velocity profile plots indicates the array element height while the horizontal bold solid lines indicate the floor and ceiling of the test section. The lower plots in figure 6 and figure 7 show locations of the elements denoted with blue squares while the pitot tube tip locations are shown as red triangles. The bold solid lines in the lower plots indicate the edge of the steel grid plate.

Velocity profile measurements from a 14x6 element array having an x and y spacing of 7.62 cm and 3.49 cm respectively are shown in figure 6. Notice that the velocity within the array drops drastically as the pitot tube is positioned further from the front of the array. Also, the profiles transition from straight vertical lines at the front of the array to highly skewed toward the rear. For the measurement location furthest from the front, the velocity within the array drops to nearly zero while the velocity above the array is approximately double that of the free-stream.

Velocity profile measurements from a 26x4 element array having an x and y spacing of 3.01 cm and 6.35 cm respectively are shown in figure 7. Much of the velocity profile trends in figure 7 are similar to those mentioned previously for the 14x6 element array. One major difference between the 14x6 and the 26x4 array is that the velocity profile in the 26x4 array is skewed much less. In other words, the velocity inside the array is much greater for the 26x4 configuration.

4.1.2. Power profiles

Using the procedure discussed in Section 3.4, smooth 3D surfaces were fitted to the average power output of each active element for all 750 data sets. Clearly it would not be practical to show the power profiles for every data set; therefore, only two profiles from array configurations similar to those mentioned previously will be presented here.

Figure 8 shows the results from one data set for the same 14x6 array discussed in Section 4.1.1 at a free-stream flow velocity of 14.3 m/s. Notice that drastic changes in power output as a function of x occur from x = 17.5 cm until approximately x = 55 cm. These results were consistent for numerous tests and agreed with the initial observations listed at the end of Section 3.2. The total estimated power output for the profile shown in figure 8 was 25.5 mW yielding an area power density of 0.148 W/m^2.

The power profile shown in figure 9 was generated from a 29x4 element array having an x and y spacing similar to the 26x4 array discussed in Section 4.1.1. The free-stream flow velocity for the 29x4 array results was set to 14.3 m/s. Again, the slope of the power profile is steep toward the front of the array then becomes more gradual toward the rear of the array. The total estimated power output for the profile shown in figure 9 was 303.9 mW yielding an area power density of 1.89 W/m^2.

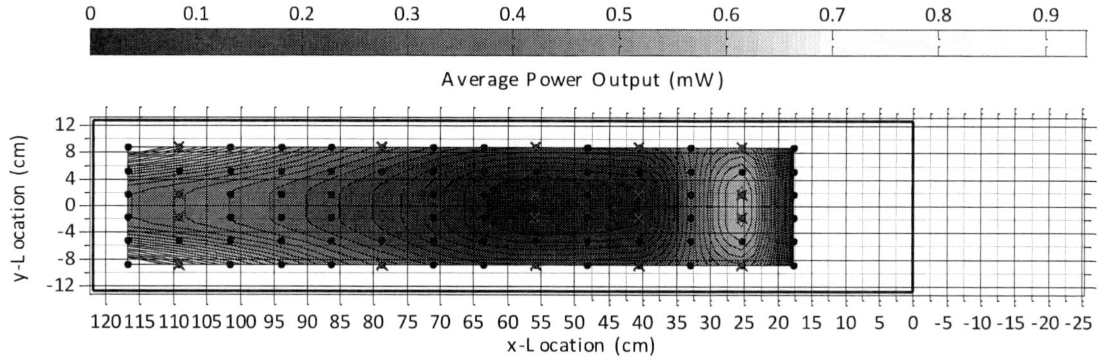

Figure 8. Typical power profile for a 14x6 element array in a mean flow velocity of 14.3 m/s showing locations of passive elements (black dots) and active elements (blue x's).

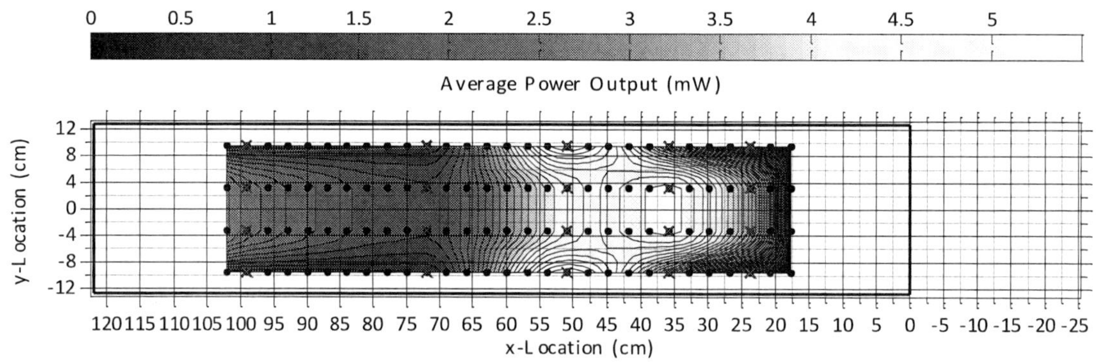

Figure 9. Typical power profile for a 29x4 element array in a mean flow velocity of 14.3 m/s showing locations of passive elements (black dots) and active elements (blue x's).

4.2. Test Cases: Multiple Array Configurations

The total average power output of each array configuration at each flow velocity was computed as described in Section 3.4. Each power profile (such as those shown in figure 8 and figure 9) was used to estimate total power that could be harvested from all elements in the array. Notice that the area of each power surface changes along with array element packing density. Packing density refers to the number of elements that are able to populate a square meter for a given configuration. In order to account for the changing area and packing density for each configuration, the total estimated power output was normalized by the area covered by each power profile. This normalized output can be thought of as an area power density of the harvester rather than the more conventional volume-based power density performance metric.

Figure 10 and figure 11 show the area power density results for arrays with varying x-spacing and having a y-spacing of 3.49 cm and 6.35 cm respectively. Note that the z-axis (out of the page) is represented in log scale. A single point on each of the power density plots was calculated from the arrays

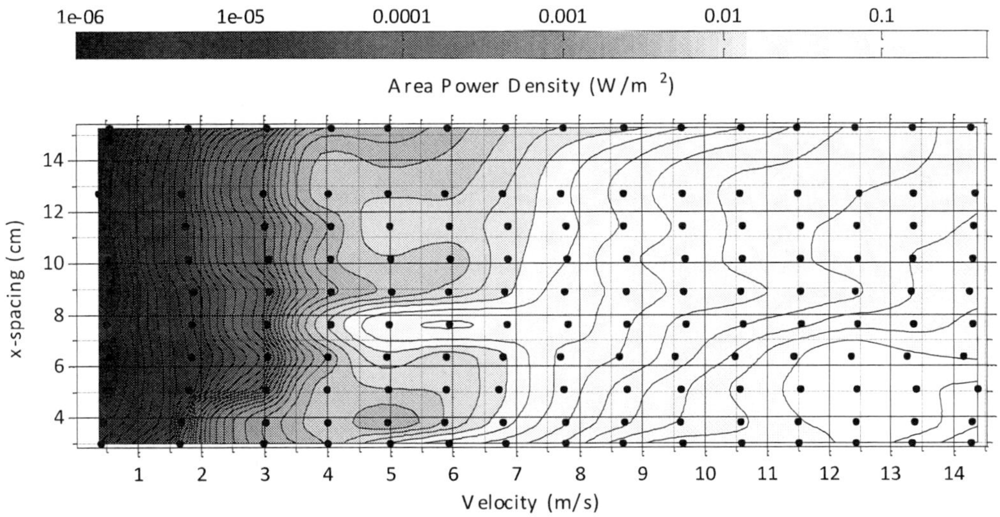

Figure 10. Total power output of array with a y-spacing of 3.49 cm

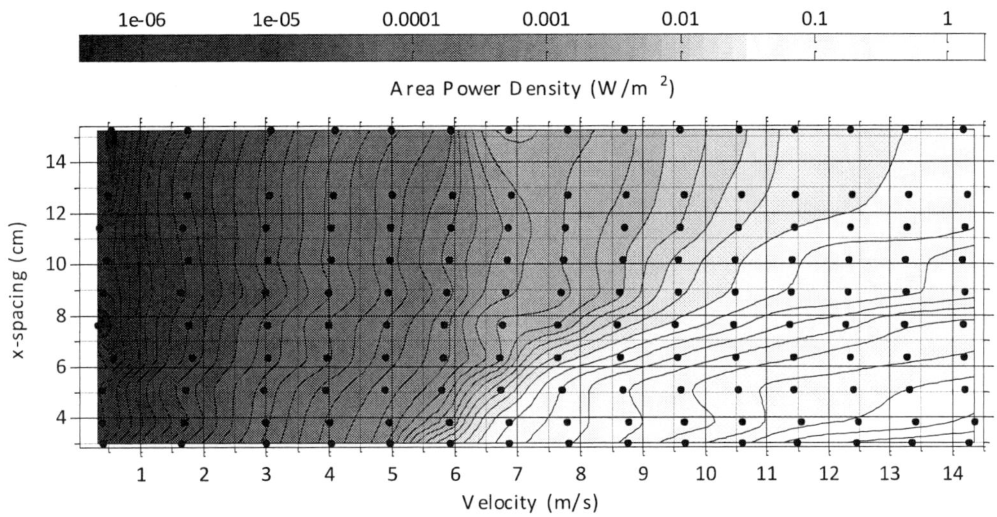

Figure 11. Total power output of array with a y-spacing of 6.35 cm

discussed in Sections 4.1.1 and 4.1.2. Each power density plot summarizes results from 10 array configurations each at 15 velocity increments.

Both power density plots show that maximum power density is attained by minimizing the spacing between array elements in the x-direction. This is not the same result; however, for the y-spacing. In fact, the opposite trend was observed where the power density increased as y-spacing increased. In other words, power density was shown to increase steadily as the number of cantilevers in the array was reduced. This is an interesting result, because it suggests that an optimum y-spacing exists. Investigating this optimum y-spacing, modeling, and scalability are currently primary focus of this work.

5. SUMMARY & CONCLUSIONS

This paper presented a brief summary of results and observations from a thorough experimental investigation of large arrays of a piezoelectric grass harvester prototype. Early investigations showed that when the flow velocity reaches a critical point, elements in the array are forced into a resonance condition where they experience large amplitude, persistent vibration. This resonance state or waving motion is called *honami* and has been found to be an extremely robust excitation mechanism for flow-induced vibration energy harvesting.

The piezoelectric grass arrays in this study were composed of passive and active elements. Passive elements were spring steel cantilevers with a length, width and thickness of 14.6 cm, 2.54 cm, and 0.254 mm respectively. Active elements were made by bonding a piezoelectric material to passive elements. A tip mass was added to the active elements in order to tune them to the same natural frequency as the passive elements.

Active elements were strategically positioned throughout the array in order to provide an average power output profile for the entire array. Total power estimates were then calculated for 50 different array configurations each for 15 velocity increments. A maximum estimated area power density of 1.89 W/m^2 was achieved. Results from the 750 tests showed trends in area power density versus velocity, x-spacing, and y-spacing. It was shown that maximum power density was achieved by minimizing the streamwise gap between elements (x-spacing) and maximizing the y-spacing. These results suggest that an optimum y-spacing exists. Future work will focus on determining this optimum spacing along with scalability of the array and modeling techniques.

Acknowledgements

The authors gratefully acknowledge the University of Michigan College of Engineering for funding this work and a special thanks to Andrew Gadway for performing numerous experiments and gathering much of the raw data which was processed and presented in the results section of this paper.

References

[1] Hobeck, J.D., and Inman, D.J., "Energy harvesting from turbulence-induced vibration in air flow: Artificial piezoelectric grass concept," *ASME 2011 Conference on Smart Materials, Adaptive Structures and Intelligent Systems, SMASIS 2011* **2**, 637–646, Scottsdale, AZ (2011).

[2] Hobeck, J.D., and Inman, D.J., "Artificial piezoelectric grass for energy harvesting from turbulence-induced vibration," *Smart Materials and Structures* **21**(10) (2012).

[3] Hobeck, J.D., and Inman, D.J., "Electromechanical and statistical modeling of turbulence-induced vibration for energy harvesting," *Proceedings of SPIE - The International Society for Optical Engineering* **8688**, 86881P1–86881P14, San Diego, CA (2013).

[4] Hobeck, J.D., and Inman, D.J., "The dynamics of large arrays of flexible structures in fluid flow with applications to energy harvesting," *Proceedings of the 25th ISMA Biennial International Conference on Noise and Vibration Engineering*, Leuven, Belgium (2012).

[5] Inoue, E., "Studies of the phenomena of waving plants (" honami") caused by wind. Part 1: Mechanism and characteristics of waving plants phenomena," *J. Agricult. Meteorol.(Japan)* **11**, 71–82 (1955).

[6] Finnigan, J., "Turbulence in Waving Wheat. I. Mean Statistics and Honami," *Boundary-Layer Meteorology* **16**(2), 181–211, Springer (1979).

[7] Finnigan, J., and Mulhearn, P., "Modelling waving crops in a wind tunnel," *Boundary-Layer Meteorology* **14**(2), 253–277, Springer (1978).

[8] Raupach, M., Finnigan, J., and Brunei, Y., "Coherent eddies and turbulence in vegetation canopies: the mixing-layer analogy," *Boundary-Layer Meteorology* **78**(3), 351–382, Springer (1996).

[9] Brunet, Y., Finnigan, J., and Raupach, M., "A wind tunnel study of air flow in waving wheat: single-point velocity statistics," *Boundary-Layer Meteorology* **70**(1), 95–132, Springer (1994).

Energy Harvesting for Sensors on using Free-Floating Flaps

Lars O. Bernhammer[1], Roeland De Breuker[1], Moti Karpel[2]

[1] Faculty of Aerospace Engineering, Delft University of Technology, Delft, The Netherlands
[2] Faculty of Aerospace Engineering, Technion, Haifa 32000, Israel

Abstract

In this paper, a novel configuration of an energy harvester for local sensing using limit cycle oscillations has been designed, modeled and tested. A wing section has been designed with two free-floating flaps. In the rotational axis of each flap a dynamo is mounted that converts the vibrational energy into electricity. It has been demonstrated numerically how a simple electronic system can be used to keep such a system at stable limit cycle oscillations by varying the resistance in the electric circuit. Additionally it was shown that the stability of the system is coupled to the charge of the battery, with increasing charge level leading to a less stable system.

1. INTRODUCTION

Fatigue is one of the main design drivers of wind turbine blades. A way to reduce these loads is by using so-called smart rotors, in which active aeroelastic devices counteract vibrations, thereby reducing fatigue damage. One of the drawbacks of this approach is that the blade dynamics need to be monitored continuously to derive the input to control algorithms. Therefore, a set of distributed sensors such as strain gauges and accelerometers is needed. Certainly for large scale offshore turbines, power supply of these devices becomes a critical design aspect. A central power system has the disadvantage that electricity needs to be transferred into the rotating blade reference frame. For this operation a system of slip rings or electric brushes is needed. Both devices are subject to heavy wear, thereby shortening the life time of the sensor system significantly. Another challenge is the cabling inside the blades. A promising solution is to power the sensors by means of local energy generation through energy harvesting.

While energy harvesting of vibrations is a well-researched topic, generating energy by exploiting aeroelastic instabilities has so far received little attention. First work in this field has been done by Bryant and Garcia [1] and Bryant, Fang and Garcia [2] who use a simple cantilever piezoelectric bender with a flap at the trailing edge to generate vibrations that can be harvested. Bryant, Wolff and Garcia [3] have further expanded this conceptual work by studying the sensitivity of the system parameters on the stability. It was found that by changing component masses and stiffnesses, the system behavior could significantly be changed. Both models exploit a two degree of freedom model coupled with a 2 dimensional unsteady, linear aerodynamic model. Park et al. [4,5] use electromagnetic energy conversion instead of piezoelectric devices. Instead of exploiting a coupled flutter mechanism, a T-shaped cantilever beam is used. A magnet is attached at the end of the cantilever beam such that it sheds vortices that lead to flutter. Coils are mounted in the non-moving reference frame.

The work presented in this paper aims to provide an implementation of an energy harvester closer to actual aerospace applications. For this purpose, the concept of a free-floating flap (FFF) further investigated. Initial work has been done by Bernhammer et al. [6], where this FFF was used as aeroelastic control effectuator that could suppress its own flutter by controlling trailing edge tabs. This work has been further evaluated by Pustilnik and Karpel [7,8], who studied the behavior of the FFF for limit cycle oscillations.

While those devices have so far been used to aerodynamically alleviate vibration loads, this paper presents a way to invert the functionality of the FFF to use them for energy harvesting. The flaps are aerodynamically balanced such that they approach marginal stability for certain critical design load cases. This ensures that the flap responds to turbulence in the wind flow and a significant amount of energy can be generated. A coil is mounted inside the flap, oscillating through a magnetic field generated by permanent magnets. The relative motion from magnet and coil due to the vibration generates electricity that can be either directly used for the sensors or stored in a battery. Details of the aeroelastic design of the wing have been presented by Bernhammer et al. [6]. The dimensions of the model were not altered and are given in Table 1. Compared to the original model, two angular potentiometers were replaced by commercially available dynamos.

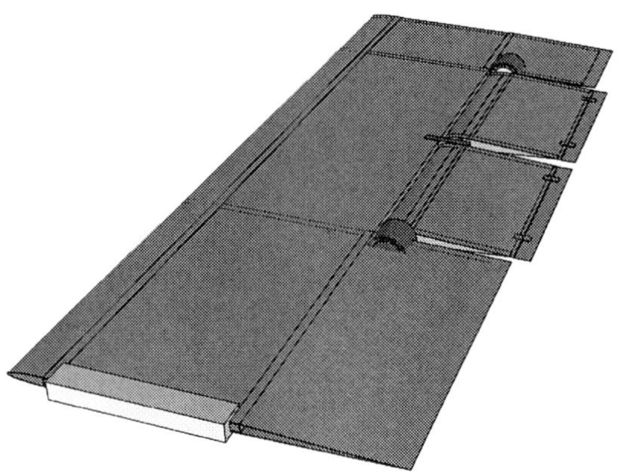

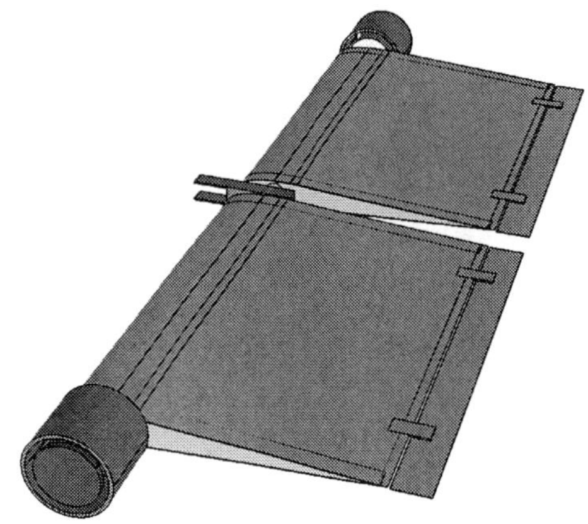

Figure 1. CAD drawing of wing (left) and close-up on flaps (right)

Table 1 Tail plane model details

Chord	Thickness	Span	Rudder chord	Rudder span	Dynamo
400 mm	11 mm	1000 mm	160 mm	250 mm	Basta Trio 3320

2. NUMERICAL MODEL

The numerical model was created in MSC/Nastran, where a modal analysis was carried out. The modes were passed to ZAERO which was used to extract the aeroelastic state-space system. Finally non-linear feedback loops were introduced in Simulink to model the effect of the structural free-play zone and the electromagnetic coupling with system.

2.1 State-Space Aeroelastic Model

The structural model was built in MSC/Nastran consisting of 914 CQUAD 4 elements. While originally being clamped at the root section, a spring was added to reflect the changes that have been made to the model during production such that it could be connected to measurement equipment. Said spring was tuned such to match the properties of the model elongation. Two additional ficticious masses [9] were added to monitor the flap motion. A modal analysis was carried out and the modal results were passed to ZAERO. While the structural equations of motion are directly taken from Nastran, ZAERO was used to construct the right-hand side of the equations of motion as given in Equation 1.

$$M_{hh}\ddot{\xi} + B_{hh}\dot{\xi} + K_{hh}\xi = F_{aero}\left(\xi,\dot{\xi},\ddot{\xi}\right) \quad \textbf{Equation 1}$$

With M_{hh}, B_{hh} and K_{hh}, being the generalized mass, damping and stiffness matrices, respectively. ξ is a vector containing the generalized degrees of freedom. The aerodynamics are modeled by a panel method. The structure is divided in 5 zones, with individual paneling, namely the root, two flaps, the section upstream of the flaps and the tip. The aerodynamics are cast into state-space format by minimum state rational function approximations in generalized coordinates including 5 lag terms. A stability analysis has been carried out by extracting the system eigenvalues to see which mode will be unstable. Figure 2 shows the first bending dominated mode that is predicted to flutter at 10.4 m/s at frequencies of 3.6 Hz in interaction with the flap deflection mode of the outboard flap.

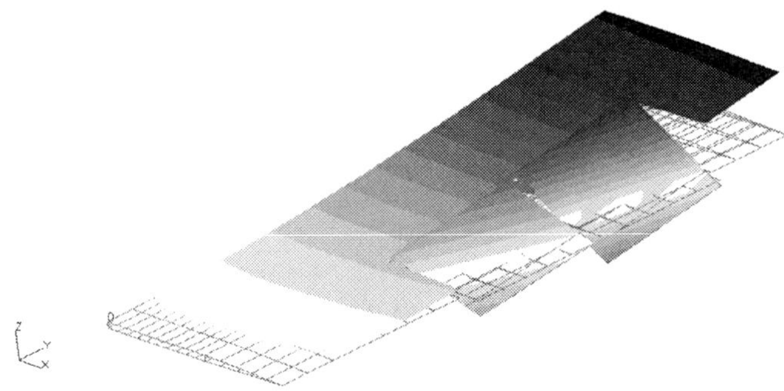

Figure 2. First bending dominated mode

2.2 Electronic circuit

The electronic circuit was kept as simple as possible. The main purpose of the flaps is to function as a power supply for sensors. Figure 3 shows schematics of a simple battery charging circuit. L1 represents the coil that is rotated in the magnetic field, DB1 is a rectifier and R1 is variable resistor.

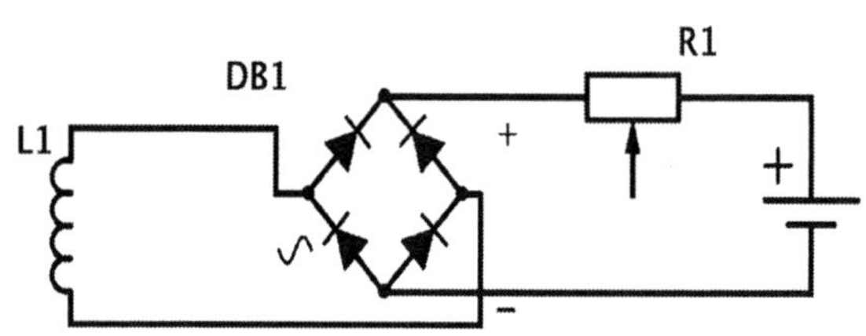

Figure 3. Schematic of modeled electronic circuit

When undergoing motion, the change in the magnetic field as seen by the coil causes a current in the electric circuit. The associated voltage created is equal to:

$$U = nl\frac{d\phi}{dt}$$ **Equation 2**

With n the number of coil turns, l the length of a coil loop and ϕ the magnetic field strength. The latter one has been experimentally determined for the dynamo used during the experiment. As only low coupling between structural motion and electronics could be observed, the magnetic flux has been increased by a factor 10 for the numerical simulations.

Table 1 Properties of dynamo

Number of coil turns	Length of coil loop	Magnetic flux divided by rotational velocity
100	94 mm	9.6e-3 Vs^2/rad

The created voltage has to overcome the voltage of the battery U_{bat} and assuming no power loss over the rectifier can be separated into two cases.

$$U = \begin{cases} nl\frac{d\phi}{dt} - U_{bat} & U > U_{bat} \\ 0 & U \leq U_{bat} \end{cases}$$ **Equation 3**

Combining Equations 2 and 3, the voltage generated can be expressed by non-linear feedback with a free-play zone with the width of the battery voltage. Furthermore, the electromagnetic harvester acts as non-linear damping to the aeroelastic system. The electrical power that is harvested adds damping with the same magnitude. Therefore, the equations of motion need to be expanded by the Equation 4:

$$P = \frac{\left(nl\frac{d\phi}{dt}\right)^2}{R_{coil} + R_{var} + j\varpi L} \approx \frac{\left(nl\phi_k\right)^2}{R_{coil} + R_{var}}\dot{\theta}^2$$ **Equation 4**

With θ as the flap deflection angle. Note that Equation 4 is simplified by approximating a constant magnetic flux ϕ_k only depending on the rotational velocity. Also, the impedance L is omitted as it is two orders of magnitudes smaller than the resistance of coil (R_{coil}) and the variable resistance (R_{var}). The damping has been implemented in generalized coordinates into the state-space system together with a structural free-play zone as shown in Figure 4.

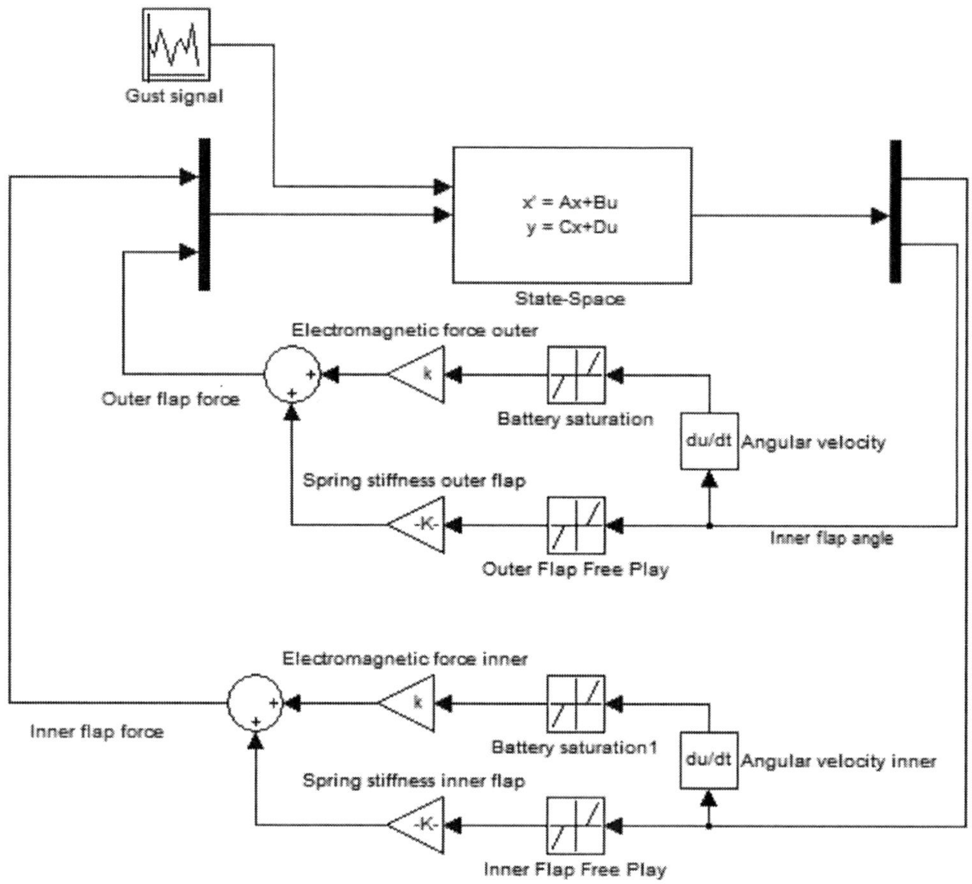

Figure 4. Simulink scheme of energy harvester model

Figure 4 also shows a gust input. The gust force history in generalized coordinates has been computed in ZAERO for a continuous gust signal with a von Karman turbulence spectrum for an altitude of 250 m above ground. Only the vertical component of the spectrum was taken into account. The turbulence intensity is 1.1%. For the gust load cases, the forces have been added to the equations of motion as given in Equation 1. The total state-space system thus reads:

$$\begin{aligned}\{\dot{x}\} &= [A]\{x\} + [B]\{u\} + \{u\}_{gust} \\ \{y\} &= [C]\{x\} + [D]\{u\}\end{aligned}$$ **Equation 5**

Where A, B, C and D are the state, the input, the output and the feed-through matrices. x is the state vector containing the modal displacements and accelerations and u is the control vector.

3. NUMERICAL RESULTS

Three sets of simulations have been performed numerically:

- 0 V battery voltage and varying resistance in the circuit
- 0.1 Ω resistance and varying battery voltage
- Gust response simulation with 0 V battery voltage and varying resistance in the circuit

All of these simulations have been done in time domain to model the non-linear structural and electromagnetical feedback. The eigenfrequencies and damping coefficients have been obtained by fitting a state-space system to the structural response.

3.1 Time-domain simulation and limit cycle oscillation

The added damping due to the power harvester has significant impact on the stability of the system. Figure 5 displays the results of a time domain simulation at 11 m/s wind speed, which, in non-harvesting case, is instable. This instability can be observed in the bottom plot of Figure 5, where only little power is generated, such that the system immediately shows an increasing oscillation. After 4 seconds a limit cycle state is reached and the voltage does not increase anymore. Decreasing the resistance leads to an increase in stability. The limit case here is a short circuit with only the resistance of the coil in the denominator of Equation 4. The associated structural response is very stable and damps out quickly.

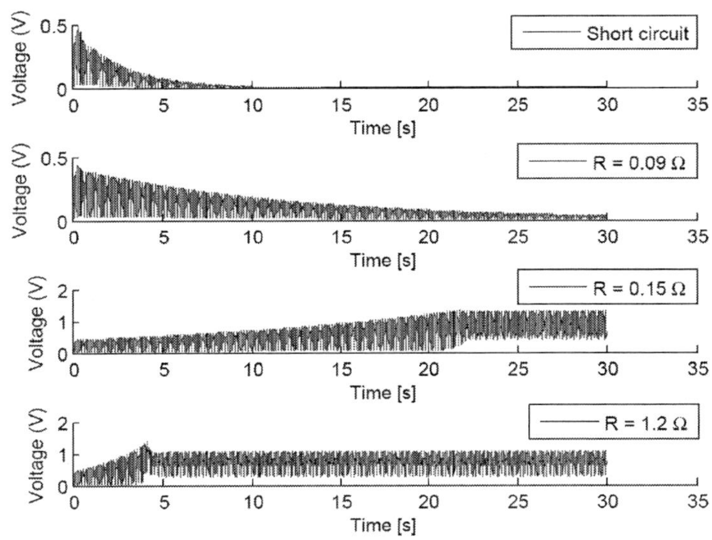

Figure 5. Time domain simulation with varying resistance at 11 m/s windspeed

A feature that is worthwhile noticing is that in the case of limit cycle oscillations as shown in Figure 5, the rectified voltage is not crossing 0 V. Before limit cycle oscillations are reached, the voltage displays sinusoidal behavior. Figure 6 gives more insight into the growth of the oscillations and the transition to limit cycles. It should be noted that the reason for the limit cycle behavior is that the amplitude of the oscillations covers the structural free-play zone. During the growth period, both flaps are oscillating in phase as shown in Figure 6. When transition to the limit cycle occurs, the smooth sinusoidal behavior disappears and ripples occur around the maximum velocity. It is

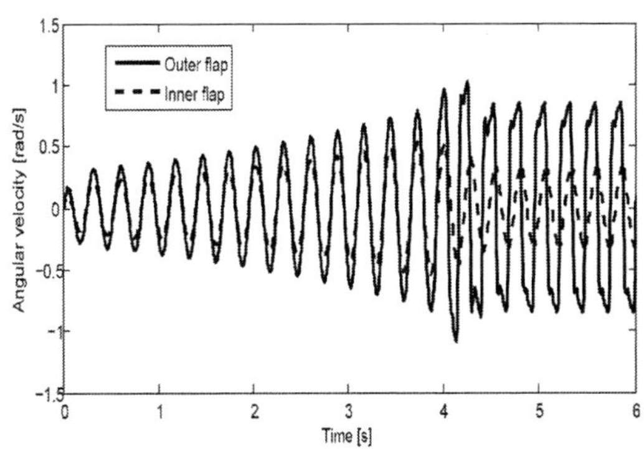

Figure 6. Transition to limit cycle oscillation

assumed that this can be attributed to the interaction between the two rudders and the phase shift that develops once a limit cycle is initialized. The phase shift in angular velocity causes the voltage always to be non-zero.

3.2 Stability evaluation

As was already noticed, the stability of the system strongly depends on the resistance in the circuit. The power generated and consequently the resistance moment on the rotational axis decreases with increasing resistance. Figure 7 shows the damping coefficient and the frequency of the oscillation as a function of circuit resistance. The baseline operation without energy harvester is equivalent to the 1.2Ω case, as the power generated is not significantly contributing to damping. The baseline flutter speed was computed to be 10.4 m/s. Adding electromagnetic damping, results in an increase of the flutter speed to 12 m/s. Controlling of the resistance in the circuit can therefore be used to keep the system marginally stable for a range of operational velocity range of more than 10 percent. Increasing the magnetic flux in the dynamo will further increase this damping such that the desired wind speed region can be spanned. Furthermore, it can be seen that the added damping is rather independent on the wind speed. While the damping is strongly influenced by the energy harvester, the frequencies stay almost unaltered. A small drop can be observed around 12 m/s, however this drop is only in the order of 3 percent.

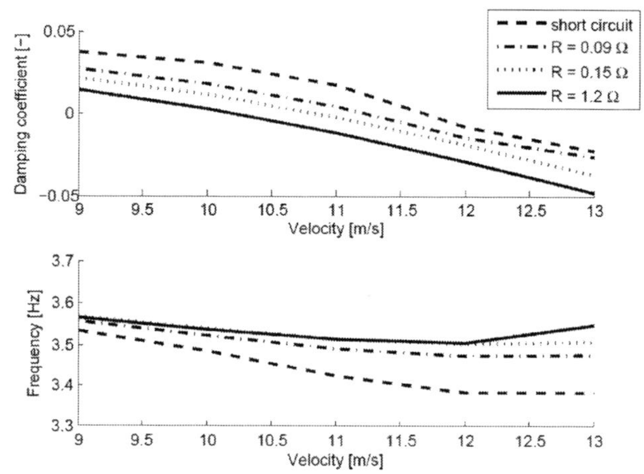

Figure 7. Frequency and Damping vs Velocity depending on resistance

The same effect can be observed when changing the battery voltage. Adding a free-play zone essentially decreases the added damping. Figure 8 shows the frequency and damping versus velocity for various voltage levels with a constant resistance of 1.2Ω. The effect is not as pronounced as in Figure 7; still an increase in flutter speed by 4 percent can be seen. This is a rather undesirable effect, as the resistance needs to be adjusted constantly when harvesting energy to charge the battery to keep the system neutrally stable. The charging effect reduces with increasing wind speed. Presumably the already instable system does not require a free-play zone to build up vibrations quickly. It

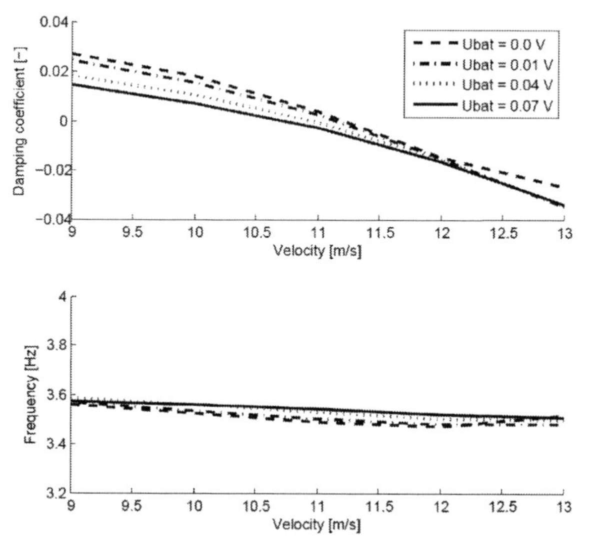

Figure 8. Frequency and Damping vs Velocity depending on battery voltage

is assumed that aerodynamic forces significantly overshadow the reduction of damping, such that the system oscillation will increase with the same speed. Similarly to the resistance variation, changing the battery voltage hardly reflects in a change in vibration frequency.

3.3 Power capture

The power capture simulation has been run for a constant inflow with an initial perturbation of the model for the first bending dominated mode as shown in Figure 2. The time history has been calculated for 30 seconds and the average power and voltage for 4 different wind speeds are plotted in Figures 9 and 10. When running the simulations for 9 m/s, the structure damps out quickly, regardless of the resistance applied. Consequently hardly any voltage is created and the system is not suitable as energy harvester. For 10 m/s the voltage is already significantly higher despite being damped, but the same trends in voltage and power still hold. Practically no power can be generated and the voltage is very low. Increasing resistance causes less damping, which means that the amplitude of the vibration due to an initial deflection decays slower, meaning that more energy can be extracted from the system.

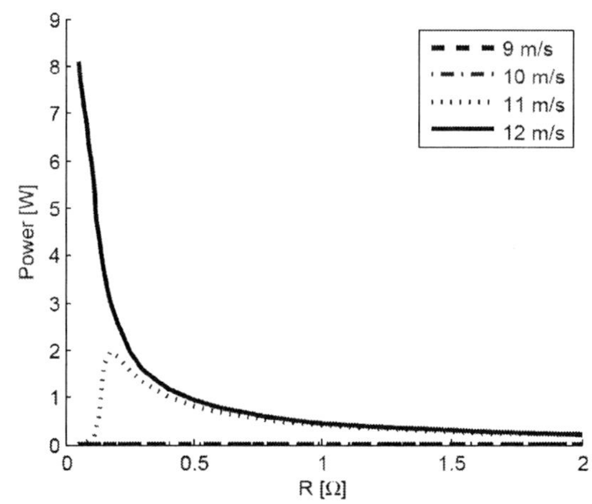

Figure 9. Power output vs resistance initial deflection

Wind speeds of 11 m/s are the first unstable velocities evaluated. Two zones can be detected in the voltage and power curves. The first part of both curves is strongly increasing until a resistance of 0.3Ω. This is the point where the wing is neutrally stable. For lower resistances, the structure is damped and not much energy can be extracted. This changes as the oscillations decay slower and slower with increasing resistance up to a point where the system becomes unstable and amplifies until it reaches limit cycle oscillations which are limited by the structural free-play zone. In this region of resistances above 0.3Ω, the generated voltage is practically constant and the power is inversely proportional with the resistance. For 12 m/s, the region of increasing voltage and power disappears, as the system is unconditionally instable regardless of the resistance as shown in Figure 7. Therefore the voltage is constant throughout the resistance range. Only a discontinuity can be observed for low resistance values, where the system is just unstable. The first low point can be attributed to the slow increase of the amplitude, while the second point, which lies above the constant curve, can be associated with the transition to limit cycle oscillations. As shown in Figure 6, the transition mechanism produces higher angular velocities yielding a higher voltage output. In the case of 0.1-0.15Ω, this transition takes place over multiple cycles, which explains the local rise in voltage. In the power production case, the inverse

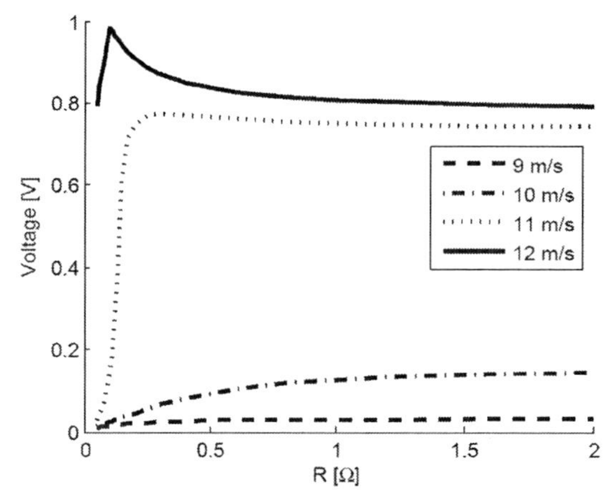

Figure 10. Voltage output vs resistance initial deflection

proportionality with the resistance is too dominant to visualize the variation of the voltage for low resistances.

3.4 Gust response

The last simulation performed is the response of the system to a gust. The gust was uniform over the wing span. The simulation length was 15 seconds. A von Karman wind spectrum was used with a turbulence intensity of 1.1 percent. Figures 11-14 show the power production in gust response. The low turbulence amplitude was chosen to not immediately have limit cycle oscillations. Overall, Figures 11-14 agree well with the simulations with an initial deflection. The higher the wind speed, the more energy can be harvested. The overall amplitude of voltages is lower, as the incremental phase contributes more. Constant voltages in the case of 11 and 12 m/s wind speeds are only reached at resistances of 1.2 and 0.5Ω. A difference to the previous simulations can be found in the comparison of the amplitudes of Figures 9 and 11. While in the limit cycle oscillation case, the generated voltage is almost equal for 11 and 12 m/s, in the gust load case, the higher wind speed shows a much higher responsiveness to the input signal, thereby tripling the output voltage. If comparing the low wind speeds to high wind speed results, a factor of 5 from 10 to 11 m/s and a factor 3 between 11 and 12 m/s are visible. For the power production this factors increase to 25 and 9, respectively. Clearly this shows the advantage of aeroelastic instabilities in energy harvesting, as the output can be increased by a factor 225, just by crossing the flutter margin and increasing the wind speed from 10 m/s to 12 m/s.

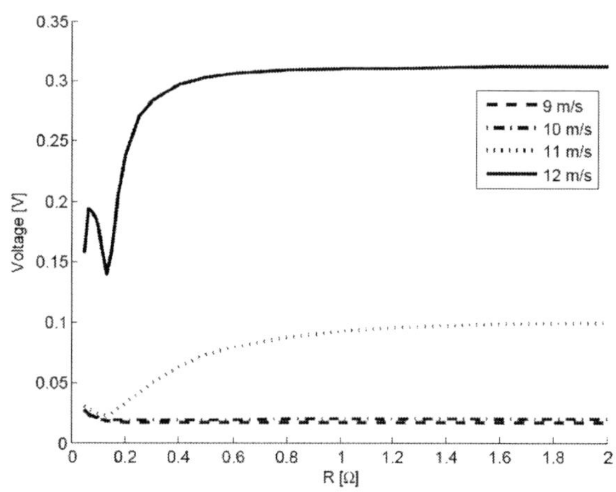

Figure 11. Voltage output vs resistance

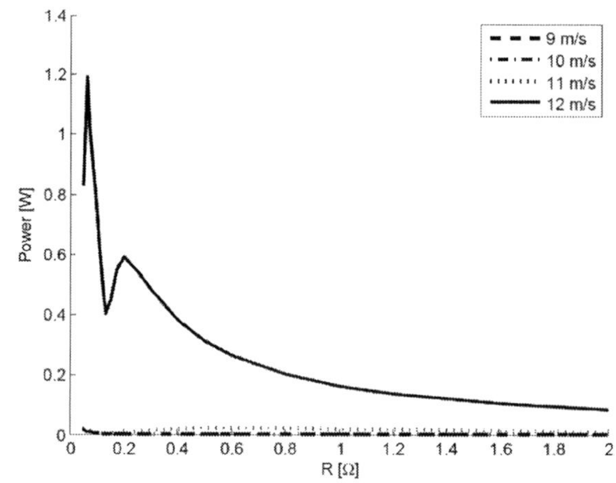

Figure 12. Power output vs resistance

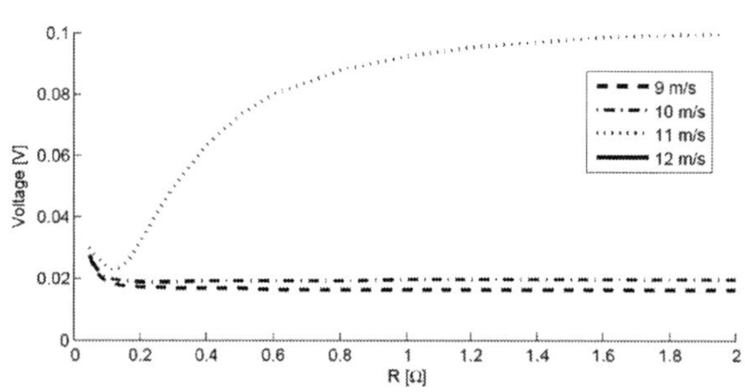

Figure 13. Voltage output vs resistance zoom in

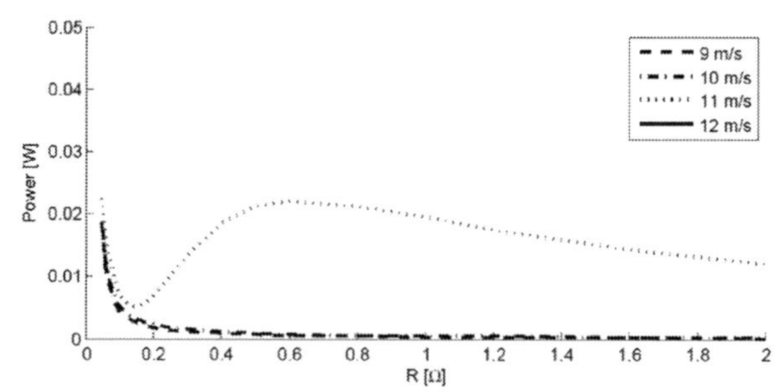

Figure 14. Power output vs resistance zoom in

In contrast to the previous simulations, a distinct peak can be observed for 0.07Ω. This peak is accompanied by a distinct dip at 0.13 Ω. The dip occurred for both 11 and 12 m/s air speeds and is particularly visible in the power production Figures 12 and 14. To explain the differences a closer look needs to be taken at the time history of the vibrations, which is plotted in Figure 15. The low gust amplitude causes a gentle increment in voltage. It is highly interesting to see that the response with varying resistance changes dramatically. Two observations can be made. The first one is that with increasing resistance, less damping is in the system and once a significant growth of the vibrations is initialized, this increase rate is getting higher as well. This falls well with what has been predicted by the stability plots in Figures 7 and 8. Eventually, the time domain simulation in the subfigures for 0.13, 0.2 and 1.2 Ω create stable limit cycles. The more remarkable observation, however, is that the point of initialization of the vibrations varies and causing the aforementioned dip. While the short circuit simulation grows slowly, but steadily from 2 seconds onwards, all cases with higher resistance postpone the initialization point of the vibrations. While for 0.13Ω, the structure does not show any response until 5 seconds, increasing the resistance also causes an increase in responsiveness.

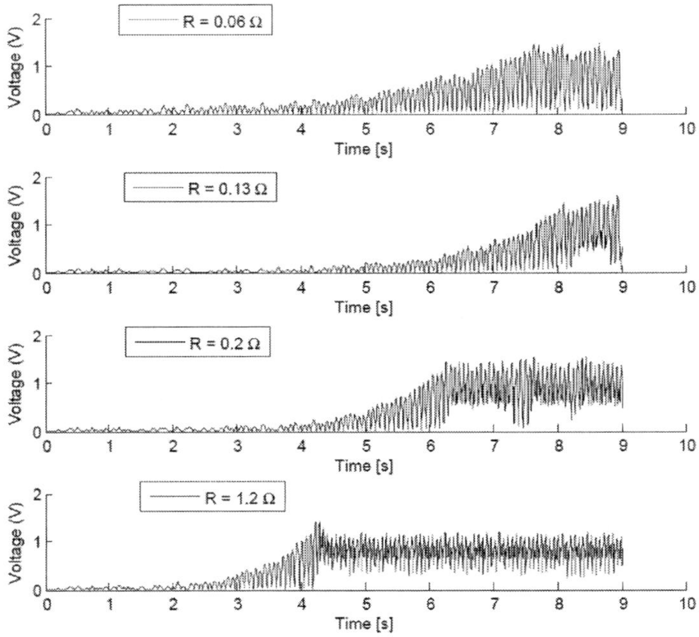

Figure 15. Voltage time histories vs resistance due to gust

This unexpected result can be attributed to the interaction of the two flaps. Figure 16 shows the limit cycle amplitude of for inner and outer flap. Both flaps behave in a rather different fashion. While the outer flap has a constant limit cycle amplitude, the inner flap decreases with increasing resistance. For low resistance values, the inner rudder becomes dominant. The electromagnetic damping has an equalizing effect on the two rudders. Most likely this is connected to the fact that the outboard flap mode is the mode that interacts with the bending mode in the flutter mechanism. Adding damping to the outboard flaps therefore has a stronger impact as the flutter mechanism is responsible for the power generation.

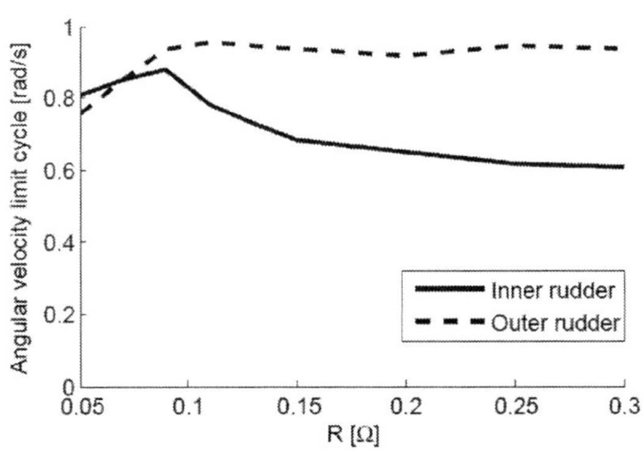

Figure 16. Power output vs resistance zoom in

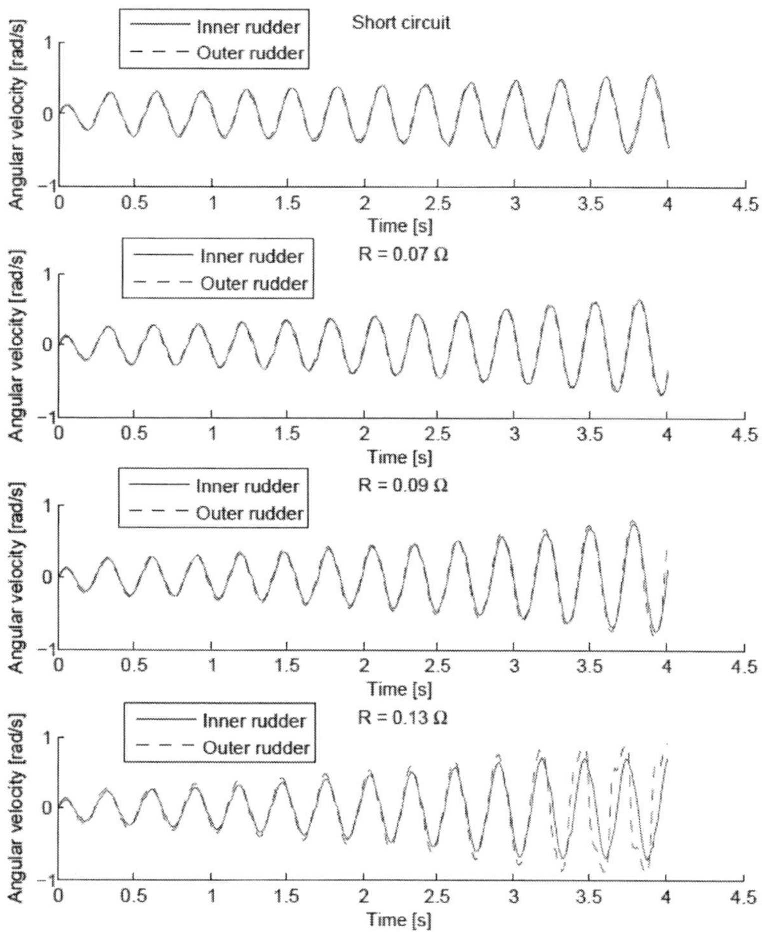

Figure 17. Time history of angular rudder velocities due to initial impulse of bending mode

Figure 17 confirms this hypothesis. Adding electromagnetic damping changes the flutter mechanism. While for high resistances, the outboard rudder is the driving mechanism behind the instabilities, the flaps are almost showing identical behavior for low resistance values. When observing closely, the inner rudder shows even a slightly higher amplitude in these cases. It can be concluded that this flutter mode with two equally participating rudders is more responsive to excitation, while the negative damping is smaller, as it takes longer to increase in amplitude over time. The flutter mode that is dominated by the outboard rudder is less responsive but the negative damping is higher, which causes the curves shown in Figure 16. Once an excitation is initialized, this flutter mode keeps increasing faster.

CONCLUSIONS

A new, practical energy harvester device has been designed and evaluated numerically. The potential of exploiting aeroelastic instabilities has been demonstrated as the increment in power production can be more than a factor 200 as compared with turbulence harvesting. Furthermore it was shown that changing the resistance is a suitable tool to control the stability of the system as it can overcome the destabilizing effect that battery charging can have on the power production and system stability. Furthermore, it can be concluded that adding non-linear damping changes the flutter mechanism. Tuning the design and controller for a successful real life application will be the next challenges for this type of energy harvester.

References

1. Bryant, M. and Garcia, E., "Modeling and Testing of a Novel Aeroelastic Flutter Energy Harvester", Journal of Vibration and Acoustics, Vol. 133, pp. 011010-1:10, 2011
2. Bryant, M., Fang, A. and Garcia, E., „Self-powered smart blade: Helicopter blade energy harvesting", Active and Passive Smart Structures and Integrated Systems, Proc. Of SPIE Vol. 7643, 764317, 2010
3. Bryant, M., Wolff, E. and Garcia, E., "Aeroelastic flutter energy harvester design: the sensitivity of the driving instability to system parameters", Smart Mater. Struct., Vol. 20, No. 125017, 2011
4. J. Park, K. Kim, S. Kwon and K. H. Law, "An aero-elastic flutter based electromagnetic energy harvester with wind speed augmenting funnel" Proceedings of Int. Conference on Advances in Wind and Structures, Seoul, Korea, 2012
5. J. Park, G. Morgenthal, K. Kim, S. Kwon and K. H. Law, "Power Evaluation for Flutter-Based Electromagnetic Energy Harvester using CFD Simulations", First International Conference on Performance-based and Life-cycle Structural Engineering, Hong Kong, China, 2012
6. Bernhammer, L.O., De Breuker, R., Karpel, M. and van der Veen, G.J., "Aeroelastic Control Using Distributed Floating Flaps Actuated by Piezoelectric Tabs", Journal of Aircraft, Vol. 50, No. 3, pp. 732-740, 2013
7. Pustilnik, M. and Karpel, M. "Dynamic Loads Alleviation Using Active Free-Floating Flaps", Proceedings of 53rd Israel Annual Conference on Aerospace Sciences, Haifa, Israel, March 2013.
8. Pustilnik, M. and Karpel, M., "Loads, Vibration and Manouver Control Using Active Floating Flaps", Proceedings of International Forum on Aeroelasticity & Structural Dynamics, Bristol, UK, June 2013
9. Karpel, M. and Presente, E., "Structural Dynamic Loads in Response to Impulsive Excitation," Journal of Aircraft, Vol. 32, No 4., 1995, pp. 853-861

PZT MEMS energy harvesting device with spiral cantilever beam

Chien-Chen Wang[2], Shun-Chiu Lin[1], Ya-Shan Shih[1], Shou-Peng Ye[2] and Wen-Jong Wu[1]

[1]Department of Engineering Science and Ocean Engineering, National Taiwan University, Taipei, Taiwan
[2]Graduate Institute of Electronic Engineering, National Taiwan University, Taipei, Taiwan

Abstract

In this paper, a PZT MEMS energy harvesting device with a spiral cantilever beam is presented. The device which is formed by laminating two piezoelectric layers is fabricated in bimorph type. A proof mass is used to adjust the resonance frequency and the accelerations. To pattern spiral shaped PZT thick film, the lift-off process uses the THB-151N negative photoresist as sacrificial layer and the PZT layer is fabricated by aerosol deposition method. Our bimorph energy harvesting device can scavenge mechanical energy from ambient vibrations and transform it into useful electrical energy at low frequencies (about 51 Hz) and low accelerations (0.1g). Finally, the performance of output voltage and power of the device in series connections are measured based on the experimental result. The experimental result shows that the output power is 2.2μW with 2.4 $V_{P\text{-}P}$ output voltages and the maximum output power of 10μW with 5.1 $V_{P\text{-}P}$ under 0.65g at its resonant frequency. The resonant frequency is kept comparably low at 51Hz.

1. INTRODUCTION

The concept of the IOT (Internet of Things) became popular in recent years. Figure 1 shows the framework of the IOT, it can be divided into three parts. The bottom part is "perception layer," which is composed by several kinds of sensor; and the middle layer is the "network layer" which includes wired and wireless network. The information data are collected and transmitted into information centers in the network layer. The top part is the "application layer". It is a data center with strong computing power and provides services to a variety of applications such as household appliances, transport, environmental monitor, house automation and healthcare.

The well-known manufacturer Cisco Systems estimates that there will be more than fifty billion smart components in the future. We briefly assume that the energy consumption of a sensor is one watt, thus the summation of fifty billion sensors are fifty billion watts. Traditionally, we would get the large amount of power by using batteries for; however, using batteries bring some drawbacks, for example: batteries have limited lifetime, and they are inconvenient to replace. Thanks to the improvement of VLSI technology, more and more low power elements has been developed and the energy consumption of each element has been reduced to even lower than one hundred micro watt. Therefore, finding a self-powered system seems to be a feasible solution to replace the usage of batteries.

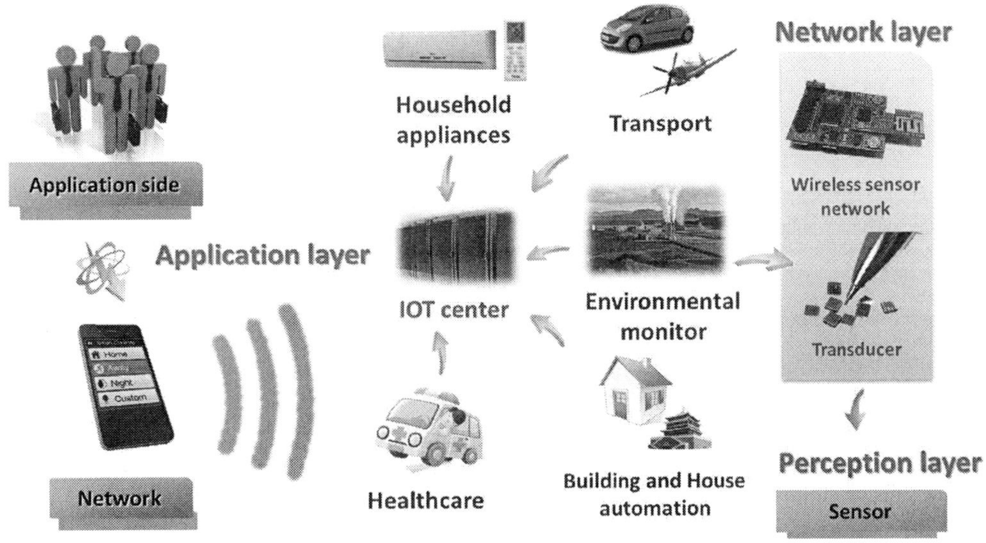

Figure 1.The framework of Internet of Thing (IOT).

Power harvesting via extracting energy from ambient has been profoundly investigated by researchers for years [1-4]. There are many kinds of energy harvesting source available such as magnetic induction energy harvester, piezoelectric energy harvester, solar cells, thermal gradients and air flow energy harvester. Compared to solar, air flow, or heat, the power source from ambient vibration carries the greatest potential. There have already been many successful vibration harvesting devices reported. Previous researchers have reported on various types of designs with the piezoelectric structures [5-8]. Most of the piezoelectric energy harvesters are designed in cantilever type. In this study, in order to create a low-frequency device, we chose the spiral structure rather than normal cantilever type and used stainless-steel instead of silicon substrate to increase flexibility. Experimental results confirm that by using the stainless steel as the device substrates, the stiffness of the beam is dramatically pulled down, and successfully lowered the device resonant frequency. In order to fabricate high-quality PZT thin-films efficiently, a self-made aerosol PZT thin-film deposition chamber based on the "jet-printing" method was developed and constructed[9]. In this paper, our devices were fabricated based on stainless-steel, which not only fortifies the robustness of the device but also lowers the resonance frequency and possesses better strength and flexibility compared to the conventional device with silicon substrate. The device is also designed with a dual oscillator structure. The dual oscillator structures are proposed to use the first resonant frequency by outer beam and second resonant frequency by inner beam to realize the broadband response. The fabricated dual oscillator device design had achieved a low resonance frequency of 51Hz. It is able to scavenge energy from ambient vibration at low frequency and low acceleration.

2. Design concept

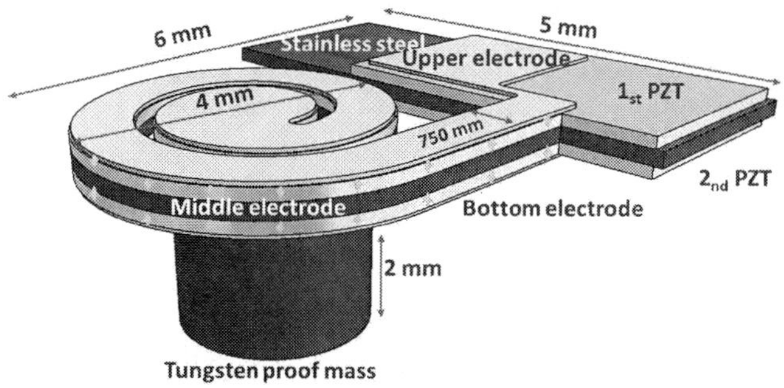

Figure 2.The schematic diagram of the spiral cantilever beam with bimorph type.

Figure 2 shows the schematic diagram of the spiral cantilever beam with bimorph type. In this study, in order to create a low-frequency device, we chose the spiral structure rather than normal cantilever type. We use stainless steel substrate instead of using silicon because the stainless steel substrate provides superior robustness allowing the micro device to withstand harsher environments and use the stainless steel to be the middle electrode, designed PZT layers of 15um on both sides, and poled the PZT layer in a serial direction.

Structure simulation

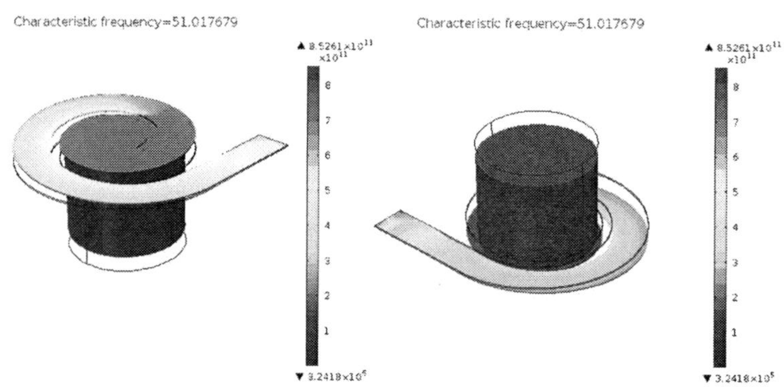

Figure 3.The simulation of the Von Mises's stress distribution of the spiral cantilever beam with Comsol© (a) The stress of upper layer (b) The Stress of bottom layer

Figure 3 (a)(b) shows the stress distribution and the resonance frequency simulation of the spiral MEMS energy harvester, which is performed by Comsol©. The resonant frequency of the device is about 51Hz. We can compare the two figures. Because of the proof mass bonding on the bottom layer, it's apparent that the stress of the root of the bottom layer is bigger than that of the upper layer.

3. Fabrication process

The fabrication process of the MEMS energy harvesting device with spiral cantilever beam is shown in figure 4. The device is fabricated on a 20μm stainless-steel substrate and the fabrication process is divided into five steps. Firstly, H_2SO_2 (sulfuric acid) and H_2O_2 (hydrogen peroxide) were used to clean the stainless-steel substrate. Secondly, the upper PZT layer is patterned by a lift-off process using photo-resist and deposited for 15μm-thickness by a self-made aerosol deposition system, which is shown in figure 5.

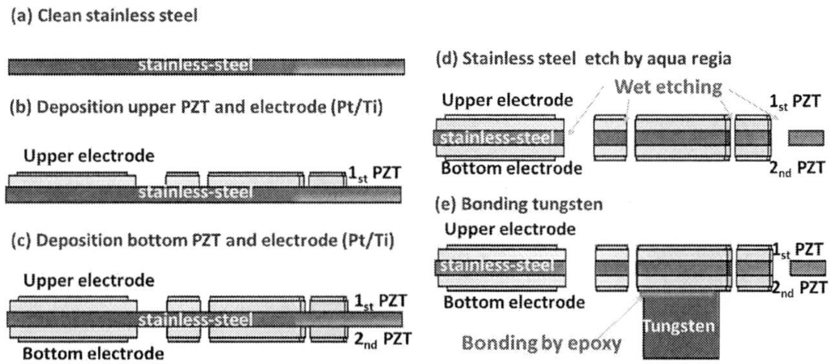

Figure 4. Schematic Diagram of the fabrication process of spiral cantilever

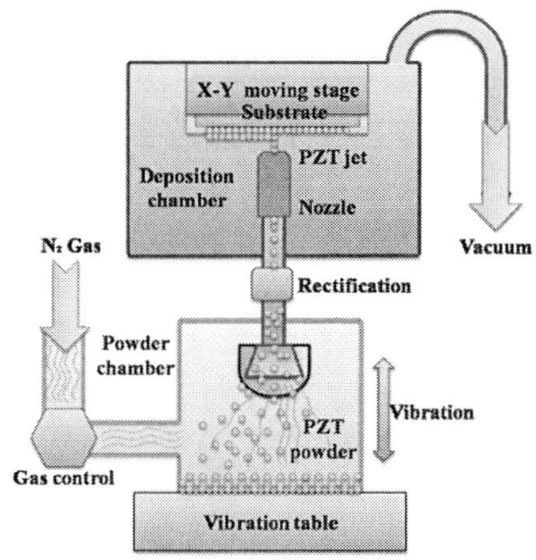

Figure 5. Schematic diagram of the aerosol PZT thin-film deposition chamber

The self-made aerosol PZT deposition chamber is able to deposit PZT film at a rate up to 0.1 micrometers per minute. As shown in the figure, the PZT powder with the particle size smaller than 1μm in diameter was put in a continuously vibrating container in order to suspend the PZT powder. Nitrogen or Helium gas was connected to the powder chamber so as to bring the PZT powder through the nozzle into the deposition chamber. With the chamber in a vacuumed condition, the PZT powder was jet out from the nozzle inside the chamber and bombarded onto the sample surface at high speed. As a result, the powder would be deposited to the sample surface. After the PZT film was deposited, the upper electrode was deposited by DC magnetron sputtering. Thirdly, the bottom PZT layer was patterned by lift-off process and deposited for 15μm-thickness, and then deposited the bottom electrode. Fourthly, and the beam shapes were then defined by aqua regia etching and the devices were annealed in a furnace and then cooled down to room temperature. Bonding of stainless-steel and the tungsten proof mass was performed afterwards. Subsequently, the PZT layer needs to be poled under a high electric field. The device was heated up to 100 °C by using a hot plate and was polled the micro piezoelectric energy harvester under 100V. The electric field was also applied during the cooling process where the device is left to cool down slowly until it reaches the room temperature.

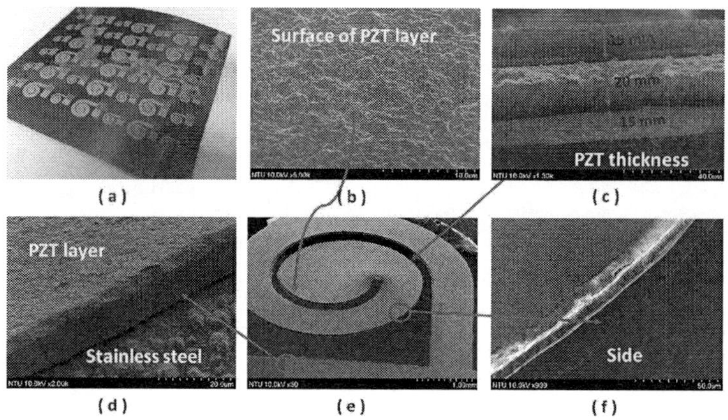

Figure 6.The photograph and SEM image of the spiral MEMS energy harvester

Figure 6 shows the photograph image and SEM image of the spiral MEMS energy harvester. We fabricated our device on an area of 80mm times 60mm stainless steel substrate. We can fabricate any shape of PZT layer by using self-made aerosol deposition system. As the figure (b) shows, the surface of PZT layer is dense on the surface of device, and we deposited the 15um-PZT layer on the both side of substrate. As the figure (c) shows, it's the lateral view of the PZT layer and stainless steel substrate. As the figure (f) shows, it's the lateral view of the arc.

5. Result and discussion

5.1 Experimental setup

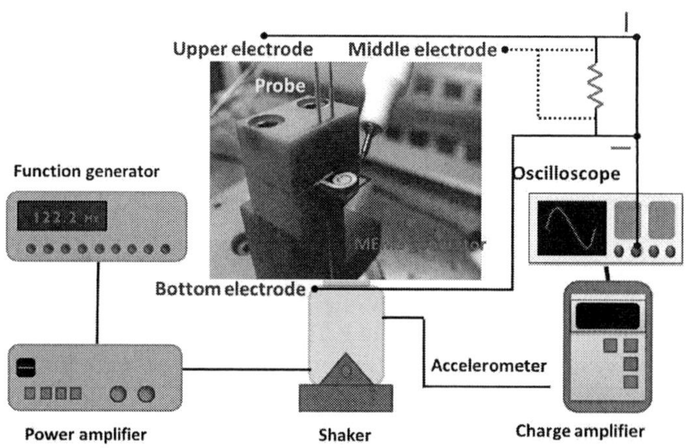

Figure 7.The schematic diagram for the experimental setup.

The schematic of the experiment setup is shown in figure 7. The piezoelectric MEMS generator was clamped by a clamper, which was then mounted on a shaker. The shaker acted as a vibration source to piezoelectric MEMS generator and was controlled by a function generator through a power amplifier which provided a sinusoidal waveform, thus we can calculate the output power by measuring the output voltage when the device is parallel to the load impedance.

5.2 Experimental result

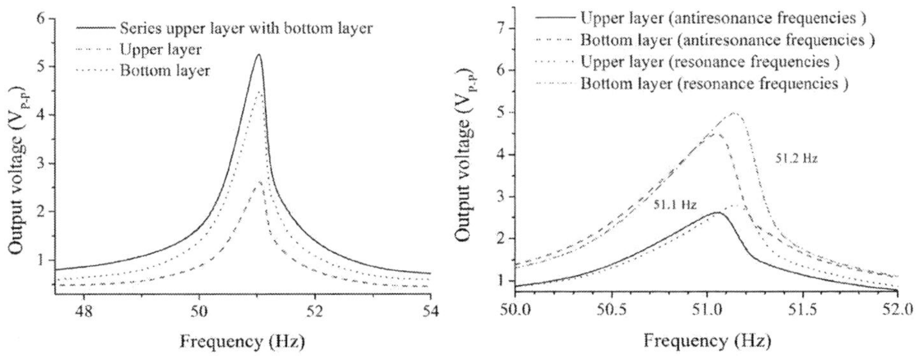

Figure 8. Experimental results of output voltage with open circuit and frequency plot under 0.1g (a) series, upper layer and bottom layer, and (b) resonant frequency and anti-resonant frequency .

Figure 8 shows the experimental results of the device. Figure 8 (a) is the output voltage – Frequency scan plot under 0.1g, we can find that maximum output voltage of bottom layer is about 4.5V, and that of the upper layer is about 2.5V. And the pick output of the series connection is 5.3V. And the resonant frequency is only 51.1Hz.Figure (b) shows the output voltage – Frequency plot of the resonant frequency and anti-resonant frequency. The resonant frequency of the top and bottom layer are both 51.2Hz and the anti-resonant frequency of both layer are both 51.1Hz

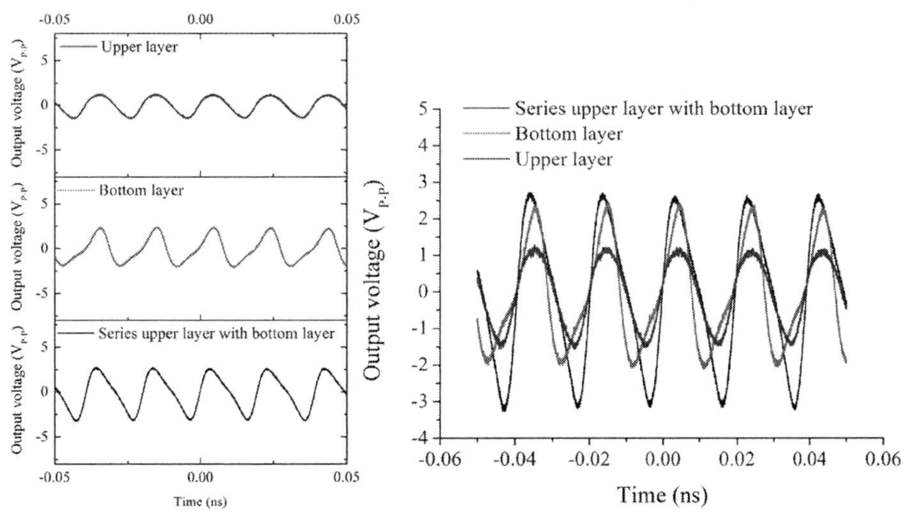

Figure 9. The waveforms of output voltage under 0.1 g acceleration.
(a) output voltage of upper, lower layer and series connect (b) combine the three figures in (a) into one.

Figure 9 shows the waveform of output voltage under 0.1g acceleration. Figure 9 (a) are the output waveforms of top layer, that of bottom layer, and that of the series connect. We combined the three figures into one, and we found that due to the different torsion of the upper and bottom layer , the waveform are not identical; therefore, the waveform of series connection is not equal to the summation of the upper and lower layer

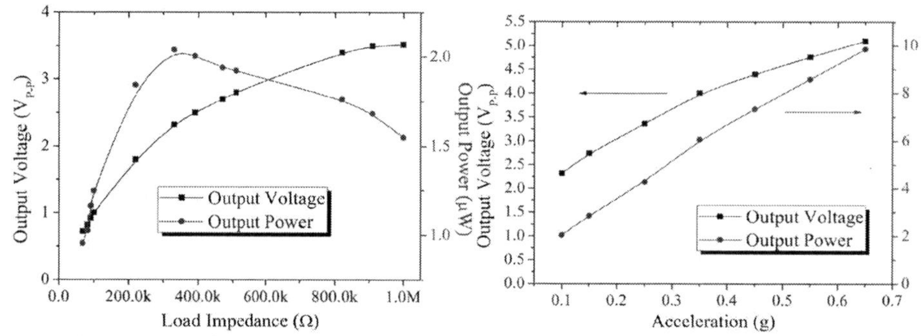

Figure 10.(a) Output voltage and output power of different loading impedances excited at 0.1 g .
(b) The maximum output power and maximum output voltages of different acceleration under 330kΩ

Figure 10 (a) shows the output voltage and output power of different load impendence plot under 0.1g acceleration. The black line represents the output voltage, and the blue line represents the output power. We found that the optimize load impendence is 330kΩ. The Maximum output voltage and Maximum output power of different accelerations shows in figure 10(b). The output voltage can be 5.1V peak to peak under0.65g acceleration.

6. Conclusions

In this paper, the PZT MEMS energy harvesting device with spiral cantilever beam has been achieved. The device is designed from the bimorphs MEMS generator that has the ability to convert ambient vibration energy to useful electrical energy. This device has resonant frequency of 51Hz, and the device has the output power of 2.2μW with 2.4 V_{P-P} output voltage. At its resonant frequency, the maximum output power of 10μW under 0.65g. In conclusion, we provided a new device which improved the stress distribution and the strength of the device. At the same time, the resonant frequency is kept comparably low at 51Hz.

References

1. C.-C. Wang, Y.-S. Shih, S.-C. Lin, J.-Y. Zeng, and W.-J. Wu, "A micro piezoelectric energy harvester based on stainless steel substrate with dual oscillators," *Active and Passive Smart Structures and Integrated Systems 2013,* vol. 8688, pp. 86880E-86880E-9, 2013.
2. J. Y. Zhang, Z. Cao, Q. Wang, and H. Kuwano, "Microstructure and piezoelectric properties of AlN thin films grown on stainless steel for the application of vibration energy harvesting," *Micro & Nano Letters, IET,* vol. 7, pp. 1170-1172, 2012.
3. Q. Wang, Z. P. Cao, and H. Kuwano, "Metal-based piezoelectric energy harvesters by direct deposition of PZT thick films on stainless steel," *Micro & Nano Letters, IET,* vol. 7, pp. 1158-1161, 2012.
4. C.-Y. Sue and N.-C. Tsai, "Human powered MEMS-based energy harvest devices," *Applied Energy,* vol. 93, pp. 390-403, 5// 2012.

5. H. Liu, C. Lee, T. Kobayashi, C. Tay, and C. Quan, "A new S-shaped MEMS PZT cantilever for energy harvesting from low frequency vibrations below 30 Hz," *Microsystem Technologies,* vol. 18, pp. 497-506, 2012/04/01 2012.
6. H. X. Hongping Hu, and Yuantai Hu, "A Spiral-Shaped Harvester with an Improved Harvesting Element and an Adaptive Storage Circuit," *IEEE transactions on ultrasonics, ferroelectrics, and frequency control,* 2007.
7. Y. Hu, H. Hu, and J. Yang, "A low frequency piezoelectric power harvester using a spiral-shaped bimorph," *Science in China Series G: Physics, Mechanics and Astronomy,* vol. 49, pp. 649-659, 2006.
8. B. A. C. Baoxing Chen, Ahmad Safari, Member, IEEE, and a. T.-W. C. Stephen C. Danforth, "Theoretical and Numerical Predictions of the Spiral-Shaped Lead Zirconate Titanate (PZT) Actuators," *IEEE transactions on ultrasonics, ferroelectrics, and frequency control,,* 2002.
9. B. Lee, S. Lin, W. Wu, X. Wang, P. Chang, and C. Lee, "Piezoelectric MEMS generators fabricated with an aerosol deposition PZT thin film," *Journal of Micromechanics and Microengineering,* vol. 19, p. 065014, 2009.

A fractional calculus based model for hysteresis in Ti_3SiC_2 and NiTi-Ti_3SiC_2 Composites

Riddhiman Bhattacharya[1,2], and Nakhiah C. Goulbourne[2,*]

1. Department of Materials Science and Engineering, H.H Dow Building, 2300 Hayward St. The University of Michigan, Ann Arbor, 48109, USA
2. Soft Materials Research Laboratory (SMRL), Department of Aerospace Engineering, FXB Building, 1320 Beal Avenue, The University of Michigan, Ann Arbor, 48109, USA

Abstract

The development of a fractional calculus based constitutive model to capture stress-strain hysteresis in Ti3SiC_2 and Ti_3SiC_2-NiTi composites over a range of temperatures, is presented. Ti_3SiC_2 is a representative member of MAX phases, which are known for unusual combination of properties in typical metals and ceramics. The high machinability combined with hysteretic response under cyclic loading is a classic example of this fact. At high temperatures, these hysteresis loops are characterized by significant visco-plastic and rate dependent effects. In this paper, a fractional calculus based approach is used to develop a phenomenological model which effectively captures the response over a wide range of temperatures (up to ~1200°C). The results indicate that only a few parameters are necessary to precisely quantify the observed phenomena over a range of temperatures, which makes it amenable for numerical implementations.
Keywords: MAX phases; Fractional calculus; Hysteresis

1. INTRODUCTION

Ti_3SiC_2 is a representative member of the $M_{n+1}AX_n$ *phases* which are thermodynamically stable ternary carbides and nitrides comprising of an early transition metal (M), a group III or IV element (A) and an interstitial element (X). The value of n typically varies from 1-3 [1,2]. These materials are often classified as Kinking Non-Linear Elastic (KNE) solids due to a non-linear elastic behavior often accompanied by microstructural mechanisms known as kink bands [2,3]. The non-linear elastic mechanical response is characterized by hysteresis loops under cyclic compression, which are reversible and rate independent at room temperature [2]. At higher temperatures, however, the hysteresis is irreversible and characterized by stress-relaxation, rate-dependence and internal stress effects [4,5]. Moreover, MAX phases are known to have excellent thermal shock resistance and thermal stability, high machinability and unusual damage tolerance under compressive loading [1,2]. Very recently, there is a growing interest among researchers to synergistically combine the unusual combination of properties existent in MAX phases, with other materials for a myriad of applications [6-8]. An example of these newer class of composites, is the NiTi-Ti_3SiC_2 hybrid composite. It is postulated that the martensitic transformation in NiTi could be used to generate compressive residual stresses within the Ti_3SiC_2 phase, which would take advantage of the damping capabilities in MAX phases [9-12].

The reversible hysteresis due to cyclic loading at room temperature has been captured using a modified Preisach-Mayergoyz (PM) model [13]. A microscale model [14] has also been developed invoking the concept of Incipient Kink Bands (IKBs), which are idealized hysteresis units responsible for fully reversible, non-linear elastic hysteresis. Lester and Lagoudas [9,10] developed a micro-mechanically motivated model for capturing residual stress effects in NiTi-Ti_3SiC_2 composites. In this paper, we report the development of a fractional calculus based models to capture room temperature and high-temperature response in Ti_3SiC_2 and NiTi-Ti_3SiC_2 composites.

The meaning of a fractional order derivative was first raised by L'Hospital in a letter to Leibniz in 1695 [15,16]. Although "fractional differentiation and integration of arbitrary order" is more mathematically appropriate, a generalized terminology "fractional calculus" (coined during L'Hospital's time [16]) is often used. In 1819, Lacroix [17] gave the first description of fractional order derivative, in which he showed that for $y=x^\alpha$ and $\alpha \in R_+$ the fractional derivative of order ½ is

$$\frac{d^{1/2}y}{dx^{1/2}} = \frac{\Gamma(\alpha+1)}{\Gamma(\alpha+\frac{1}{2})} x^{\alpha-1/2}, \quad \ldots\ldots (1)$$

where Γ is the gamma function (for details, see [15]) and R_+ represents the set of positive real numbers. This results obtained from this definition was consistent with the Reimann-Liouville definition, which is explained in Section 2. Subsequently several mathematicians (such as Fourier, Abel, etc.) led the research in formalizing the theory underlying the broad definition of fractional calculus until 1941 [16]. In the meanwhile, Nutting (1921, [18]) proposed a general deformation law to model stress-relaxation where he pointed out that fractional powers of time are crucial to capture the response. Gemant (1936, [19]) showed that fractional powers of frequency can be used to model viscoelastic response from which he postulated that fractional order derivatives can effectively capture time and frequency dependent effects simultaneously. More recently, Bagley and Torvik [20-23] formulated constitutive relationships for materials which exhibit damping and are especially attractive for engineering structural components [22]. The key advantages of a fractional calculus based formulation are fewer number of parameters which leads to well-posed problems for Finite Element (FE) (or any numerical) implementations [22, 24, 25, 30]. Such implementations are cumbersome (and may even be impossible) for hysteresis models like the PM model due to very high computation time [26] and heavy dependence on weighting functions of hysteresis operators (for further details regarding the PM model, see [27, 28]). The physical significance of a fractional calculus based model can be realized using rheological representations consisting of springs and dashpots [25, 29]. The generalized constitutive law for such systems is

$$\sigma = C\frac{d^a}{dt^a}\varepsilon, \quad \ldots\ldots (2)$$

where σ is stress, ε is strain and C is a material parameter and a is the fractional derivative order lying between 0 and 1. For zero order derivative (a=0), the behavior is linear elastic and Eqn (2) reduces to Hooke's law with C as the Young's modulus. For a first order derivative (a=1), the behavior is completely viscous where the stress is linearly dependent on the strain-rate in the material. Therefore, a fractional order derivative ($0<a<1$) represents a viscoelastic type response which can be represented as a spring-pot [25,26,29].

The paper is organized into 5 sections. In section 2, the fractional calculus based analytical equations and numerical operators are defined. A description of the hysteresis model formulation in MAX phases and results are presented in section 3. A critical analysis of the model, along with its scope and limitations are discussed in Section 4. The key conclusions from this analysis are summarized in Section 5. An appendix is provided which outlines the optimization scheme (developed at SMRL) used for parameter determination of the model(s).

2. The Fractional Calculus (FC) method

The fundamental definitions and equations of fractional calculus operators and numerical algorithms are presented in this section. The most classical and widely accepted form of fractional time derivative is the Reimann-Liouville definition given by:

$$ {}^{m}_{ti}D^{\alpha}_{tf}\, f(t) = \frac{1}{\Gamma(m-\alpha)}\frac{d^m}{dt^m}\int_{t_i}^{t_f}\frac{f(\tau)}{(t-\tau)^{\alpha+1-m}}d\tau, \quad\text{.......................................(3)} $$

for $m \in N$, *and* $m > \alpha > m\text{-}1$.

In Eqn. 3, D^{α} is the derivative operator of order α (usually lying between 0 and 1), t_i and t_f are the lower and upper bounds of time between which the derivative is evaluated. The factor m is the range of fractional derivative order. For 1-D models, typically $m=1$ and the derivative operator reduces to $(1 > \alpha > 0)$

$$ {}^{1}_{0}D^{\alpha}_{t}\, f(t) = \frac{1}{\Gamma(1-\alpha)}\frac{d}{dt}\int_{0}^{t}\frac{f(\tau)}{(t-\tau)^{\alpha}}d\tau, \quad\text{..(4)} $$

Similarly, a fractional integral operator of order 1 can be defined as [8,19]:

$$ {}^{1}_{0}J^{\alpha}_{t}\, f(t) = \frac{1}{\Gamma(1+\alpha)}\frac{d}{dt}\int_{0}^{t}\frac{f(\tau)}{(t-\tau)^{-\alpha}}d\tau. \quad\text{..(5)} $$

For the models presented here, the initial time is assumed to be at $\tau=0$ and final time $\tau=T$, where T is the period of the cyclic stress input. Henceforth, these scripts will be dropped and fractional calculus operators in Eqns. 4 and 5 will be designated as $D^{\alpha}(.)$ and $J^{\alpha}(.)$ respectively.

Numerical algorithms have been developed by several researchers to evaluate these operators using different approaches based on basic definitions (for details see [15]). The operators used in this paper are based on the Grünwald-Letnikov definition of fractional calculus, which can be summarized as [26]:

$$ D^{\alpha}\big(f(t)\big) = \frac{\Delta t^{-\alpha}}{\Gamma(2-\alpha)}\Big[\frac{(1-\alpha)}{n^{\alpha}}f(0) + \textstyle\sum_{j=0}^{n-1}\big((j+1)^{(1-\alpha)} - j^{(1-\alpha)}\big)\,(f_{n-j} - f_{n-j-1})\Big], \quad\text{.........(6)} $$

$$ J^{\alpha}\big(f(t)\big) = \frac{\Delta t^{\alpha}}{\Gamma(2+\alpha)}\Big[\frac{(1+\alpha)}{n^{-\alpha}}f(0) + \textstyle\sum_{j=0}^{n-1}\big((j+1)^{(1+\alpha)} - j^{(1+\alpha)}\big)(f_{n-j} - f_{n-j-1})\Big], \quad\text{...........(7)} $$

where the analysis time [0, t] is discretized into n equally sized steps with size of each step Δt which is equal to step size of the time integral, and $f_j = f(j\Delta t)$. Fractional derivatives and integrals of periodic functions causes a phase shift. For example, a sinusoidal function phase shifts by an amount $\pi\alpha/2$ when operated by a derivative operator of order α. Similarly, an integral operator of order α causes backward phase shifting by the same amount. Either operations lead to a hysteresis loop formation [26]. Although the end results look similar, the difference between the two operators can be visualized by plotting the derivative/integral versus function at different time steps. Figures 1 and 2 show the derivative and integral plots for a sine function respectively at different time steps for $\alpha=0.5$. The clockwise and anticlockwise evolution of the two operators is clearly revealed here (indicated by arrows).

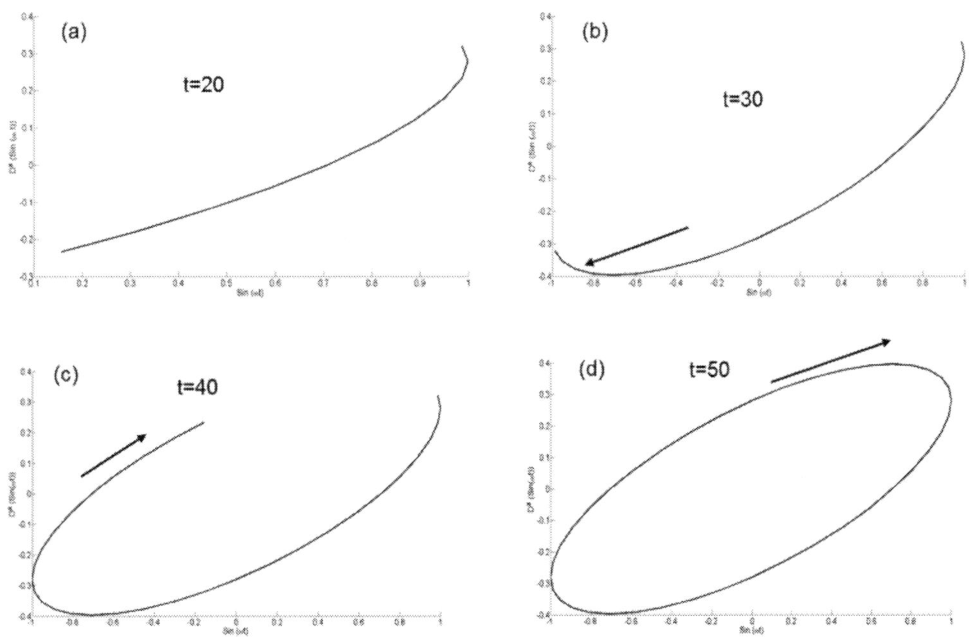

Figure 1 (a)-(d). Derivative operator (of order 0.5) versus periodic function (sine) at different time steps, showing the clockwise evolution of the loop formed due to phase shifting. The arrows indicate the direction of movement with respect to previous time step.

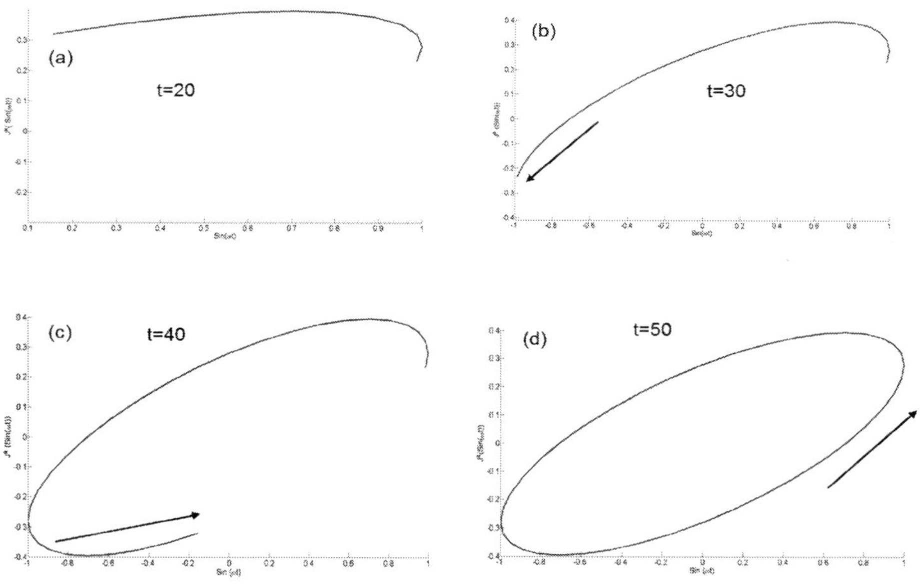

Figure 2 (a)-(d). Integral operator (of order 0.5) versus periodic function (sine) at different time steps, showing evolution of the loop formed due to phase shifting. The arrows indicate the counter-clockwise nature of the integral operator.

The algorithm used to derive Eqn.6 and 7 is also known as the L1 algorithm [15]. Another algorithm commonly used in the literature is the G1 algorithm in which a recursive relationship is used to calculate quadrature weights for the time integral (for details, see [15,25,29]). A detailed analysis on the derivations and efficiencies of various methods and discretization schemes is presented in [22, 23, 25] and is beyond the scope of this paper.

3. Hysteresis model development

3.1 Pure Ti_3SiC_2

The cyclic responses of MAX phase materials are characterized by a) fully reversible, strain-rate independent hysteresis at room temperature and b) irreversible, rate dependent response at higher temperatures with anelastic and internal stress effects. A phenomenological model is developed which effectively captures these features for both Coarse Grained (CG) and Fine Grained (FG) Ti_3SiC_2 using only 3 parameters. The parameters are determined by numerically evaluating a fractional integral at each time step and minimizing the deviation (or residual) between experimental results [from [4] and [5]] and model calculations using a Monte Carlo based optimization scheme (see section 3.3 and Appendix).

A stress-controlled experimental protocol was used for these experiments and hence the model is designed to calculate the strain for a given cyclic stress-input. For this type of experimental protocol, the hysteresis loop moves counter-clockwise (strain with respect to stress) which is captured using a fractional integral operator (J^{α}). A triangular waveform (stress versus time, not shown here) is input to the model which is similar to the stress controlled loading protocol used for the experiments [5]. The strain is assumed to be a combination of a) non-linear function of stress and b) fractional integral operator. This is defined as:

$$\epsilon(t) = C_o\sqrt{[\sigma(t)]} + C_1 J^a \sigma(t), \ldots\ldots (8)$$

where C_o, C_I and a are the parameters of the model which are optimized to fit the experimental data.

Using the expression for J^a, Eqn. 8 expands to

$$\epsilon(t) = C_o\sqrt{[\sigma(t)]} + C_1 \frac{\Delta t^a}{\Gamma(2+a)} \left[\frac{(1+a)}{n^{-a}} \square(0) + \sum_{j=0}^{n-1}\left((j+1)^{(1+a)} - j^{(1+a)}\right)\left(\square(t_{n-j}) - \square(t_{n-j-1})\right)\right] \ldots\ldots (9)$$

The total time (or period of the cyclic loading) is discretized into n divisions of equal size with time increments of $\Delta t = 0.5$. To capture the phenomenology of a strain-controlled experiment, the same formulations can be repeated using a derivative operator (D^{α}) due to the clockwise movement of the response with respect to its input.

3.1.1 Room Temperature Response

The room temperature response of coarse grained Ti_3SiC_2 for the stress-controlled tests (of up to ~160 MPa, see [5]) is presented in this section. The fine grained counterparts show a near elastic response for temperatures of up to ~900°C with insignificant hysteretic effects. Therefore, the stress-strain hysteresis of fine grained Ti_3SiC_2 is discussed in the subsequent section, which deals specifically with the high temperature response.

The stress-strain response for coarse grained Ti_3SiC_2 is shown in Figure 3. The experimental data is obtained from the literature [4,5] and smoothed using the *ExponentialMovingAverage (EMA)* function in Mathematica® to remove sharp corners, which when operated by the integral operator to calculate strains, cause numerical instabilities. In contrast to simple moving average, in this method a weighted smoothing factor is used which takes values between 0 and 1. For all the data reported in this paper, the smoothing factors lie in the range 1/8-1/15. Figure 3(a) shows a comparison between the originally published data [5] and the smoothed data used to compare the model calculations. It is observed that abrupt changes in the response are eliminated with minimal information loss. Figures 3(b) shows the stress-strain behavior obtained from experimental measurements and model calculations. As mentioned earlier, the strains are calculated using Eqn. 9 in which the parameters (C_o, C_I and a) are determined by minimizing the residual between experimental data and model prediction using the strain versus time behavior (not shown here). It must be noted that this stress input function has a symmetric loading for all the loading cases presented in Figs. 3-5. Due to the phase shifting nature (see section 2) of the fractional integral operator, the time at which peak strain is attained gets shifted (with respect to stress). This phase shifting effect is due to the phenomenon of stress relaxation (or viscoelastic effects) which causes a "lag" in strain measurements. This effect is relatively less pronounced at room temperatures but becomes more significant at higher temperatures, as outlined in section 3.1.2.

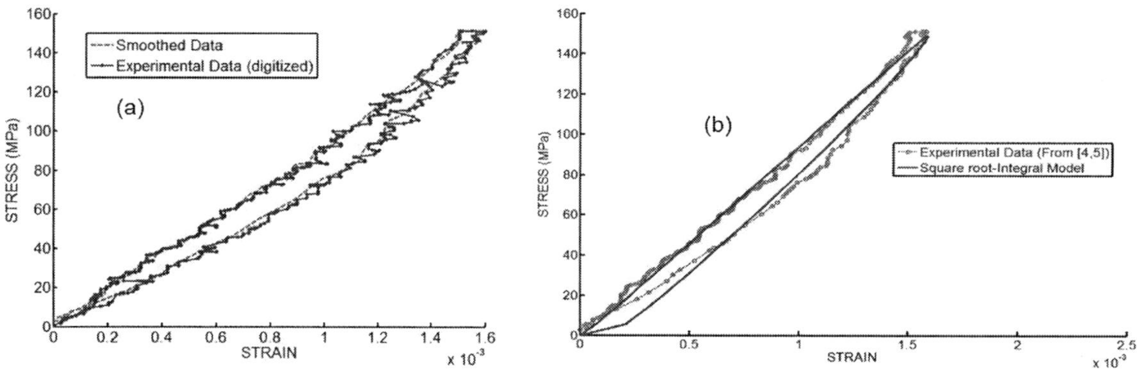

Figure 3. Stress-strain response of Ti_3SiC_2 under room temperature cyclic loading showing a) digitized versus smoothed data used for model fit, and b) comparison between model calculations and experimental data.

3.1.2 High Temperature Response

The response of Ti_3SiC_2 (and MAX phase materials, in general) consists of a linear-elastic response at very low stresses followed by non-linear elastic behavior and hysteresis over a range of temperatures. At higher temperatures, the deviation from the linear elastic behavior to non-linear response begins at relatively lower stresses (or strains) for both fine grained and coarse grained microstructures [5]. It is also observed that the area under the hysteresis loops increase with temperature and at a certain temperature the hysteresis loops become open and rate dependent. This is due to a temperature dependent brittle-ductile transition [5]. Figure 4 shows the stress-strain response obtained (from model and experiments) for coarse grained material at temperatures greater than room temperature. The response for fine grained samples is shown in Figure 5. The results suggest that the model is able to effectively capture the damping, stress-relaxation, brittle-ductile transitions and residual stress phenomena in both coarse grained and fine-grained materials over a range of temperatures.

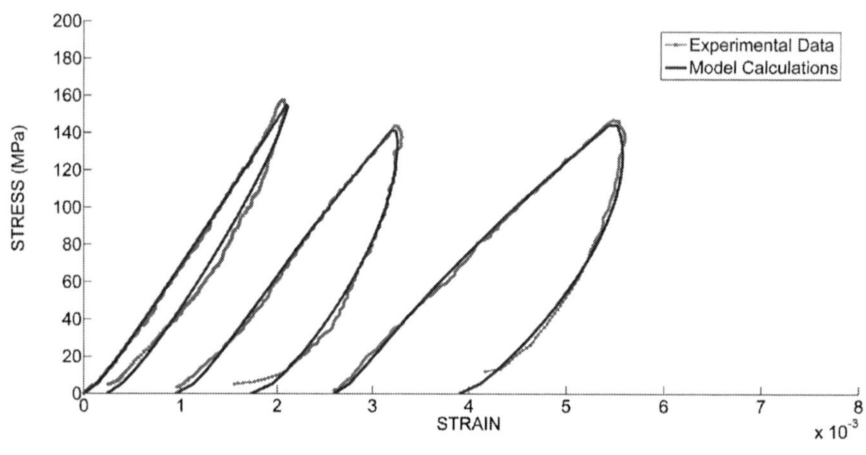

Figure 4. Experimental results and fractional-integral model calculations for coarse grained Ti_3SiC_2 at temperatures of 500°C, 1100°C and 1200°C, showing the stress-strain response.

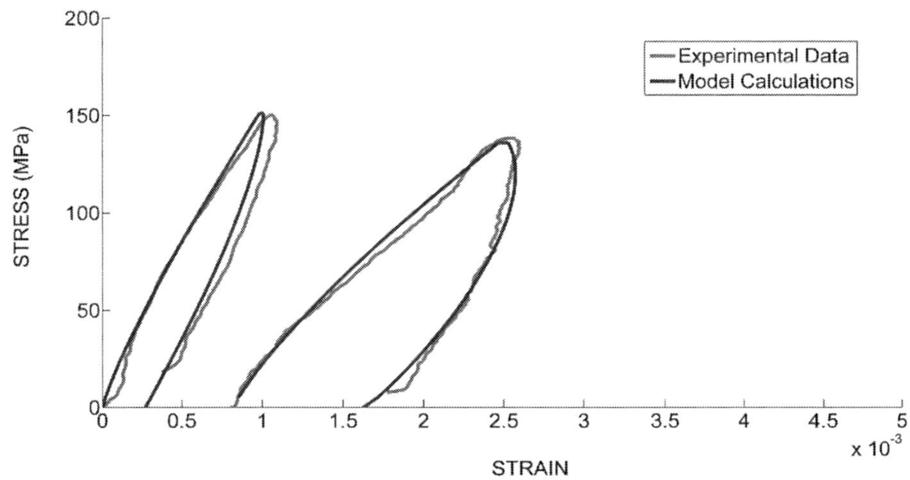

Figure 5. Stress-strain response showing experimental results and fractional integral model for fine grained Ti_3SiC_2 at temperatures of 1100°C and 1200°C.

The capabilities of the model in predicting the phenomenology accurately is evaluated by calculating the area enclosed under the stress-strain hysteresis loop. This quantity physically represents the damping of the material under a given loading condition and is not used in calculating the residuals, which are minimized with respect to experimental data. The result from this analysis is summarized in Table I. It is observed that the average relative error between area measurements from experimental data and model calculations is ~ 6 %. From this analysis, it is inferred that the three-parameter model is able to capture the hysteretic response effectively over a range of temperatures for different grain sizes.

Table I: Areas under hysteresis loops for coarse grained and fine grained Ti_3SiC_2

Coarse-grained Ti_3SiC_2		
Temperature (°C)	Enclosed Area (MJ/m^3)	
	Experimental	Model
Room Temperature	0.0173	0.0178
500	0.0393	0.0363
1100	0.0928	0.0947
1200	0.1577	0.1597
Fine-grained Ti_3SiC_2		
1100	0.0415	0.0328
1200	0.0951	0.0973

3.2 NiTi-Ti_3SiC_2 Composites

The NiTi-Ti_3SiC_2 is an example of the SMA-MAX phase composites, which are currently being developed to generate residual stresses, thereby leading to a pre-stress in the ceramic (or MAX phase) layer. The experimental data, used for hysteresis model development is obtained from the literature; unlike the pure MAX phases, prior to cyclic loading, the composites are pre-stressed to a value of 5 MPa. Subsequently, the composite samples are loaded cyclically to different stress levels in the range 50-200 MPa (for further details, see [12]). It is observed that with increasing stress levels at a given temperature, the damping increases. To capture this response, a 4-parameter fractional-integral model is used – which is a modified version of the 3-parameter model for pure MAX phases, presented in Section 3.1. The stress-strain relationship for the composite model is given by:

$$\epsilon(t) = C_o\sigma^{\mu}(t) + C_1J^{a}\sigma(t) \qquad (10)$$

Where μ is the fourth parameter introduced in this model.

The stress-strain response comparison between experimental data and model calculations for NiTi-Ti_3SiC_2 composites is shown in Figure 6. It is observed that for lower stress, the response is nearly linear elastic but becomes more non-linear with increasing maximum stresses.

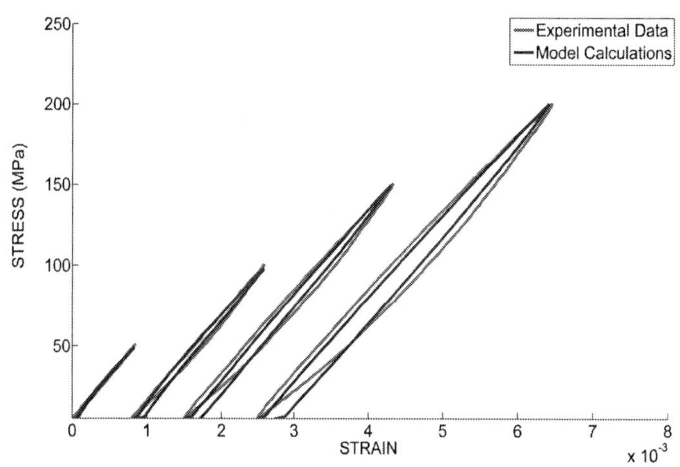

Figure 5. Stress-strain response comparison between experimental data and model calculations for NiTi-Ti_3SiC_2 composites, tested at different maximum stress levels at ambient temperatures.

Similar to the analysis for pure Ti_3SiC_2 data, the area enclosed by the hysteresis loops is used a metric to determine the accuracy of model calculations. The average relative error for the different stress levels were found to be within ~ 15%. The numerical values for areas under the loop from experimental data and model calculations for composites are summarized in Table II.

Table II: Areas under hysteresis loops for NiTi-Ti_3SiC_2 Composite

Maximum Stress (MPa)	Experimental Data (MJ/m^3)	Model Calculation (MJ/m^3)
50	0.00012	0.000425
100	0.008	0.0078
150	0.027	0.0188
200	0.0669	0.0528

4. Discussion

In this paper, the high-temperature hysteresis in Ti_3SiC_2 and room temperature response to different stress levels NiTi-Ti_3SiC_2 composite has been effectively captured for the first time using a phenomenological fractional calculus approach. The hysteretic stress-strain response in Ti_3SiC_2 over a range of temperatures (Room Temperature -1200°C) and for different grain sizes (coarse and fine) could be captured using only 3 parameters. The parametric values were optimized with respect to experimental data ([4], [5]). From the characteristic output of the model for coarse and fine grained material, it is observed that the model is able to predict the stress relaxation and internal stress effects. The key advantage of this model is that the constitutive law governing the stress-strain response uses fewer parameters, thereby making it convenient for numerical implementations which has been reported in the literature for viscoelastic systems [24,30]. Moreover, the parameters can be determined using a computationally inexpensive optimization scheme. To effectively capture the response of NiTi-Ti_3SiC_2 composite at room temperature and different stress levels, an additional parameter was introduced to the 3 –parameter model. For both pure Ti_3SiC_2 and the composite, it was observed that the model was able to predict the damping reasonably well (max. relative error $<\sim$ 15%). Current and future research efforts are focused at validating the model formulation with experimental data using different rates and loading protocols, and numerically implementing this constitutive law for different loading conditions.

5. Conclusions

The stress-strain hysteresis in a representative member of the MAX phase family (Ti_3SiC_2) over a range of temperatures, and MAX phase based composite for different stress levels is captured using a fractional calculus approach. The fractional integral based formulation uses only 3-4 parameters to capture the cyclic response for temperatures of up to ~1200°C and is able to reasonably predict the overall response (such as stress relaxation, internal stress effects and damping). The fewer number of parameters makes it highly suitable for implementation in numerical analysis.

Acknowledgements

The authors would like to gratefully acknowledge the financial support by the United States Air Force Office of Scientific Research (AFOSR), Award No. FA9550-09-1-0686. R.B. acknowledges the support and help from lab member, Mr. J.D. Davidson, who developed the Monte Carlo fitting algorithm.

References

[1] Barsoum MW, El-Raghy T, Radovic M, 2000, "Ti_3SiC_2: A Layered Machinable Ductile Carbide", Interceram, 49 (4), 226-233.
[2] Barsoum MW, Radovic M, 2011, "Elastic and Mechanical properties of the MAX Phases", Annu. Rev Mater Res 41: 195-227.

[3] Barsoum MW, Farber L., El-Raghy T., 1999," Dislocations, Kink Bands and Room-Temperature Plasticity of Ti_3SiC_2", Met. Mat. Trans. A, 30A, 1727-1738.

[4] Barsoum MW, Zhen T., Kalidindi S.R., Radovic M., Murugaiah A., 2003 "Fully reversible, dislocation-based compressive deformation of Ti_3SiC_2 to 1 GPa", Nat. Mat., 2, 107-111.

[5] Zhen T., Barsoum M.W., Kalidindi S.R., Radovic M., Murugaiah A., 2005 "Effects of temperature, strain rate and grain size on the compressive properties of Ti_3SiC_2, Acta Mater., 53, 4163-4171.

[6] Radovic, M., & Barsoum, M. W. (2013). MAX phases: Bridging the gap between metals and ceramics. *AMERICAN CERAMIC SOCIETY BULLETIN*, *92*(3), 20-27.

[7] Wang, J., Li, A. J., Wang, S. M., & Zeng, H. X. (2012). Dependence of the microstructure and properties of TiC/Ti_3SiC_2 composites on extra C addition. *Ceramics International*, *38*(7), 5967-5971.

[8] Wang, W. J., Gauthier-Brunet, V., Bei, G. P., Laplanche, G., Bonneville, J., Joulain, A., & Dubois, S. (2011). Powder metallurgy processing and compressive properties of Ti_3AlC_2 / Al composites. *Materials Science and Engineering: A*, *530*, 168-173.

[9] Lester, B. T., & Lagoudas, D. C. (2013). Modeling of Residual Stresses in Shape Memory Alloy-Ceramic Composites.

[10] Lester, B. T., & Lagoudas, D. C. (2012). Computational Micromechanical Modeling of Ceramic-SMA Composites.

[11] Kothalkar, A., Hu, L., Schaff, F., Basu, S., Radovic, M., & Karaman, I. (2012). Processing and Characterization of NiTi-MAX Phase Composites Prepared by Spark Plasma Sintering. In *AIP Conference Proceedings*. American Institute of Physics, Ste. 1 NO 1 Melville NY 11747-4502 United States.

[12] Kothalkar A., Benitez. R, Hu L., Radovic M., and Karaman I., (2013) "Thermo-mechanical response and damping behavior of shape memory alloy/MAX phase composites", submitted to Metallurgical and Materials Transactions A.

[13] Zhou A.G, Basu S., Friedman G., Finkel P., Yeheskel O., and Barsoum MW, 2010 "Hysteresis in kinking nonlinear elastic solids and the Preisach-Mayergoyz model",Phy. Rev. B, 82, 094105-1-10.

[14] Barsoum MW, Zhen T., Zhou A., Basu S., and Kalidindi S.R., 2005, "Microscale modeling of kinking nonlinear elastic solids", Phy. Rev. B, 71, 134101-1-8.

[15] Oldham K.B., and Spanier J., 1974, "The Fractional Calculus", Academic Press, New York and London, ISBN: 0-12-525550-0.

[16] Hilfer R., 2000, "Applications of Fractional Calculus in Physics", World Scientific Publishing, ISBN: 9810234570.

[17] Lacroix, S.L.,1819, “Traite du calcul differential et du Calcul Integral”, Courcier, Paris, Vol 3., 409-10.

[18] Nutting, P. G. "A new general law of deformation.",1921, Journal of the Franklin Institute 191.5: 679-685.

[19] Gemant, A., 1936 "A Method of Analyzing Experimental Results Obtained from Elasto-Viscous Bodies." Physics 7.8: 311-317.

[20] Bagley, Ronald L., and P. J. Torvik, 1983 "A theoretical basis for the application of fractional calculus to viscoelasticity." Journal of Rheology 27: 201.

[21] Bagley, Ronald L., and Peter J. Torvik, 1986, "On the fractional calculus model of viscoelastic behavior." Journal of Rheology 30 : 133.

[22] Diethelm, Kai, et al., 2005, "Algorithms for the fractional calculus: a selection of numerical methods." Computer Methods in Applied Mechanics and Engineering194.6: 743-773.

[23] Adolfsson, Klas, Mikael Enelund, and Stig Larsson, 2004, "Adaptive discretization of fractional order viscoelasticity using sparse time history." Computer methods in applied mechanics and Engineering 193.42: 4567-4590.

[24] Cortés, Fernando, and María Jesús Elejabarrieta, 2007, "Finite element formulations for transient dynamic analysis in structural systems with viscoelastic treatments containing fractional derivative models." International journal for numerical methods in engineering 69.10: 2173-2195.

[25] Schmidt, André, and Lothar Gaul., 2006, "On the numerical evaluation of fractional derivatives in multi-degree-of-freedom systems." Signal processing 86.10 : 2592-2601.

[26] Sunny, M. R., Kapania, R. K., Moffitt, R. D., Mishra, A., & Goulbourne, N., 2010, "A modified fractional calculus approach to model hysteresis",. Journal of applied mechanics, 77(3).

[27] Mayergoyz, Isaak D, 2003, Mathematical models of hysteresis and their applications. Academic Press.

[28] Guyer, R. A., McCall, K. R., Boitnott, G. N., Hilbert Jr, L. B., & Plona, T. J.,1997, "Quantitative implementation of Preisach-Mayergoyz space to find static and dynamic elastic moduli in rock." Journal of geophysical research 102.B3: 5281-5293.

[29] Mendiguren, Joseba, Fernando Cortés, and Lander Galdos., 2012, "A generalised fractional derivative model to represent elastoplastic behaviour of metals."International Journal of Mechanical Sciences.

[30] Schmidt, André, and Lothar Gaul. "Finite element formulation of viscoelastic constitutive equations using fractional time derivatives." *Nonlinear Dynamics*29.1-4 (2002): 37-55.

Modeling the Stress-Strain-Resistance Behaviour of Shape Memory Alloy for Sensorless Actuator Position Control

Brian Lynch[1], Xin-Xiang Jiang[2], Fred Nitzsche[3], Alex Ellery[4]

[1] Mining Systems Laboratory, Dept. of Mining Eng., Queen's University, Kingston, Canada
[2] Engineering Laboratory, Transportation Safety Board of Canada, Ottawa, Canada
[3] Rotorcraft Research Group, Dept. of Mech. & Aero. Eng., Carleton University, Ottawa, Canada
[4] Space Exploration Engineering Group, Dept. of Mech. & Aero. Eng., Carleton University, Ottawa, Canada

Abstract

The application of shape memory alloys in actuator systems has been well studied and continues to be an important topic of research in various fields of engineering including aerospace, biomedical, and robotics. Of particular concern is precision control of the actuator position. Many approaches have been investigated for position control, but most of these depend upon feedback of the actuator position as measured by a sensor. Research into the material properties of shape memory alloys has revealed a significant relationship between the electrical resistance and strain. This correlation has been used to model the strain as a function of the electrical resistance in order to predict the actuator position and remove the need for sensor feedback. However, the presence of R-phase in the microstructure of the alloy as well as the influence of applied stress make modeling difficult. The work presented here focuses on the development of a new predictive model of the relationships between electrical resistance, applied stress, and strain in Ni-Ti shape memory alloy. Characterization of the material was performed and the results used in order to develop the model based on empirical relationships. The model was then simulated for verification and successfully used to replace sensor feedback in a simple position control experiment.

1. INTRODUCTION

Shape memory alloy (SMA) has the phenomenal ability to remember and recover a previous geometry simply through the application of heat. This phenomenon, called the shape memory effect, allows an SMA component (such as a wire or plate) to be used as an actuator. Although various shape memory alloys have been studied, the most popular and useful alloy is Nitinol due to its superior properties [1]. This alloy is so named due to its nearly equiatomic composition of nickel and titanium. The shape memory effect is caused by transformations in the crystal microstructure of the material as the applied stress and temperature are changed [2]. This transformation occurs between two primary phases, martensite and austenite, but may also involve the formation of R-phase depending upon the manufacturing processes used [3]. As stress is applied to an SMA component, a large amount of plastic strain occurs. Although this deformation appears to be permanent, it is in fact due to "de-twinning" of the martensite crystals as shown in Figure 1 [4]. Raising the temperature of the material above a critical temperature causes the martensite to transform into austenite, which upon cooling retains the original geometry of the component. Note that the application of a bias load is required in order to ensure that the actuator returns to the extended position during cooling. An important feature of the phase transformation process is the hysteresis effect that occurs due to differences between the critical transformation temperatures during heating and cooling, which makes modeling and control more difficult. There also exists a dependence between the transformation temperatures and applied stress which adds further complexity.

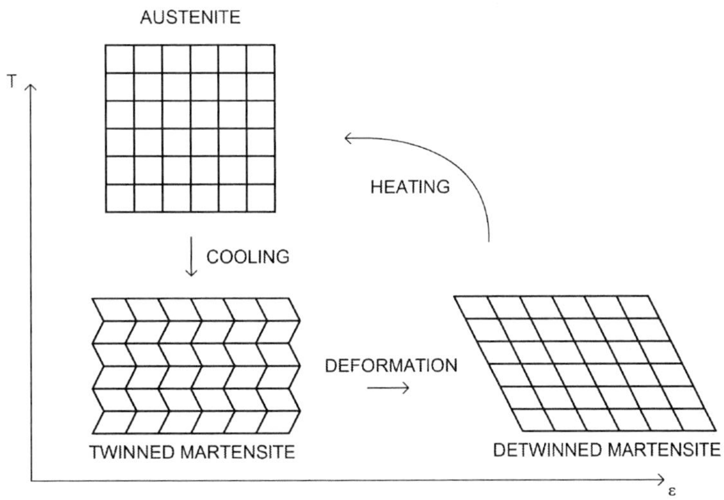

Figure 1. Phase transformations which cause the shape memory effect.

Recent research has demonstrated remarkable relationships between the temperature, applied stress, strain, and electrical resistance of SMA [5]. Each of the material phases (martensite, austenite and R-phase) possess a different electrical resistivity. Thus, changes in the phase composition cause changes in the electrical resistance. Additionally, the electrical resistance also changes due to elongation and contraction of the wire. Therefore, there are significant and repeatable correlations between the electrical resistance and stress, strain, and temperature of the material. Given a model of these correlations, the strain of an SMA component may be predicted from measurements of the electrical resistance. Feedback of the resistance is more convenient compared to feedback of the actuator position since resistive heating is typically used and the applied voltage and current may be easily determined. Shape memory alloy may therefore be considered both an actuator and a sensor.

The self-sensing ability of SMA actuators has been studied in various applications. Overheating of actuators may be avoided by using resistance feedback to confirm complete actuation [6], which allows for faster actuation speeds and reduces the effects of fatigue. Pre-heating of actuators has also been studied using resistance feedback in order to prime the actuator for faster response under varying ambient conditions [7]. Various types of active structures using SMA wires with resistance feedback have been investigated, such as transducers for monitoring the effects of earthquakes [8] and SMA-embedded composites [9]. Precision position control of SMA actuators has been studied using resistance feedback to predict position, but with only limited success depending upon the complexity of the predictive model. Most often, the applied stress conditions are fixed and the relationship between resistance and strain is modeled using linear or polynomial functions [10][11][5][12]. Compensation of hysteresis effects has also been investigated using various approaches including separate heating and cooling functions [13] and artificial neural networks [14]. However, none of these solutions are able to account for both the presence of R-phase during transformation and the effects of varying applied stress.

This paper presents a new predictive model of the correlation between stress, strain, and electrical resistance in Nitinol shape memory alloy. The model is based upon empirical relationships which are derived from the results of material characterization experiments. Both electrical resistance and applied stress are measured and used as feedback in order to predict the actuator position. Precision position control of SMA actuator wire is then demonstrated using the predicted position as feedback within a simple proportional-integral-derivative (PID) control law.

2. MATERIAL CHARACTERIZATION

Experimental studies were performed using a laboratory bench-top setup built and arranged at the Canadian Space Agency headquarters. This setup consisted of a vertical frame from which an SMA wire was suspended and loaded using either bias masses or springs. Measurement of the applied load was provided by a load cell located between the frame and wire connection point. The actuator position was measured using a linear variable displacement transducer (LDVT) mounted on the frame with a probe attached to the free end of the SMA wire. Feedback of the applied voltage and current was used to determine the electrical resistance of the actuator wire. Data acquisition was performed using primarily National Instruments hardware and LabView software. The SMA wire itself is a Nitinol alloy called Flexinol that is commercially available from Dynalloy. Figure 2 shows the experimental setup used for both characterization and control.

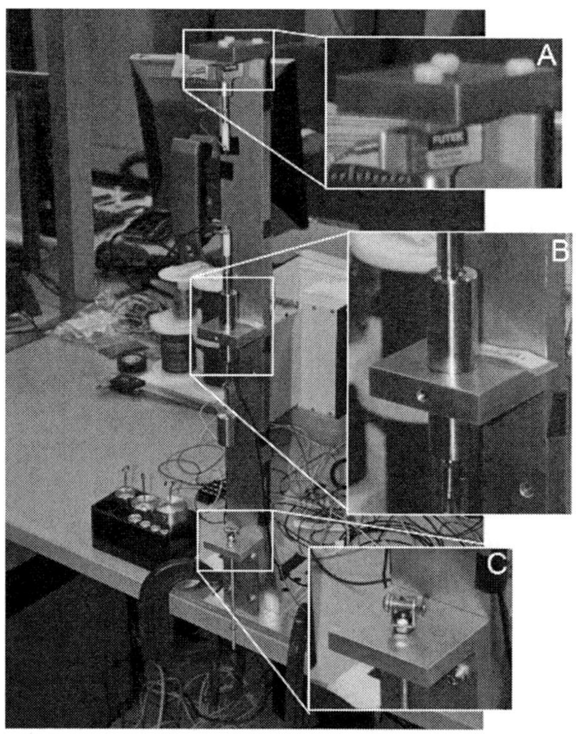

Figure 2. Setup used for material characterization and control experiments.

Characterization of the material behaviour was performed in order to observe and evaluate the fundamental correlations between electrical resistance, applied stress, and strain. The characterization experiments focused on important aspects of the material behaviour in the context of an actuator application: complete heating and cooling behaviour, incomplete actuation, and the influence of applied stress. The overall relationship between resistance and displacement during both complete heating and cooling is shown in figure 3. Starting at point a, the actuator is initially heated to point b, at which point the material begins transformation from martensite into austenite until reaching point c at full actuation. Heating is then stopped and the actuator cools from point c to point d, at which point transformation from austenite to R-phase begins. Transformation from austenite to R-phase occurs from point d to point e, at which point the R-phase and any remaining austenite begin transformation back into martensite.

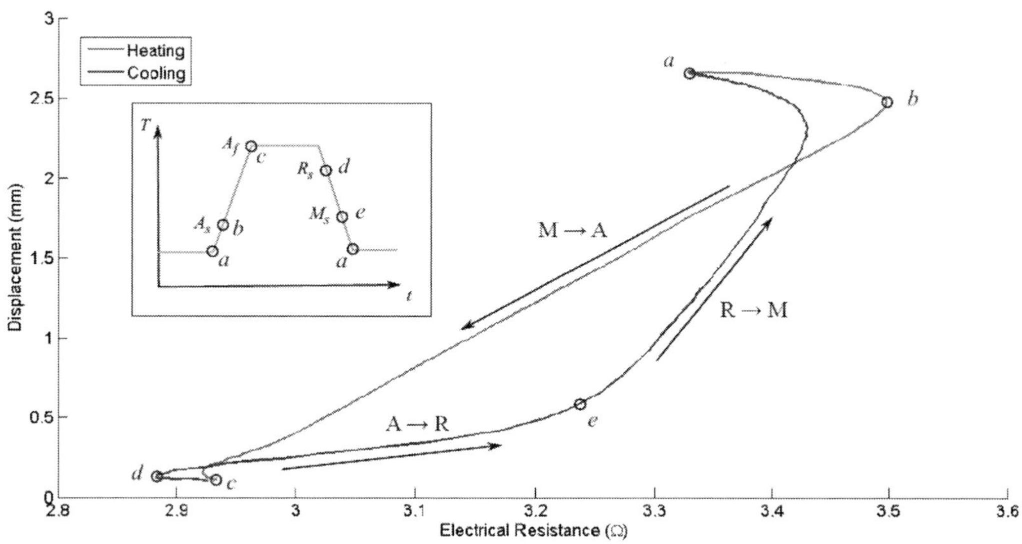

Figure 3. Complete heating and cooling resistance-displacement curves at an applied load of approximately 0.96 N (approximate temperature profile over time shown in inset).

The hysteresis effect observed in the complete heating and cooling curves of the resistance-displacement behaviour is due to a combination of various thermomechanical effects. The electrical resistivity is different for each phase of the material, and depends linearly upon temperature as in typical metals. Electrical resistance is also dependent upon the length and cross-section area of the wire. As the actuator is heated and cooled, both the length and phase composition change and the electrical resistance follows a generally non-linear and hysteretic pathway, but with some linear portions during both heating and cooling.

Repeating the complete heating and cooling experiment under a series of varying applied loads resulted in the family of curves shown in Figure 4, where the colour changes from blue to red with increasing applied stress. The most prominent feature of the resulting behaviour is the decrease in hysteresis width as applied stress is increased. In addition, both the initial and final values of resistance and displacement increase with increasing applied stress.

The behaviour of the SMA actuator wire under incomplete actuation was also studied, both during heating and cooling. Partial actuation was achieved by progressively decreasing the heating duration for a series of actuation cycles. This process was repeated similarly for the cooling duration, resulting in the curves shown in Figure 5. When the actuator is partially heated and subsequently cooled, there is a transition curve branching from the heating curve and approaching the cooling curve asymptotically. Similar behaviour is also seen when the actuator is partially cooled and subsequently heated.

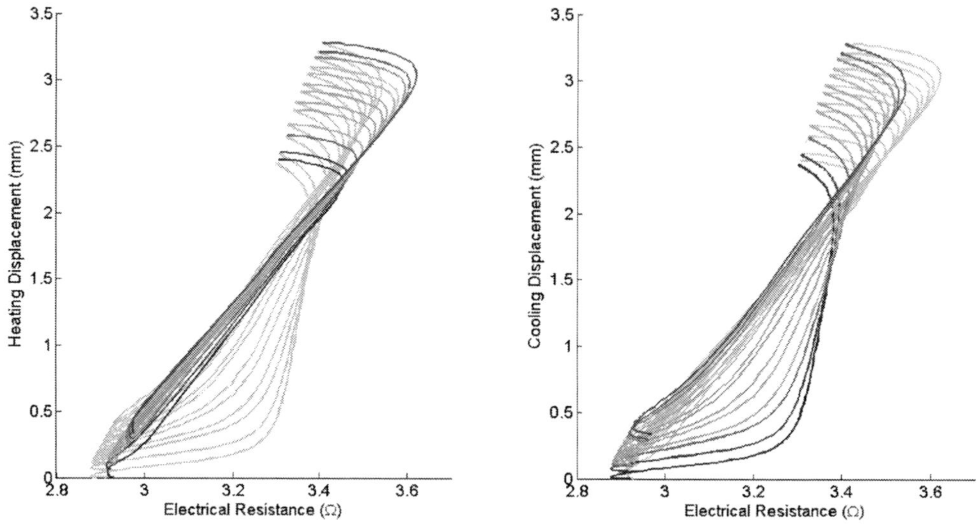

Figure 4. Family of heating (left) and cooling (right) resistance-displacement curves for various levels of applied stress. Note the remainder of the full curves are shown in gray.

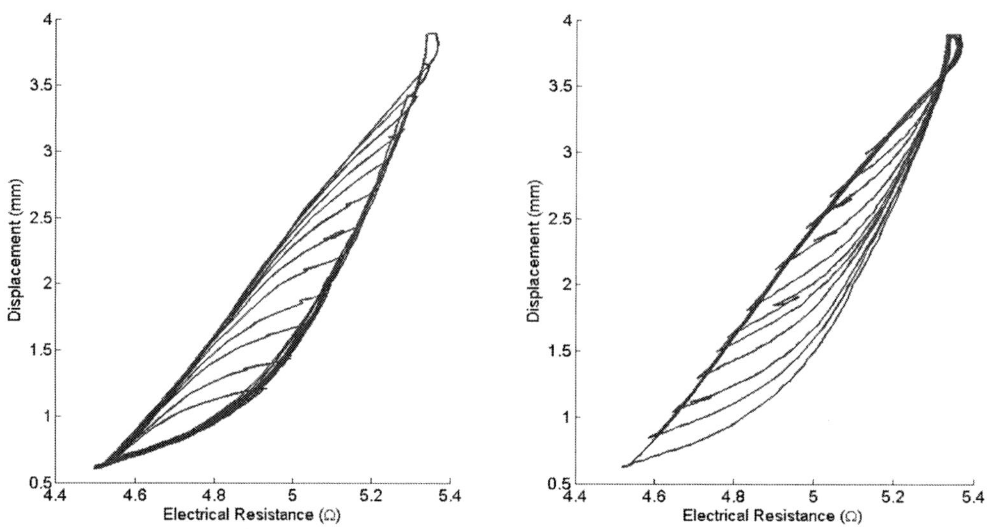

Figure 5. Family of resistance-displacement curves for partial heating (right) and cooling (left).

3. MODELING

The relationships between electrical resistance, applied stress, and strain were modeled for the purpose of predicting position in actuator control algorithms. The model was developed with careful consideration of the overall heating and cooling curves, incomplete actuation curves, hysteresis effects, and the influence of applied stress. Although the model depends upon the applied stress as an input, it may be measured or estimated depending upon the loading conditions. Temperature was not used as a parameter within the model even though it is the fundamental driving parameter for the shape memory effect. Typically, the phase composition of the material is determined from the temperature and applied stress. Instead, the electrical resistance was used as an indicator of the phase composition.

3.1 Heating & Cooling Curves

The heating and cooling curves during complete actuation form the basis of the model. From the results of material characterization, it can be seen that much of the heating and cooling curves appear to be linear. Although there exist non-linear and indeterminate regions near the upper and lower bounds, linear functions were chosen and fitted to the experimental results for both heating and cooling curves. Equations 1 and 2 give the linear functions for heating and cooling, where δ and R are the displacement and resistance, and a and b are the slope and offset. The subscripts H and C refer to heating and cooling.

$$\delta_H = a_H R + b_H \tag{1}$$

$$\delta_C = a_C R + b_C \tag{2}$$

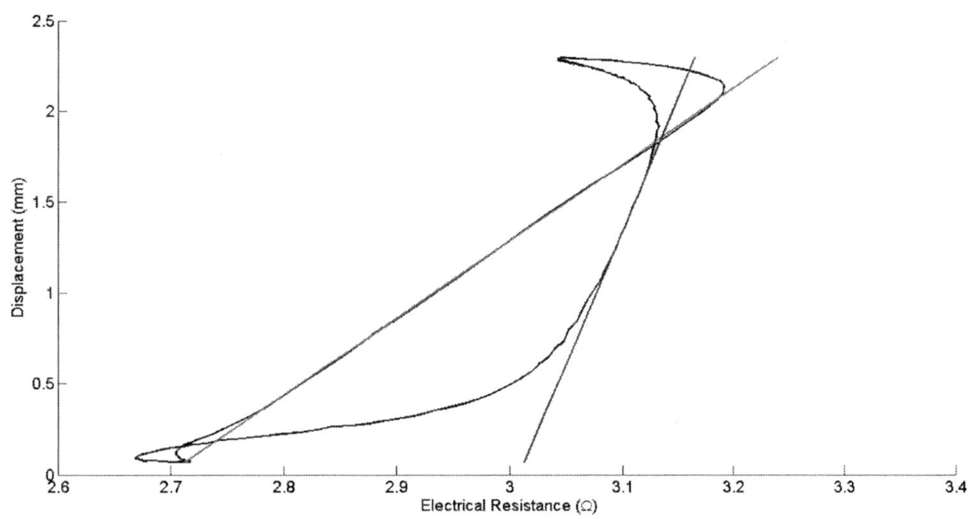

Figure 6. Linear heating and cooling asymptotes fitted to experimental data.

Figure 6 shows the results of fitting linear functions to the heating and cooling curves at a constant applied load of 0.47 N. The heating curve has a larger linear region compared to the cooling curve, which resulted in a better fit. Note that the cooling curve was considered to be the region of martensite transformation since the R-phase transformation was considered separately. The non-linear and indeterminate regions are ignored since they are relatively small compared with the overall actuation range. These linear heating and cooling curves are henceforth referred to as asymptotes since the actuator resistance and displacement tend to approach the heating and cooling curves during transition from one to the other.

3.2 Influence of Applied Stress

The experimental results of material characterization demonstrated a significant influence of the applied stress upon the resistance-displacement curves during heating and cooling. Both the slope and offset of the heating and cooling asymptotes vary, although more significantly for the latter. Analyzing the experimental data resulted in a family of heating and cooling asymptotes for which the slope and offset were functions of applied stress. These functions were modeled using quadratic polynomials as shown in equations 3 to 6, and fitted to the experimental data. Figure 7 shows the resulting fits compared to experimental data for the slope and offset parameters.

$$a_H = \beta_{aH2}\sigma^2 + \beta_{aH1}\sigma + \beta_{aH0} \tag{3}$$

$$a_C = \beta_{aC2}\sigma^2 + \beta_{aC1}\sigma + \beta_{aC0} \tag{4}$$

$$b_H = \beta_{bH2}\sigma^2 + \beta_{bH1}\sigma + \beta_{bH0} \tag{5}$$

$$b_C = \beta_{bC2}\sigma^2 + \beta_{bC1}\sigma + \beta_{bC0} \tag{6}$$

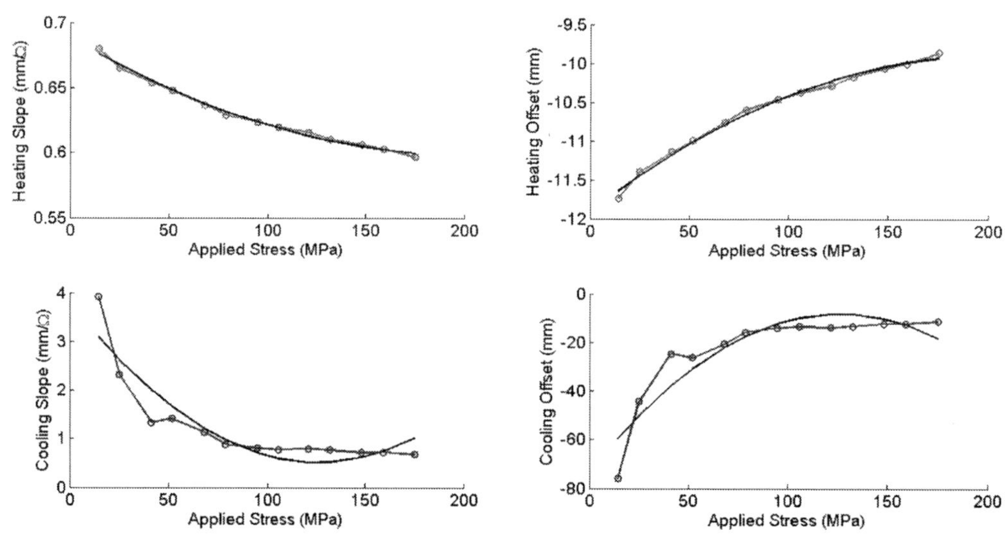

Figure 7. Polynomial fitting of the heating and cooling slope and offset as a function of applied stress.

In practice, the applied stress must be continuously updated in order to track the varying resistance-displacement curves. If sensor measurement of the applied load is unavailable, the stress may be estimated either from known loading conditions or as a function of the strain through a stiffness model. Note that higher order polynomials or other functions may be used to improve the accuracy of the slope and offset fits.

3.3 Hysteresis Effects

The heating and cooling asymptotes account for the overall changes in resistance and displacement during full actuation. However, the hysteresis effects which occur during transition from heating to cooling must be included within the model as well. Both complete and incomplete actuation during heating or cooling result in such transitions, and therefore a general rule for transition was developed based upon the experimental results of material characterization. During transition from heating to cooling, the resistance-displacement curve of the material tends to approach the heating curve asymptotically (and vice versa). Therefore, an exponential function was chosen to model the transition curve. This choice is appropriate since thermodynamic processes are characterized by differential equations which have solutions based upon exponential functions.

A differential approach was used in order to generalize the transition function for any initial point on or between the heating and cooling asymptotes. The slope of the resistance-displacement transition curve, a, is modeled as an exponential function of the distance between the current point on the curve and the projected point on the relevant asymptote, z, as clarified in Figure 8.

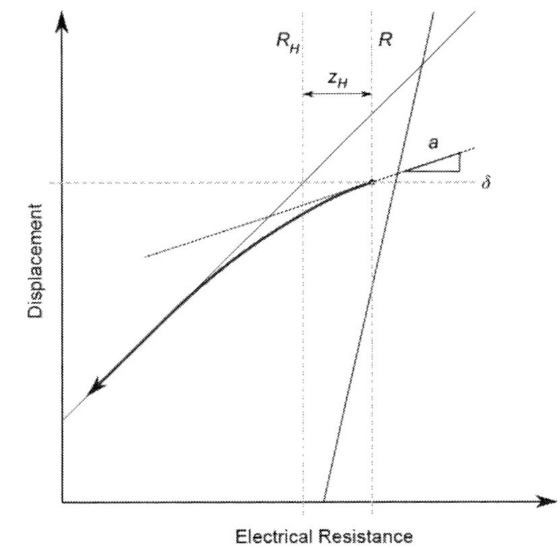

Figure 8. Diagram of the transition curve during heating with relevant parameters (heating begins at displacement d and resistance R).

Equations 7 to 10 describe the exponential differential transition curve equations during heating and cooling with initial slope a_0 and transition parameter τ. Integration of the differential transition equation then yields the actuator displacement.

$$a = (a_H - a_0)e^{-\tau z_H} + a_0 \tag{7}$$

$$z_H = R - \left(\frac{\delta - b_H}{a_H}\right) \tag{8}$$

$$a = (a_C - a_0)e^{-\tau z_C} + a_0 \tag{9}$$

$$z_C = \left(\frac{\delta - b_C}{a_C}\right) - R \tag{10}$$

3.4 Normalization

The relationships presented so far describe the behaviour of SMA in terms of electrical resistance and displacement. Although model coefficients for wires of various length and diameter may be experimentally determined for each particular application, the model may also be normalized to model strain as a function of electrical resistivity and stress. Normalization of the model is accomplished by comparing the resistance-displacement curves of a series of actuators of varying lengths and correlating electrical resistance and resistivity according to the well established relationship in equation 11, where ρ is the electrical resistivity, L is the length, and A is the cross-section area.

$$R = \frac{\rho L}{A} \tag{11}$$

The normalized resistance, λ, is computed from the initial resistance, R_0, which is a function of the martensite resistivity, ρ_M, and initial length and area, L_0 and A_0, as shown in equation 12.

$$\lambda = \frac{R}{R_0} = \frac{A_0 R}{\rho_M L_0} \tag{12}$$

The normalized displacement is simply the strain, and figure 9 shows the overlapping normalized resistance-strain curves for three actuator wires of different lengths. Equation 12 is then substituted within the model equations for transition curves, slope and offset coefficients, and heating and cooling asymptotes.

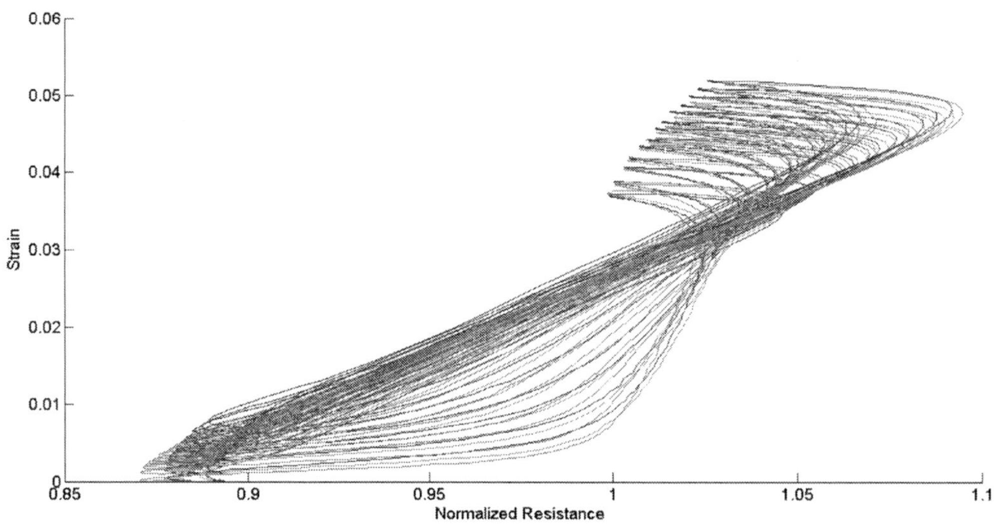

Figure 9. Overlapping normalized resistance-strain curves for actuator wires with initial lengths of 59 mm, 64 mm, and 76 mm.

3.5 Model Correlations

The model developed in this research was intended to be used in position control applications with resistance feedback in order to simplify the system and eliminate the need for position sensors. The use of theoretical models, such as the Liang and Rogers model, were deemed inappropriate since temperature feedback was unavailable. However, there is some correlation that may be made between this new model and previously established relationships.

In order to make such a comparison, a model which extends the relationships developed by Liang and Rogers which included the effects of R-phase was analyzed [15]. Equations for the slope and offset of the heating and cooling asymptotes were determined as functions of the applied stress, material properties, and actuator geometry. Since the true material properties were unavailable, values from a similar alloy were used to compare the extended theoretical model to experimental results. Figure 10 shows the slope and offset parameters as functions of applied stress for both heating and cooling.

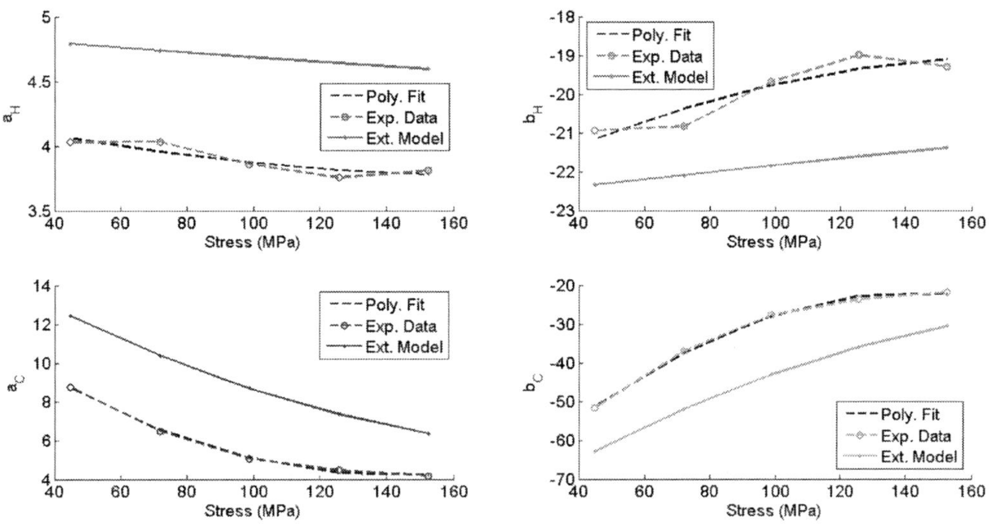

Figure 10. Comparison of slope and offset parameters determined empirically and using the extended theoretical model.

Although there is some deviation between the empirical values and theoretical values, the trends in the slope and offset functions are similar for each parameter. The correlation between the two analyses would be expected to improve significantly if the true material properties were known. Furthermore, it may be possible to use the extended theoretical model to determine the slope and offset of the heating and cooling asymptotes instead of polynomial functions, greatly improving the quality of the model.

4. POSITION CONTROL

The new model is applied within a simple PID control law in order to predict the actuator position instead of using position feedback. Electrical resistance and applied load are measured and used as inputs to the model, which predicts the actuator position and compares it to the desired position. The error then drives the PID control law that outputs the desired electrical current (which is saturated at upper and lower thresholds).

Gains for the PID control law are coarsely tuned by trial and error in order to provide a quick but stable response. Note that the position control experiment is performed to verify the performance of the model and not the control law itself. Therefore, while the difference between the desired position and the actual position is the error which concerns the end-user, the important measure for verification is the difference between the predicted position and actual position. Aspects of the control law performance such as overshoot and settling time are not of concern.

Figure 11 shows the results of the position control experiment using a bias spring to provide varying load conditions. The target jumps randomly to a new position periodically and the actuator position can be seen track the target. Since the model position is used for feedback, error between the model and actual positions arises due to inaccuracies within the model fitting process. However, the overall performance is good, with model error remaining approximately 0.1 mm throughout the experiment.

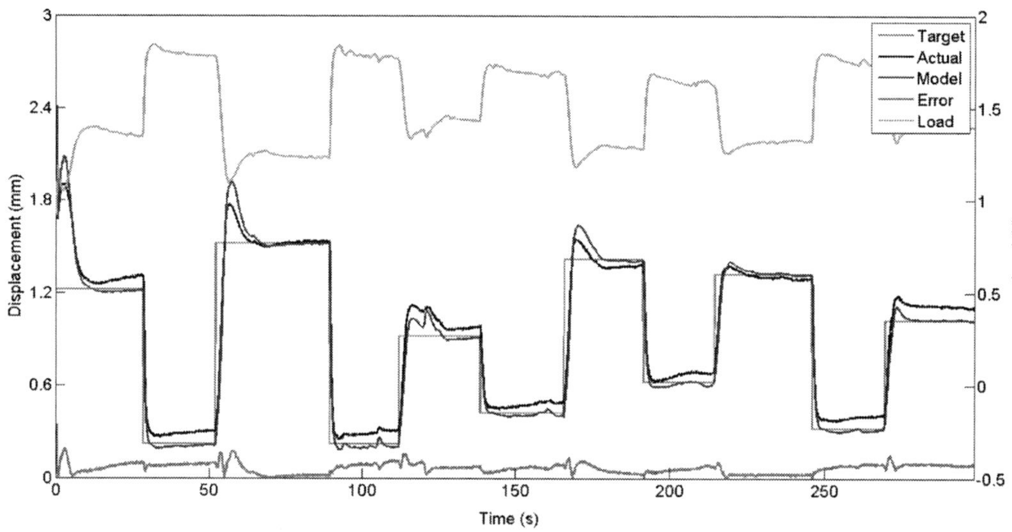

Figure 11. Results of position control based upon model feedback under bias spring load conditions. The error, shown in red, is the absolute difference between the model and actual positions.

5. CONCLUSIONS

The research presented in this paper is a significant contribution to modeling and control of SMA actuators. Measurement of electrical resistance and applied load were successfully used as feedback for a predictive model of the actuator displacement which was developed based upon the results of material characterization. The performance of the model was verified in a simple PID control experiment, demonstrating the ability to capture the overall heating and cooling behaviour, influence of applied stress, and incomplete actuation. Previous research has investigated the use of resistance feedback for position control of SMA actuators, but no solutions have addressed the presence of R-phase, influence of varying applied stress, and incomplete actuation behaviour in tandem. This model overcame all of those issues and has also been normalized for actuator wires of any length.

Further development of the model may investigate more comprehensive non-linear functions for the overall heating and cooling curves as well as the fitting functions for the influence of applied stress. Determination of the material properties may allow the model coefficients to be determined analytically, as shown in section 3.5. It may also be possible to include the effects of fatigue upon the model parameters, and predict the actuator temperature based upon the extended theoretical model. Although the model performance was verified using a PID control law, it is possible to use the predicted actuator displacement to replace sensor feedback in other types of control laws as well. Therefore, this model is easily adapted to many applications.

References

1. Es-Souni, M., et al., "On the transformation behaviour, mechanical properties and biocompatibility of two Ni-Ti-based shape memory alloys: Ni-Ti-42 and Ni-Ti-42-Cu-7," *Biomaterials*, 2001, Vol. 22, pp. 2153-2161.
2. Huang, X., et al., "Crystal structures and shape-memory behaviour of Ni-Ti," *Nature Materials*, 2003, Vol. 2, pp. 307-311.
3. Liu, Y., McCormick, P., "Criteria of transformation sequences in Ni-Ti shape memory alloys," *JIM Materials Transactions*, 1996, Vol. 37, pp. 691-696.
4. Ng, K., Sun, Q., "Stress-induced phase transformation and detwinning in Ni-Ti polycrystalline shape memory alloy tubes," *Mechanics of Materials*, 2006, Vol. 38, pp. 41-56.
5. Lan, C.-C., Fan, C.-H., "An accurate self-sensing method for the control of shape memory alloy actuated flexures," Sensors and Actuators A, 2010, Vol. 163, pp. 323-332.
6. Featherstone, R., Teh, Y. H., "Improving the speed of shape memory alloy actuators by faster electrical heating," *Proceedings of the International Symposium of Experimental Robotics*, 2004, pp. 18-21.
7. Dib, M. E., et al., "Adaptive SMA actuator priming using resistance feedback," *Smart Materials and Structures*, 2011, Vol. 20, pp. 1-14.
8. Mao, C., et al., "Experimental study on shape memory alloy displacement transducer," Earth and Space Conference, 2008, Vol. 323, pp 1-8.
9. Triantafyllou, G., Psarras, G., "Probing the reverse martensitic transformation in constrained shape memory alloys via electrical resistance," *Journal of Intelligent Material Systems and Structures*, 2010, Vol. 21, pp. 975-982.
10. Ikuta, K., et al., "Shape memory alloy servo actuator system with electric resistance feedback and application for active endoscope," *Proceedings of the 1988 IEEE International Conference on Robotics and Automation*, 1988, Vol. 1, pp 427-430.
11. Yang, K., Gu., C., "A novel robot hand with embedded shape memory alloy actuators," *Proceedings of the Institution of Mechanical Engineers*, 2002, Vol. 216, pp. 737-746.
12. Shi, Z., et al., "Accurate controlled shape memory alloy actuator for minimally invasive surgery," *Proceedings of the 2011 IEEE International Conference on Mechatronics and Automation*, 2011, pp. 817-822.
13. Meier, H., et al., "Control loops with detection of inner electrical resistance and fatigue behaviour by activation of Ni-Ti shape memory alloys," *Proceedings of the 2009 European Symposium on Martensitic Transformations*, 2009, Vol. 05006, pp. 1-8.
14. Asua, E., et al., "Sensorless control of SMA-based actuators using neural networks," Journal of Intelligent Material Systems and Structures, 2010, Vol. 21, pp. 1809-1818.
15. Brammajyosula, R., et al., "Model for resistance evolution in shape memory alloys including R-phase," *Smart Materials and Structures*, 2011, Vol. 20, pp. 1-11.

Load Monitoring using a Rayleigh Backscattering Fibre Optic System

Jérémi Bussières[1,2], Marcias Martinez[1,*], Darun Barazanchy[1], Dimitri Debruyne[3], Pascal Lava[3]

[1] Faculty of Aerospace Engineering, Delft University of Technology, Delft, Building 62, Kluyverweg 1, 2629HS Delft, P.O. Box 5058, 2600 GB Delft, The Netherlands

[2] Department of Mechanical Engineering, École de Technologie Supérieure, 1100 Rue Notre-Dame Ouest, Montreal, QC H3C 1K3, Canada

[3] Department of Mechanical Engineering, Catholic University College Ghent, KU Leuven Assoc, B-9000 Ghent, Belgium

Abstract

Optical Fiber System can be divided into three categories; discrete point sensing; integrated and distributed sensing. The present study will make use of a distributed sensing system for load monitoring of thermoplastic composite panels. The distributed sensing system utilized in this research study is based on the physical principle of Rayleigh Backscattering. Rayleigh Backscattering is produced by the interaction of the light wave with the silica impurities in the fiber. This study concentrates on the evaluation of an optical fiber distributed sensing technologies that can measure strain and temperature along the entire length of a fiber at a resolution of 5 mm over a 10 meter length at a frequency of up to 10 Hz. The primary objective of this research paper is to demonstrate the use of an embedded fiber optic line to measure strain distribution on the entire section of a thermoplastic composite panel, manufactured in accordance with the aerospace industry requirements in term of size, materials and process. In addition, the strain results were verified with respect to standard commercially off the shelf strain gauges. Detail information on the procedures used to embed the optical fiber in the panel will be described as part of this paper. The newly manufactured smart thermoplastic composite panel will be subjected to static pure bending loads. In addition, a Finite Element Analysis (FEA) model of the thermoplastic composite has been developed to verify and validate the experimental results.

1. Introduction

The aerospace industry has relied on the use of strain gauges and accelerometers for monitoring the loads of aircraft structures [1,2]. The use of these sensors has allowed the industry to acquire typical load spectrums, which are then used in full-scale test applications for fatigue life calculations, thus allowing scientist and engineers to determine remaining useful structural life of critical components. However, in military operational environments, the use of aircraft outside their established missions creates a level of uncertainty in the lifing of the structure. It is not uncommon to identify aircrafts that were initially intended for search and rescue operations in a maritime environment, operating in desert regions fulfilling a completely different mission profile. In addition, newer aircraft

* Corresponding author. Tel.: +31 (0) 152784414;
E-mail address: m.j.martinez@tudelft.nl (Marcias Martinez).

types such as Unmanned Air Vehicles (UAV) also face a complete different challenge due to the lack of historical load data, thus creating a level of uncertainty in their design usage. In flight test aircraft, many strain gauges are utilized to generate the response equations. Load monitoring systems based on strain gauge sensors present, however, certain problems such as: the gauges themselves become damaged and fatigued; they require complex wiring networks to be installed and maintained [3]. In the case of flight load monitoring systems based on accelerometers, the three dimensional components of acceleration are integrated twice to determine position, and from the position (deformation), loads can be calculated. However, the integration process is prone to numerical errors. From the generated data and the known loading geometry, predictive loading equations are generated that are sufficient to calculate the bending moments, shear loads and torsion moments on the structure of interest [4]. Thus, the need for a reliable load monitoring system that one day could be installed on every aircraft for individual aircraft load monitoring. In order to overcome many of these challenges, the use of a Rayleigh Backscattering Fiber Optic Sensor (FOS) is being considered as a potential solution for individual load monitoring of critical aerospace components.

Optical Fiber Sensors can be divided into three categories, discrete point sensing, integrated sensing and distributed sensing [5]. Rayleigh Backscattering belongs to the subfield of distributed sensing. In discrete point sensing systems, the use of Fiber Bragg Grating (FBG) has been heavily used as potential replacement to standard strain gauges [6,7]. Many researchers have also been able to lay down multiple FBGs into a single continuous line thus providing a network of sensors [8].

In Rayleigh Backscattering (distributed sensing), the backscattering phenomenon is based on a specific form of light wave reflection, were the light wave is reflected back from its source in a diffuse manner. In contrast, a specular reflection, like that one observed with a mirror, the light wave is reflected from a single incoming direction into a single outgoing direction [5]. However, depending on the wavelength of the backscattering, this one can be subdivided into three specific peaks known as the Raman, the Brillouin and the Rayleigh as shown in Figure 1. In addition, the backscatter spectrum can also be subdivided into stoke and anti-stoke components. As the light wave interacts with the medium through which it is travelling, the molecules that compose the medium absorb energy due to this interaction, which is known as stoke scattering. Conversely, if the molecules in the medium lose energy due to the interaction with the light wave, it is known as the anti-stoke scattering [5].

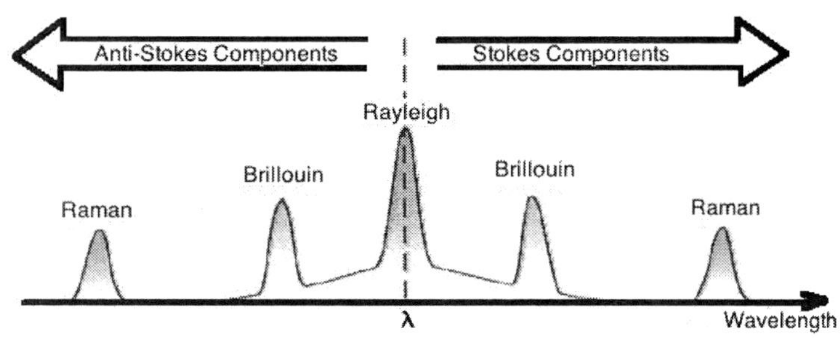

Figure 1: Backscattering Spectrum [5]

Raman scattering occurs due to the interaction of the light wave between the photons (basic unit of light) and the thermal vibration of the silica molecules in the fiber material. Raman scattering is thus, very susceptible to temperature variations. Brillouin scatter occurs due to photon-photon interaction, thus making it susceptible to fiber geometry, fiber density, temperature and strain [5]. Although many researchers have studied the Brillouin and the Raman scatter, the primary objective of this study is to evaluate the effects of Rayleigh Backscattering and its potential use in load monitoring applications. In contrast to the Raman and Brillouin, Rayleigh Backscattering is produced by the interaction of the light wave with the silica impurities of the fiber. This interaction produces a reflection in the range of 10^{-7} m^{-1}, which can be detected and analyzed with an Optical Time Domain Reflectometry (OTDR), Optical Frequency Domain Reflectometry (OFDR) or a Polarization OTDR (POTDR) [5,8]. In our study, an OFDR was used which allowed for the measurement of an optical fiber deformation at a constant temperature with a very high spatial resolution (δz).

An OFDR consist of the following items, as depicted in Figure 2.

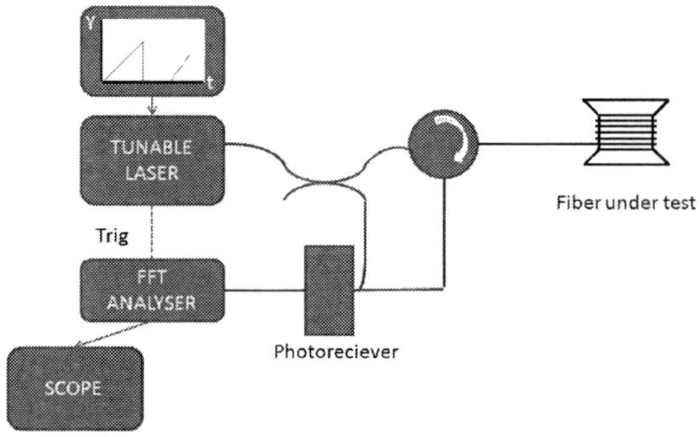

Figure 2: Main components of an OFDR [5]

The special resolution, mathematically expressed in Eq. 1, of an OFDR is a function of: *c,* the speed of light, Δv, the bandwidth of the tunable laser depicted in Figure 2, and the group index n_g. The group index is defined as the ratio of the group velocities of light traveling through vacuum to that of the light traveling through a medium. The group velocity is defined as the apparent velocity of the envelope containing a group of waves propagating through a medium [9]. Group velocity can be independent of phase velocity, i.e., the distance it takes for a single wave cycle to propagate at a given frequency is not exactly the same distance travelled by the entire group of waves. The concept of group velocity relates a group of waves of different phase speeds that travel as a group. However, it is important to note that the bandwidth of the tunable laser (Δv) can be very high allowing for really small special resolutions [10].

$$\delta z = \frac{c}{2n_g(\Delta v_{max})} \tag{1}$$

The maximum measurement range d_{max}, in OFDR systems is a function of the frequency resolution of the laser, δv, as expressed in Eq. 2:

$$d_{max} = \frac{c}{2n_g(\delta v)} \tag{2}$$

The OFDR used in this study allows for the operator to measure strain and temperature over a fiber up to 10 meter in length at a data acquisition frequency of 100 Hz for a strain gauge length of 5 mm. In addition, the OFDR can be setup to measure at a lower data acquisition frequency of 23 Hz with a gauge length of 1.25 mm. Regardless of the type of backscatter equipment used for strain sensing, the principal objective of measuring distributed strain or temperature is of interest in the aerospace industry. Much of the technology utilized to date provides discrete points, which present a major challenge in obtaining the full strain field and load distribution along a fuselage or a wing of an aircraft, thus the need for a system that can accurately monitor strain distribution in critical components. The following section discusses how an optical fiber embedded in a composite panel, in combination with a Rayleigh Backscattering system was utilized to measure the strain distribution of the panel subjected to small bending loads.

2. Experimental Setup

In order to verify the capability of a Rayleigh Backscattering system (LUNA ODISI-B) to measure bending loads, a composite panel was manufactured with an embedded FOS. The embedding of fiber on the panel was a point of concern, in that the presence of the fiber could lead to the initiation of a delamination site. However, the approach followed as part of this research, was to over design the composite panel and incorporates the fiber at a layer that does not play a significant role on the integrity and load carrying capabilities of the panel. In addition, only small bending loads were applied to determine the sensitivity of the technology at low loads. The next section, provide a detail explanation on the composite panel manufacturing and the test bench setup.

2.1 Laminate Manufacturing (Step 1)

A small laminate of 370 mm x 170 mm x 1.6 mm, was manufactured from semi-preg T300JB 5-harness satin carbon fiber fabric with a poly-phenylene-sulfide matrix (5H-CF/PPS) from TenCate Advanced Composites. The laminate was made in a two-step operation: first a 4-plies laminate was first developed followed by a second set of plies co-cured with the fiber sensor placed in between the composite laminates. The 6 plies of semi-preg were cut using a Gerber Technologies DCS 2500 cutter with a 25.4 mm cutting wheel. The laminate was cut to final dimensions using a Unitom Water-Jet cutter.

Four plies were assembled in a symmetrical stack sequence and held together with an ultrasonic welder. The periphery of the panel was enclosed with an aluminum tape to prevent the matrix from flowing out. The 4-plies laminate was placed between two molds and manufactured by inserting it into a JOOS hot-platen press at 320°C with a consolidation pressure of 10 bars.

2.1.1 Mold preparation

The molds are two 600 mm x 600 mm x 1 mm steel plates. Both surfaces were free of scratches, cleaned with a degreaser (PFQD by PT Technologies Europe) and treated with a release agent (227-CEE by Marbocote Ltd). Three layers of Marbocote 227-CEE were applied in different directions, with a drying time of five (5) minutes between each layer. Manufacturers' application instructions were followed as shown in [11, 12].

2.1.2 Fiber Optic Installation

The fiber optic was bonded to the top surface of the cured laminate by following this procedure, based on Luna Innovations Fiber Bonding Tutorial found in [13]. A summary of the installation process is described next and graphically depicted in Figure 3:

1. The aluminum tape was removed from the cured laminate and cleaned with PFQD followed by a secondary clean with isopropyl alcohol.
2. For maximal mechanical adhesion, a sanding of the surface with #300 sandpaper was performed in all directions using silica carbide where the fiber was bonded.
3. The bonding surface was cleaned with a wipe impregnated with alcohol, followed by drying the surface with a dry wipe.
4. The routing of the fiber on the panel was marked using a graphite pencil. The x-axis was set parallel to the warp direction.
5. The fiber was held in place using small pieces of Kapton tape (approximately 3mm x 3mm in size). In order to ensure good contact between the fiber and the laminate surface, the fiber was pulled tight. A 20 mm diameter punch was used to ensure the fiber followed a 10mm radius, thus avoiding potential damage or breaking of the fiber.
6. Small drops of adhesive (2-Ethyl Cyanoacrylate) were used to adhesively bond the fiber to the laminate. The adhesive was gently spread along the length of the fiber using a clean swab.
7. In the regions were the fiber was held in place with Kapton tape, the tape was removed and small drops of adhesive were used to ensure fixing the fiber to the laminate.
8. Excess adhesive was gently removed with a sharp tool and ensuring that no damage is introduce into the panel.

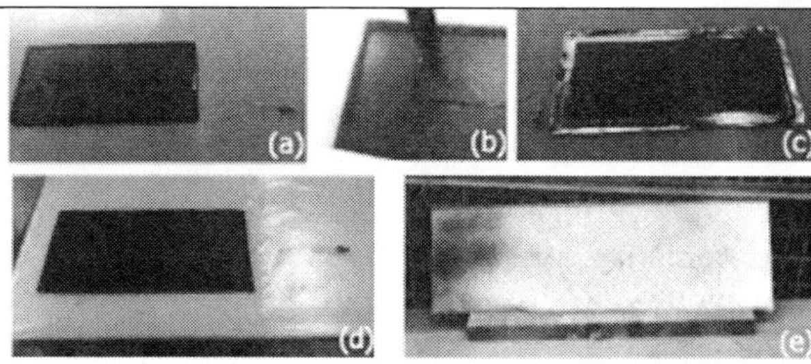

Figure 3: (a) Bonding of the fibre optic line onto the 4-plies laminate (b) 20 mm punch ensure the fibre to follow the desired radius (c) the instrumented laminate, enclosed between two plies, ready to be manufactured (d) Final 6-plies laminate with an embedded fibre optic (e) Preparation of the laminate for Digital Image Correlation.

The final fiber placement within the laminate is shown in Figure 4.

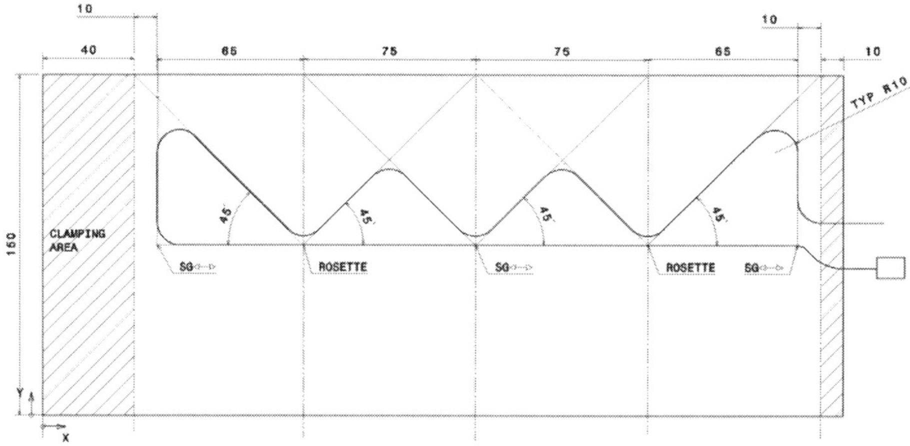

Figure 4 - Location of the strain gauge and rosettes, plus the path of the embedded fibre optic. Laminate is clamped on the left and the load is applied on the right.

2.1.1 Laminate Manufacturing (step 2)

The instrumented laminate was enclosed between two additional plies. The egress and ingress of the fiber optic was placed on the side of the laminate, between plies 5 and 6. Thus, the fiber optic was protected and out of the way of the aluminum tape. This ensured it wouldn't be enclosed in any potential matrix flow. The laminate was then placed in the press. Luna Technologies, recommends a maximum temperature of 80 °C for their connector. Thus, the connector was placed outside of the hot area of the press and enclosed in an insulating fabric (vacuum breather) in order to prevent damage to the connector. The entire cycle was monitored with the Luna ODISI-B DAQ system to ensure the integrity of the fiber optic, while it was subjected to high temperature and pressure. After the curing process, the aluminum tape was removed and the laminate was cut to final size. The fiber optic enclosed within the laminate created no visible bump or thickness variation on the upper surface of the laminate. Only a print-through of the fiber optic was visible.

3. Test Bench

The test bench was manufactured to meet the dimensions of the instrumented laminate. All parts of the fixture were made of aluminum, with the exception of the depth micrometer and the fasteners. The laminate is clamped on one end, between two 25 mm x 5 mm plates as shown in Figure 5. On the opposite side a 25 mm x 10 mm plate with three ports to place a micrometer for point load displacement is attached (shown in Figure 5). Those ports are used to install a depth micrometer, which is used to apply a deflection on the laminate. However for the bending test cases, the micrometer is used on the central port, while the side ports are offset by 60 mm and can be used for torsion and combined loading deflections. When the laminate is installed, the longitudinal direction of the fiber is aligned with the center of the plates. The test bench was setup by aligning the laminate thickness in the direction gravity in order to minimize the effect of gravity in the measurement.

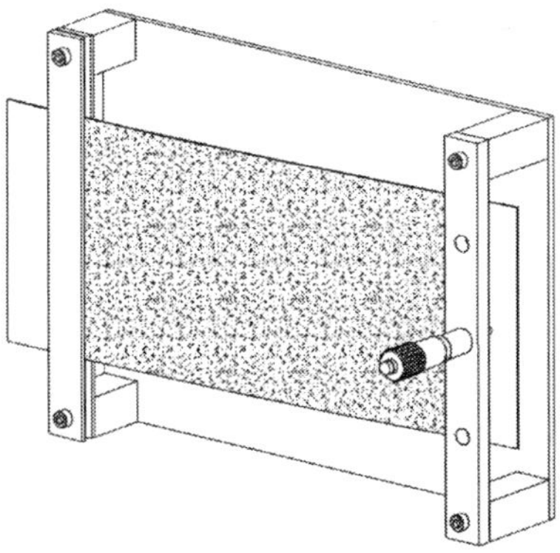

Figure 5: Test Bench Setup

4. Results

The results presented in this study consist on applying a 10 and 20 mm tip deflection in bending on the composite laminate. The result of the strain field measured by the Rayleigh Backscattering system is shown on **Figure 6**. As seen in this figure, the strain distribution matches those measured with standard commercially off the shelf strain gauges that were strategically located on the surface of the panel. However, it is also visible that for a 10 and 20 mm the strain level has an unacceptable noise. The extreme ends of the panel (at 0 mm and 300 mm) as shown in **Figure 6** could not be verified since the fiber did not cover the end points of the plate. However, the initial results demonstrate the ability of the system to capture with great level of accuracy the strain on a composite plate.

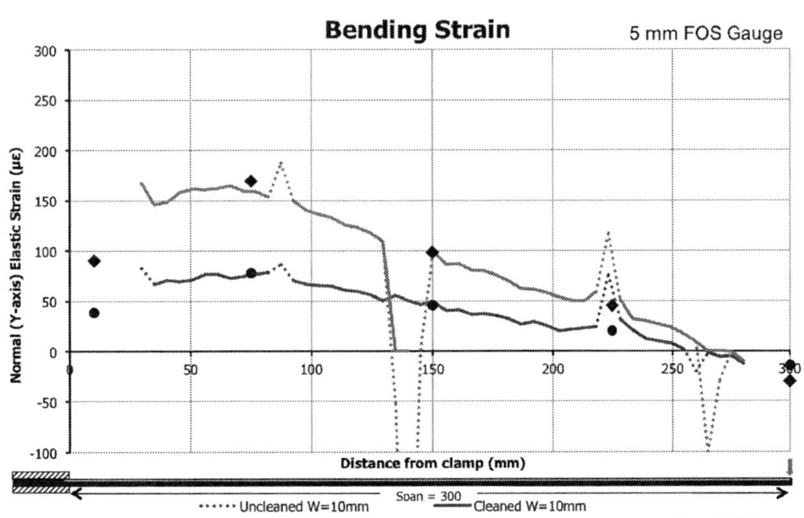

Figure 6: Bending Strain vs. Distance (5mm Fibre Optic Gauge)

4.1 Finite Element Verification

As a verification of the experiment, a Finite Element Analysis (FEA) model was developed using the commercially software package Abaqus/CAE 6.11. Due to the symmetry of the composite plate, the model size was reduced to 340 mm by 75 mm by 1.6 mm. The symmetric boundary condition was applied at the top surface of the plate as shown in Figure 7, in order to reduce the number of elements needed to perform the analysis. A uniform displacement field was applied on one extreme edge of the plate. The clamping of the plate was modeled with two rigid bodies as shown in Figure 7. These two rigid bodies were set as un-deformable with the same approximate dimensions to those used during the experimental setup (40 mm by 100 mm by 5 mm).

The composite laminate plate was meshed using 20-node reduced integration quadratic brick elements (C3D20R). The meshed plate was assigned material properties as shown in Table 1, which corresponded to the carbon fabric material used during the experiments. The composite layup option in Abaqus was used to model the plate; the element width and length were set 5 mm and the number of elements through the thickness was set to one. Due to the composite layup method used in the model, the strains could be extracted from the layer without increasing the number of elements through the thickness. Furthermore, changing the element size in width and length did not affect the results. The clamps were meshed using 4-node 3D bilinear rigid quadrilateral elements (R3D4). The element mesh size for the clamps was set to 5 mm in all directions. The FEA simulations were executed using a computational cluster HPC12 at Delft University of Technology. Each simulation was set with four Intel XEON E 5520 (quad core) and 24 GB 1333 ECC RAM.

Table 1 Material properties used for FEA model

Material property	Values
E_1[GPa]	55.49
E_2[GPa]	55.49
E_3[GPa]	13.70
ν_{12}[-]	0.08
ν_{13}[-]	0.08
ν_{23}[-]	0.41
G_{12}[GPa]	4.17
G_{13}[GPa]	4.17
G_{23}[GPa]	3.48

The strain in the x-direction (E11) was extracted at the fifth layer from the bottom (z=1.33 mm) starting at the clamp (x=40 mm) and ending on the right boundary (x=340 mm) of the plate. The extracted strain was then compared to the experimental results, as shown in Figure 7 and Figure 8 for a deflection of 10 mm and a deflection of 20 mm respectively.

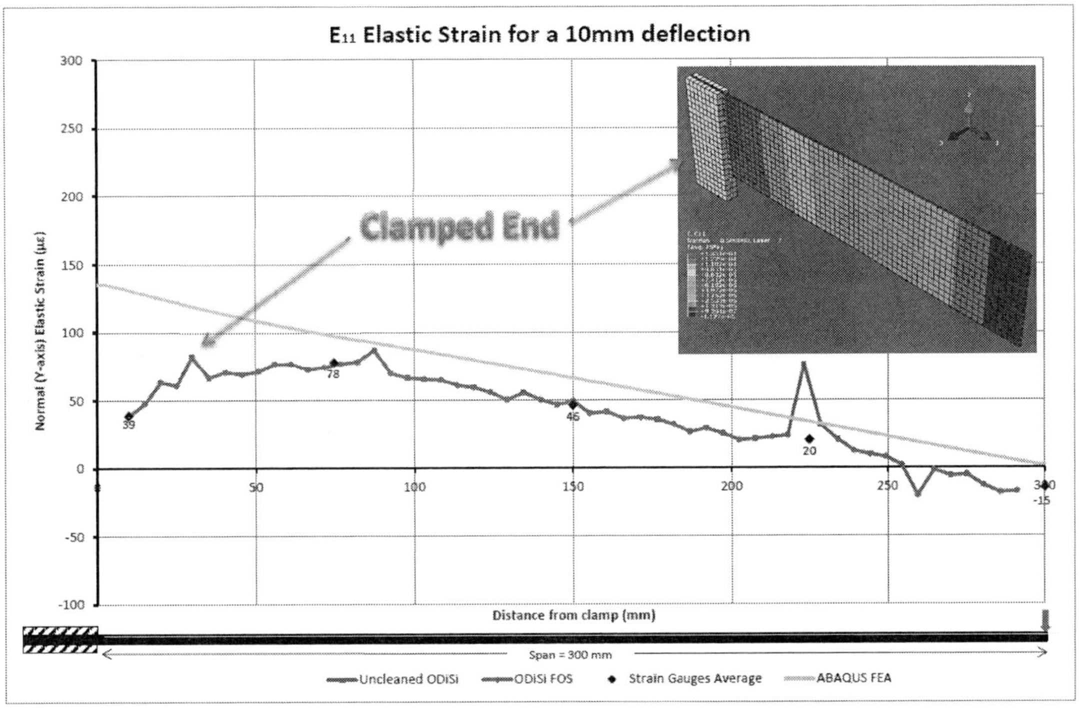

Figure 7: E11 Elastic Strain for 10 mm deflection at the tip.

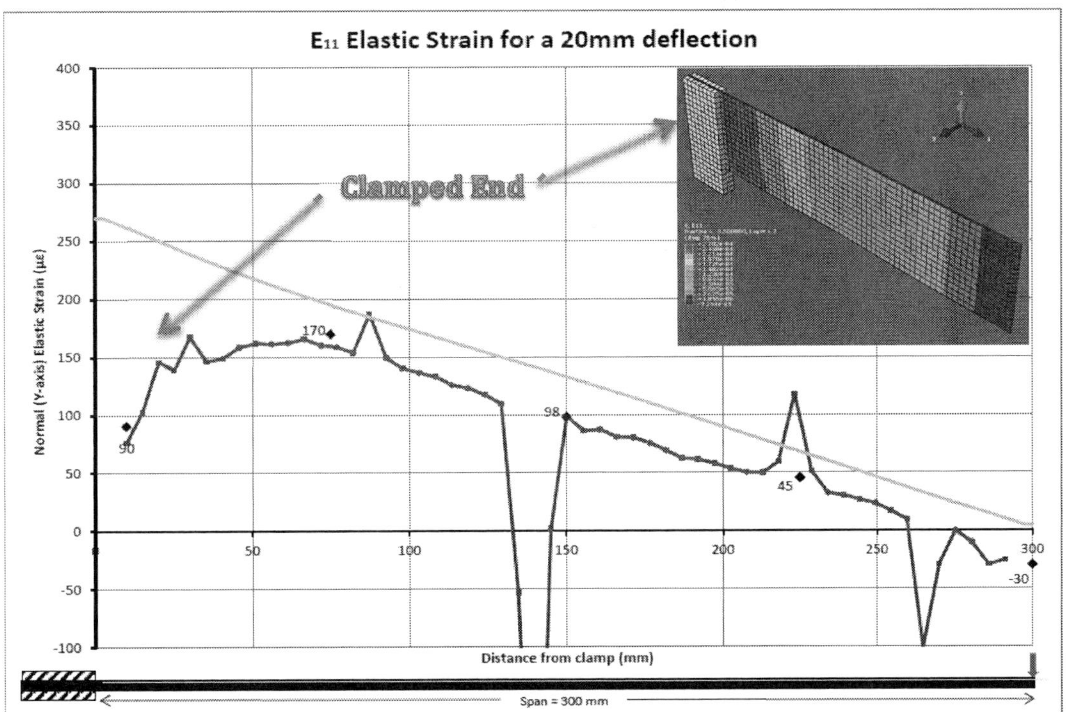

Figure 8: E11 Elastic Strain for a 20 mm deflection at the tip

5. Discussion

As shown in Figure 7 and Figure 8 the results measured by the Rayleigh Backscattering system matches very closely those of the commercially off the shelf strain gauges. The ability of a Rayleigh Backscattering system to capture higher strain field had been previously documented in [10]. The present test was primarily designed in order to study the ability of an embedded fiber optic to capture lower strain field values of a composite laminated subjected to bending loads. The finite element model results matched those of the Rayleigh Backscattering and the commercially off the shelf strain gauges. However, the model is unable to duplicate the lower strain values close the clamped tip. A potential cause for this behavior could be the lack of rigidity on the clamping plates in comparison to those modeled in the finite element model. During the experiments it was observed that the plate showed signs of movement and warping. Finally, the method of applying the deflection to the plate was not exactly the same for the experiments as in the FEA analysis. For the experiments the deflection was concentrated at one point, while for the FEA analysis the deflection was uniform over the boundary of the plate. It is likely that this also contributed to the difference between the obtained results.

Nevertheless, the FEA analysis proved a useful tool for a first order estimation of the strains in the plate. It is important to note that the slope of strain through out the panel matches that of the FEA. Therefore, improving the model and experiments can results in more accurate results in future scenarios where the composite plate will be subjected to bending, torsion and dynamic load spectrum. In order to simulate these conditions, a new model will be constructed that may includes: (i) a better clamping method for the experiments such that the plate is not able to warp, (ii) replacing the point deflection in the experiments for a uniform displacement at the boundary such that loading case do not differ.

It is important to note that the strain field (250 microstrain) produced by a 10 or 20 mm deflection of the composite plate is small in comparison to those experienced by composite panels found on wing panels. These results are part of a very preliminary study that envisions future development of embedded sensors in composite panels. In addition, the techniques require the operator to develop an algorithm to minimize or discard unnecessary data. The amount of data that may be obtained with a 10 meter fiber optic, at a data acquisition rate of 100 Hz may present some major data management challenges.

6. Conclusions

The use of Rayleigh Backscattering for strain measurement on a composite panel with an embedded composite fiber has been demonstrated. The current results demonstrate the ability of this technique to measure strain accurately. However, an unacceptable level of noise in the measurement makes the technique cumbersome to use for in-field applications. In addition, a FEM was created to verify the experimental results and the results demonstrated that the values measured by the Rayleigh Backscattering system match those of the FE model.

Although the Rayleigh Backscattering system has a primary advantage of providing data along the entire length of the fiber, this may represent a challenge during the post processing of the signal in real-time applications. There is a requirement of an algorithm that may store only the peak and valley values of the data during in-field operations. This would provide the operator of the technology the necessary information on the loads suffered by the structure for useful life estimation calculation.

7. References

[1] Reed, S., "Encyclopedia of Structural health Monitoring", Edited by Christian Boller, Fu-Kuo Chang and Yozo Fijino, 2009, John Wiley & Sons, Ltd. ISBM: 978-0-470-05822-0.

[2] Engineering Sciences Data Unit, "Estimation of the endurance of civil aircraft wing structures", IHS Eng. Sciences Data Unit Data Items #79024, October 1979.

[3] Wu J, Yuan S, Shang Y and Wang Z 2009 Strain distribution monitoring wireless sensor network design and its evaluation research on aircraft wing box Int. J. Appl. Electromagn. Mech. 31 17–28

[4] Martinez, M., Rocha, B., Li, M., Shi, G., Beltempo, A., Rutledge, R., Yanishevsky M., "Load monitoring of aerospace structures utilizing micro-electro-mechanical systems for static and quasi-static loading conditions", Smart Materials and Structures, Vol 21, Number 11, doi:10.1088/0964-1726/21/11/115001, 2012.

[5] http://www.rp-photonics.com/group_index.html (Last Accessed Nov 7th, 2011).

[6] Foote, P. D. and Read, I., "Optical sensors for aerospace structural monitoring", paper presented at the *IEEE Colloquium (Digest),* 1995, 87, 2/1-2/6.

[7] Honglei, G., Gaozhi, X., Mrad, N. and Yao, J., "Fiber Optic Sensors for Structural Health Monitoring of Air Platforms", Journal of Sensors, 2011, 11, 3687-3705; doi: 10.3390/s110403687.

[8] Mrad, N., "Optical Fiber Sensor Technology: Introduction and Evaluation and Application", Published by John Wiley & Sons, Inc., New York, 2002, *Encyclopedia of Smart Materials*, 2, 715-737.

[9] http:/en.wikipedia.org/wiki/Group_velocity (Last Accessed Nov 7th, 2011).

[10] M. Martinez, A. Beltempo, M. Yanishevsky, R. Rutledge, B. Rocha, "Load Monitoring on SHM Platform 1A using LUNA Technologies, Distributed Sensing Fibre Optic System", National Research Council of Canada, Unclassified Lab Technical Report, LTR-SMPL-2012-0047, March 2012.

[11] http://www.pttechnologies.com/images/stories/PDF/pf-qd brochure_11.06.12.pdf (Last Accessed August 2013).

[12] http://www.ecfibreglasssupplies.co.uk/images/TechnicalDataSheet/1819.pdf (Last Accessed August 2013).

[13] http://www.youtube.com/watch?v=EW9U98N7xh4 (Last Accessed August 2013).

Time Reversal of Lamb Waves for Damage Detection in Thermoplastic Composites

Maria Barroso-Romero[1,2*], Darun Barazanchy[2], Marcias Martinez[2], Roger M. Groves[1], Rinze Benedictus[2]

[1] Aerospace Non-Destructive Testing Laboratory, Faculty of Aerospace Engineering, Delft University of Technology, Kluyverweg 1, P.O. Box 5058, 2600 GB Delft, The Netherlands
[2] Faculty of Aerospace Engineering, Delft University of Technology, Kluyverweg 1, P.O. Box 5058, 2600 GB Delft, The Netherlands

Abstract

The objective of this study is to evaluate the use of correlation methods as a damage detection criterion in the application of the Time Reversal of Lamb Waves (TRLW) approach. The dispersion and attenuation characteristics of the waves are used for determining the maximum area to be inspected in a two transducers configuration. The Finite Element (FE) Model and experimental data are compared for verification of the method. The experimental setup consisted of a pair of piezoelectric transducers, working both as actuator and sensors, separated by a predetermined distance. Different defect sizes and positions with respect to the transducers were introduced on both aluminum and thermoplastic composite plates. A waveform generator was used to excite the piezoelectric actuators. An oscilloscope was used for acquiring the signals from the transducers. Those signals were processed in Matlab™ for the application of time reversal and a correlation analysis. The damage consisted of a hole through the thickness varying its diameter and location. The FEM simulates the propagation of Lamb waves in aluminum plates in order to analyze TRLW for damage detection. Lamb waves were generated using a 5 cycle Hanning window at different frequencies. The FEM consisted of standard 3D elements (C3D8I) for the plates and PZT elements (C3D8E) for the transducers. In order to simplify the model the PZTs were assumed to be perfectly bonded to the plates. The FEM analysis was mainly focused on the S_0-wave. The indication of the presence of damage is performed using the correlation between the original actuation and the time reversed signal.

1. INTRODUCTION

The use of Time Reversal of Lamb Waves (TRLW) for damage detection is by definition a baseline free technique, and especially suited for composite materials [1, 2]. Before it can be implemented as part of an inspection system, the limitations of the method must be understood. The use of correlation in combination with TRLW is discussed in this research.

In order to guarantee the integrity of a structure, effective Structural Health Monitoring (SHM) systems for real-time application are needed [3]. Such systems should be able to detect the presence of damage and

* Corresponding author, M.BarrosoRomero@tudelft.nl

evaluate its severity along the operational life of the structure, preferably without the need of a baseline from the undamaged structure.

In damage detection several parameters influence the capabilities of a Non Destructive Inspection (NDI) technique. For instance considering the material and geometry of the structure being inspected and the size and type of the defects being analyzed allows the user to select suitable NDI from the broad range available.

We consider the use of ultrasonic guided waves propagating in thin walled structures, or Lamb waves. Their main difference with respect to the bulk waves is their dispersive and multimodal characteristics, which makes their analysis highly complex for high frequencies [4]. The material and inspection requirements define the frequency range, in which only the desired modes are actuated, in order to make the analysis less complex [5]. The use of Lamb waves with their propagation characteristics, dispersion and attenuation, allows the inspection of relatively wide areas without the need of a point-by-point scan [6], reducing the inspection time. Lamb waves can propagate long distances with relatively low energy loss. The velocity dispersion and attenuation of the waves provide critical information on possible scatters in the material [7].

In the classical application of Lamb waves for damage detection, the behavior of the wave on the undamaged structure is first captured, in order to obtain a baseline response. In presence of damage in the structure, the Lamb wave behavior changes, thus providing a different response for the same actuation signal [8, 9]. Both baseline and damaged responses are then compared for damage detection. The same process needs to be repeated from at least three different locations in order to determine the position of the damage [9].

Mostly, two configurations are used for damage detection based on Lamb waves: pulse-echo and pitch catch. In the pulse-echo configuration the actuation signal is sent from the location of the actuators. This signal propagates through the structure and is reflected by defects and boundaries along the wave propagation path. These reflections are then captured back at the actuation location for comparison with the baseline. Reflections from the edges may complicate the analysis. In the case of pitch-catch configuration, the response is captured from a different location from which it originated, thus registering the wave changes after the propagation. Both pulse-echo and pitch-catch techniques can be used with Time Reversal of Lamb Waves (TRLW) analysis.

TRLW applied to damage detection represents a promising technique for SHM applications in both metallic [2], and composite materials [10]. One of the primary advantages of the time reversal technique is that it does not require a baseline of the signal, since the actuation signal is compared to its own time reversal in each interrogation process [2]. The simplest time reversal process consists of the comparison of the time-reversed Lamb wave signal, with the original sent signal. That is an actuation signal is sent, then captured at a second transducer, reversed in time and sent back to the first transducer. This last received signal is then compared to the original actuation. This technique uses the transducers as both actuator and sensor. In this technique the time-reversed signal can be sent from the same sensor from which it has been recorded [11]. Nevertheless, a MTRM (Modified Time Reversal Method) was proposed by Watkins [10], consisting of one transducer acting as actuator and another 4 as sensors. The signal received at each sensor is time-reversed and then sent, not from the position of the sensor, but from the original actuator. They show the similarity of the results from the MTRM with respect to the classical approach.

In the classical use of Lamb waves, the defects are more easily detected in homogeneous materials, such as metals, while for composite materials the propagation velocity of the waves strongly depends on the direction of propagation, as well as on the layup of the structure [6], as the propagation is faster in the direction of the fibers. It is in this aspect where time reversal offers its major advantage for damage detection in composites, as the inhomogeneities are compensated in the time reversal process [1]. As stated by the reciprocity theorem, the wave experiences the same changes in both directions along the same path [12]. This property implies that the velocity dispersion is totally compensated by the time reversal process, and this applies even for inhomogeneous materials [7]. Additionally, the short time needed for the interrogation (time between the original actuation is sent and the time reversal response is received, is of the order of seconds) makes the process also ambient-independent, as the temperature and load variations and their influences can be neglected. Only the post-processing can take longer depending on the selected approach, i.e. considering time domain, frequency or wave number in the analysis will require different computing times.

The number of transducers needed for the damage detection in a structure depends on the structure itself, on the size and type of the damage, and on the frequency of excitation, which determines the wavelength and velocity of the propagating wave, thus defining the area that can be efficiently inspected. The optimization of the number of transducers of a network needed for a certain area depends directly on the one-to-one transducer capability of the system, that is the area covered by each pair of transducers, given a structure and an actuation pulse. The objective of this research is to analyze the use of the TRLW with correlation as criteria for damage detection, in this one-to-one configuration, in its application for thermoplastic materials.

The next section gives an introduction to TRLW, and the references to the correlation methods used in the study. Section 3 includes a description of the experimental procedure that was used to assess the ability of TRLW for damage detection for homogeneous and inhomogeneous materials, in a pitch-catch configuration. Two transducers were used for both actuation and sensing, and the classical time reversal process mentioned before was applied. The damage consisted of a hole through the thickness varying its diameter and location. Presence of damage was determined by cross correlating the actuation and reconstructed signals. In order to exclude the influences of the composite inhomogeneities in the correlation values, an aluminum plate was also tested and in this case the experimental results were validated with a Finite Element Model (FEM), as explained in Section 4. The experimental results for the thermoplastic composite, and the comparison between numerical and experimental results for aluminum are gathered in Section 5. These results are discussed in Section 6. Finally some conclusions from this study are given in Section 7.

2. TIME REVERSAL OF LAMB WAVES FOR DAMAGE DETECTION

2.1 Lamb Waves

In this section the propagation of elastic waves in a thin wall or plate-like structure is considered. These guided waves are named Lamb waves, which were explained for the first time in 1917 by the mathematician Horace Lamb [13]. The complexity of these waves comes from their dispersive and multimodal propagation. This means that the waves propagate as different modes, and in each mode the velocity depends on the frequency, phenomenon that is known as group velocity dispersion, shown by the dispersion curves. These curves, are obtained from the derivation of the equations of motion in the absence of the body forces, Eq. 1, imposing the boundary conditions for plane strain [4] in Eq. 2, with $\vec{u}$ representing displacement field, ρ density, σ stress field, t the variable time, and x_i the space variables in the 3 orthogonal directions,

$$(\lambda+\mu)\nabla(\nabla\cdot\vec{u})+\mu\nabla^2\vec{u}=\rho\ddot{\vec{u}} \tag{1}$$

$$u_1=u_1(x_1,x_3,t),\ u_2=0,\ u_3=u_3(x_1,x_3,t),\ \sigma_{13}(x_3=h)=\sigma_{33}(x_3=h)=0 \tag{2}$$

Using Eq. 1 and 2, the dispersion curves of the Lamb waves in Eq. 3, provide the variation of phase or group velocity as a function of the product of frequency and plate thickness

$$\frac{\tan(qh)}{\tan(ph)}=\frac{4k^2qp\mu}{(\lambda k^2+\lambda p^2+2\mu p^2)(k^2-q^2)} \tag{3}$$

in terms of Lamé constants λ, μ, wave number k, thickness h, and the parameters p, q, defined as

$$p^2 = \frac{\omega^2}{c_L^2} - k^2, \quad q^2 = \frac{\omega^2}{c_T^2} - k^2 \tag{4}$$

with ω, c_L and c_T being the angular frequency and the longitudinal and transverse velocities, respectively. Eq. 3 was coded in Matlab™ and the dispersion curves were calculated for aluminum Al-6082-T6 and carbon reinforced polyphenylene sulfide, CF/PPS, taking the approximation of homogeneous material for the composite plate case. These curves were experimentally validated, as shown in Fig. 1. The dispersion curves allow the choice of the range of frequencies in which only the first modes S_0 and A_0 are excited and where the variation of the velocity with the frequency is as low as possible in order to reduce the velocity dispersion. In this case we chose the range between 125-350 kHz for a 2mm thick plate.

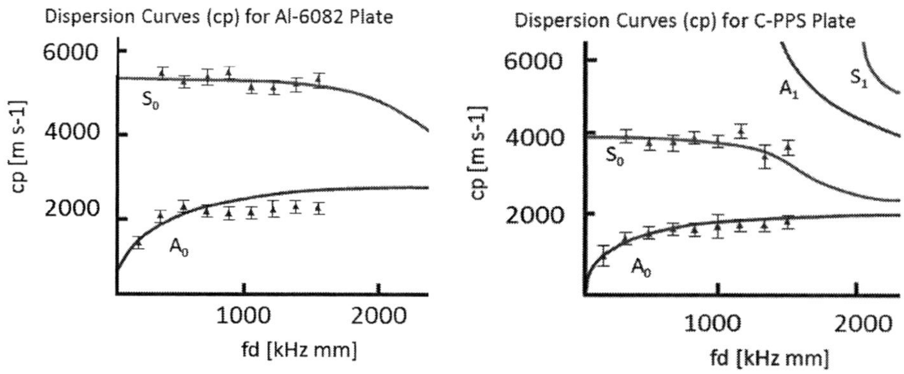

Figure 1. Dispersion curves for Al-6082 (left), and CF/PPS (right), and experimental validation of S_0 and A_0 modes, for 2mm thick plates

2.2 Time Reversal of Lamb Waves

The Time Reversal of any process consists on reversing the variable time in the equation defining the process [14]. In the general form of the wave propagation equation $\vec{u} = \vec{a} \cdot e^{i(\omega t - kx)}$, where x is the space variable in the direction of propagation and $\vec{a}$ proportional to the amplitude of the displacement, it can be observed that the time reversal operation is equivalent to the complex conjugation in the frequency domain.

The experimental setup for the time reversal process applied in this paper is shown in Fig. 2.

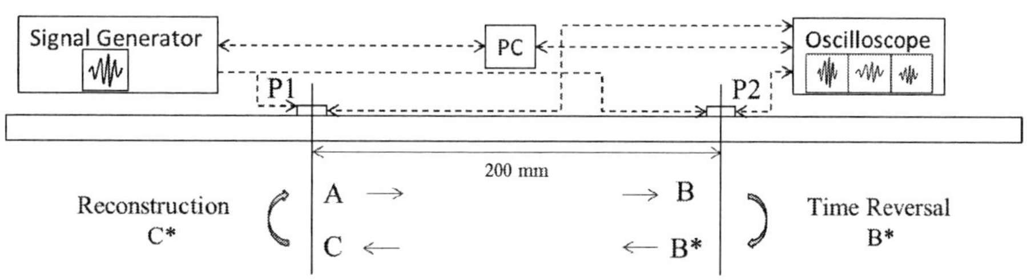

Figure 2. Actuators P1 and P2 are bonded to the plate. The dimensions of the plates are 580x580x2mm for the CF/PPS plate, and 750x750x2mm for the aluminum plate

The actuation signal A is propagated from P1 and received at P2 as response B. B is time reversed into B^* and back-propagated from P2 to P1, where C is received and time reversed for the comparison with A. The response B received at transducer P2 is related to A as in Eq. 5, where K is the transfer function relating an excitation at P1 to the response at P2. K can be expressed as an exponential function as shown in Eq. 6 [7], where γ is the amplitude dispersion function, with the amplitude decreasing in proportion to the inverse of the square root of the propagation distance [6, 15].

$$\mathrm{B} = \mathrm{K} \times \mathrm{A} \tag{5}$$

$$\mathrm{K} = \gamma \cdot \exp(-\mathrm{ikx}) \tag{6}$$

B is then time reversed, or conjugated in the frequency domain, B^*, and back-propagated from P2. As mentioned in the introduction, the reciprocity theorem applies, for which the response at P1 to an excitation from P2 has the same transfer function K [12]. With this property the response at P1 is then

$$\mathrm{C} = \mathrm{K} \cdot \mathrm{B}^* = \mathrm{K} \cdot \mathrm{K}^* \cdot \mathrm{A}^* \tag{7}$$

The group velocity dispersion exp(-ikx) is compensated in the time reversal process, and C, the response received at P1, is proportional to the time reversed original excitation A^* [7]. Both signals, C with A^* are compared at the end of the process, which is the same as comparing A and C^*, or time reversed C. Two correlation procedures, explained in Subsection 2.3, are evaluated as a comparison method. The presence of damage in the wave propagation path will cause a drop in the signal reconstruction level, which is proportional to the damage size and the proximity of the damage to the transducers.

2.3 Correlation

The degree of signal reconstruction on the time reversal process is evaluated by correlation between the reconstructed signal C^* and the original actuation A, as found in [11]. For the correlation method, different approaches are examined in order to evaluate the influence of the methods used on the comparison of the two signals. Hence the correlation coefficient and the maximum cross-correlation between the signals after normalization are calculated using the functions *corrcoef* and *xcorr* available in the Matlab™ software library.

2.3.1 Correlation coefficient (*corrcoef*)

The correlation coefficient ρ_{C^*A} between the two time series C^* and A, is calculated as shown in Eq. 8.a [16], based on the covariance of the series C^* and A, σ_{C^*A}, and the square of the standard deviations or their variance, $\sigma_{C^*}^2$ and σ_A^2, as calculated in Eq. 8.b.

$$\text{(a)} \quad \rho_{C^*A} = \frac{\sigma_{C^*A}}{\sqrt{\sigma_{C^*C^*}\sigma_{AA}}} = \frac{\sigma_{C^*A}}{\sqrt{\sigma_{C^*}^2\sigma_A^2}} = \frac{\sigma_{C^*A}}{\sigma_{C^*}\sigma_A} \qquad \text{(b)} \quad \sigma_{C^*A} = E[(C^*-\bar{C}^*)(A-\bar{A})] \tag{8}$$

2.3.2 Cross correlation (*xcorr*)

This correlation takes into account the time lag for correlation of the series, that is the series are time-shifted one with respect to the other, and the correlation coefficient is calculated for each time lag, evaluating the time shift needed for maximum correlation of the series [16]. The calculation is expressed in Eq. 9, where m is the time lag expressed in number of data points. The maximum cross-correlation coefficient is then considered in order to evaluate the degree of signal reconstruction.

$$\rho_{C^*A}(m) = \frac{E[(C^*_m - \bar{C}^*)(A_{n+m} - \bar{A})]}{\sigma_{C^*}\sigma_A} \tag{9}$$

3. EXPERIMENTAL PROCEDURE

One Al-6082-T6 plate 750x750x2 mm and 3 CF/PPS plates 580x580x2 mm were used for the experiments. The composite plates were manufactured by hot-pressing 8 plies of woven CF/PPS prepreg, with a final thickness of approximately 2.1 mm. The aluminum plate was used to test the correlation method in homogeneous materials. On one surface of each plate two APC850 piezoelectric transducers, 12.6mm diameter, P1 and P2, were bonded with cyanoacrylate. The transducers were separated by a distance of 200 mm and aligned parallel to the edges and in the middle of the plate in order to reduce the influence of the wave reflections with the edges of the plates. In the case of the composite plate the transducers were aligned with the 0/90 direction of the fibers. As an actuation signal, a 5-cycle Hanning window pulse was chosen as a trade-off between short time duration and narrow band spectrum [15]. The maximum amplitude of the pulse was 20 Vpp. As shown in Fig. 2, this nominal pulse is sent from a 30 MHz signal generator (Agilent-33522B) to the transducer P1, and recorded by an oscilloscope (PicoScope-6402A) at 250 MHz sampling rate, as *A(t)*. The response *B(t)* captured at P2 is the incident wave from the propagation of the original wave through the plate, and thus shows the dispersion effects and the reflections from the defects along the propagation path. The time-reversed signal B^* is then sent from P2 and received at P1 as *C(t)*. The reconstructed signal C^* obtained by time reversing *C*, is finally correlated with *A*.

External noise was filtered out with a low-pass filter, at a cut-off frequency of 900 kHz. The central frequencies of the actuation signal, between 125 and 350 kHz, were chosen so that the S_0 and the A_0 waves arriving at the second transducer do not overlap, and that these are the only 2 modes excited, as can be inferred by the dispersion curves in Fig. 1. Four central frequencies were used: 125, 180, 245 and 315kHz, where the response to the actuation showed these two modes included their reflections from defects and the plate edges. The reflections from the possible defects may overlap with the reflections from the edges. In such cases these reflections cannot be distinguished. Therefore, only the incident waves until the first reflections from the edges, were considered for the analysis.

In order to obtain the best reconstruction of the wave it is necessary to time-reverse the entire propagated field [1]. Nevertheless, in order to simplify the analysis, a limited time interval including only the incident waves was selected from *B(t)* to be time-reversed, and only the propagated field at location P2 was considered. For that reason the reconstruction of the wave is not perfect, not even for the case of the undamaged structure. The correlation level for the reconstruction of the wave in an undamaged plate defines a threshold for the correlation value, which also depends on this time interval. The waveform indicates the presence of a defect, and depends on its type and size as well as on the distance to the line defining the direct path between the transducers.

The time interval from response *B* chosen to be time-reversed and back-propagated may include only mode S_0, or both S_0 and A_0 modes, as shown in Fig. 3 (left). This last time interval including modes S_0 and A_0 may provide some information not present if only the S_0 mode was selected.

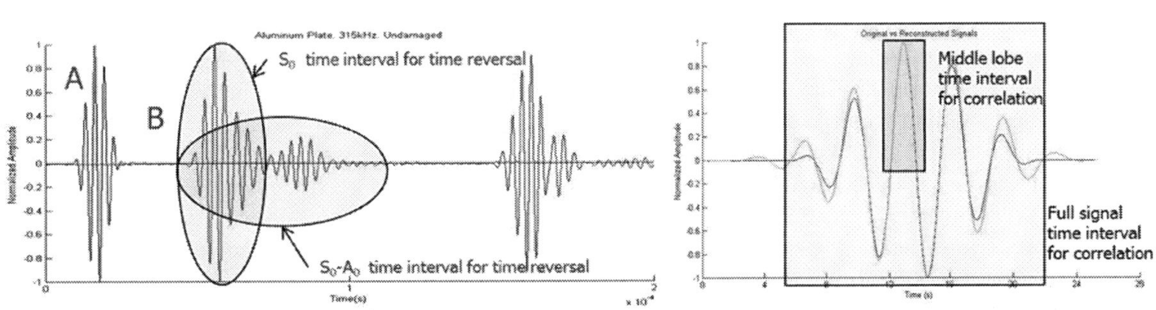

Figure 3. On the left, normalized actuation signal *A* and response *B*, showing the time reversal intervals including mode S_0 and modes S_0 and A_0 for time reversal. On the right, normalized actuation signal *A* and reconstructed signal C^*, showing the time intervals for full-signal and middle-lobe correlation

From C, another limited time interval selection was needed in order to define the reconstructed signal C^*. This time interval selection also affects the correlation value, as will be shown in Section 5. This influence was analyzed by selecting two different time intervals: (i) the full-signal with a time interval equivalent to the corresponding length of signal A for each frequency; and (ii) the main or middle lobe crossing at an amplitude of zero, as shown in Fig. 3 (right). In both cases the analysis was performed after normalization of the series with respect to their corresponding maximums.

The 3 CF/PPS undamaged plates were tested first. Then a 2 mm hole was applied in each plate, at different distances offset from the P1-P2 line, as depicted in Fig. 4. The hole was then widened from 2 to 5 mm in each plate, and retested. The same procedure was followed with the aluminum plate. In Section 5 the experimental results are compared with the numerical simulations for the undamaged (D0) and the 5mm hole (D5) case.

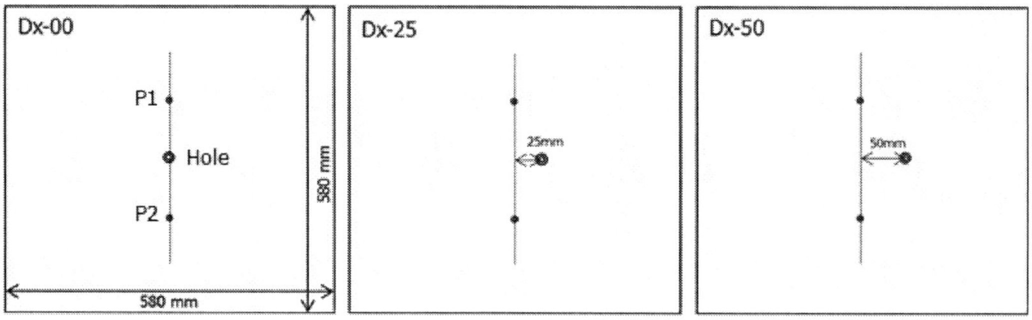

Figure 4. Three CF/PPS plates tested for hole detection in three different locations. From left to right, hole with diameter x (x = 0, 2, 5 mm) in line with the transducers P1-P2, hole 25 mm and 50 mm from P1-P2 line. P1 and P2 separated 200 mm, along the symmetry line of each plate

4. FEA: TIME REVERSAL PROCESS FOR ALUMINUM

A Finite Element Analysis (FEA) for verification of the experimental tests described in Section 3, was conducted using the commercially available software package ABAQUS™. The FE model was constructed using standard 3D elements, C3D8I, for an aluminum plate as described in Fig. 2. The properties required for these elements were $E = 69\,GPa$, $\rho = 2700\,kg \cdot m^{-3}$, and $\nu = 0.33$.The actuators/sensors, P1 and P2 were modeled using C3D8E piezoelectric elements, requiring the following material properties: density ($7600\,kg \cdot m^{-3}$), piezoelectric charge constants (d_{33}=400, d_{13}=-175, d_{15}=590 $\cdot\, 10^{-12}\,C \cdot N^{-1}$), and the dielectric constant (K^T=1900), which were provided by the manufacturer, and the stiffness matrix, which was obtained from the literature [17]. As they were bonded along the middle line of the plate, a symmetric boundary condition was used, reducing the number of elements needed for the plate by a factor of two, and thus reducing the computational time. The other boundaries were modeled as free boundaries. The defect considered was through the thickness hole with different diameters, in the middle of the plate.

The contact between the plate and the piezoelectric transducers was assumed to be perfectly bonded, using a **TIE*-constraint in the ABAQUS™ model. In order to simulate a voltage input, the piezoelectric transducers were actuated via an electric potential boundary condition available under the dynamic implicit solver. The electric potential for the piezoelectric transducer surface connected to the plate was set to zero throughout the simulation. The free top surface of the piezoelectric actuator was set with a boundary condition as to have an electric potential and amplitude variation in time, which matched that of the experimental actuation signal. The piezoelectric transducers used in the experiments were polarized through the thickness.

The actuation signal used in the FE model consisted of a 5-cycle Hanning window. Three different frequencies (125 kHz, 180 kHz and 245 kHz) were verified with the model. The piezoelectric transducer P1 was actuated such that a Lamb wave was generated, and subsequently acquired by the piezoelectric transducer P2, by analyzing the *electric potential* output as a function of time.

The convergence of the FEA simulation depended on the integration time and maximum size of the elements. A good rule of thumb is given by Chen et al. [18], which used at least twenty points per cycle at the highest actuation frequency, f, in order to determine the FE integration time step, $\Delta t = (20 \cdot f)^{-1}$. For the element size, Chen et al. [18] state that the largest element size can be determined by: $L_e = v/(10 \cdot f)$, where v is the lowest propagation velocity of the wave. In order to ensure that the integration time step and element size would not affect the results, the integration time steps of 0.1 μs and 1mm element size were used for all numerical simulations. The numerical simulations were executed using the computational cluster HPC12 at Delft University of Technology. For each simulation one node equipped with four Intel XEON E5520 (quad cores) and 24 GB 1333 ECC RAM was used, resulting in a computational time of approximately fifteen hours per simulation.

5. RESULTS

5.1 Composite Plate

The CF/PPS analysis consisted of evaluating a time interval that included only the S_0 mode and a time interval that included the S_0 and A_0 modes. The longer time interval was expected to include more information on the propagation and have a larger potential for the interaction of the wave with the damage. The coefficient of correlation showed a strong dependency on the time interval selected for the correlation. In the correlation of the middle lobe, the coefficient varied between 93.84 and 99.98%. In all the cases at 125 kHz, damaged and undamaged, the coefficients are lower than 99%, and for the rest of the frequencies, damaged and undamaged cases, the coefficients remain above 99%. Hence the middle lobe cannot be considered for the correlation criteria.

In contrast, Table 1 shows the correlation coefficient calculated from the full signal correlation, where the length of the time interval selected corresponds to the time interval of signal A for each frequency.

In order to discard the influence of the reinforcing fibers on the propagating waves, and then on the correlation coefficient calculated, the homogeneous case represented by an aluminum plate was analyzed and presented in the next subsection, in conjunction with a FEM.

Table 1 Full signal correlation between the original actuation signal A, and the reconstructed signal C^* obtained from the time-reversal of the time interval including S_0-A_0 modes

CorrCoeff (%)/ X-Corr (%)							
Frequency (kHz)	Undamaged	Hole 2mm diameter			Hole 5mm diameter		
		50mm off the line P1-P2	25mm off the line P1-P2	In line with P1 and P2	50mm off the line P1-P2	25mm off the line P1-P2	In line with P1 and P2
125	53.3/68.1	51.5/66.7	54.6/71.9	52.9/67.1	51.5/66.4	53.9/71.6	63.8/63.9
180	61.4/63.2	61.7/63.0	61.6/63.3	12.0/13.6	60.8/62.5	60.9/62.6	13.4/15.2
245	86.3/86.3	83.9/84.2	88.6/88.5	86.2/86.3	83.8/84.1	88.3/88.2	86.2/86.3
315	30.9/34.3	29.2/33.2	34.0/37.3	28.5/32.1	29.0/33.0	33.4/36.7	28.6/32.1

5.2 Aluminum plate

For the aluminum plate, both experimental and numerical results were compared. Two cases were considered for the comparison: (i) undamaged plate (D0), and (ii) a hole through the thickness, 5 mm in diameter (D5) placed in the middle of P1-P2 line.

In Fig. 5 it can be observed how the FEA results matched the experimental in case of the direct response (signal B), received in P2. For all the frequencies of study the numerical waveform coincided with the experimental one, although a slight delay is observed on the FEA result with respect to the experimental, which caused the wave's spread in time. This mismatch may be caused by an error in the estimation of the material properties used as an input in the model.

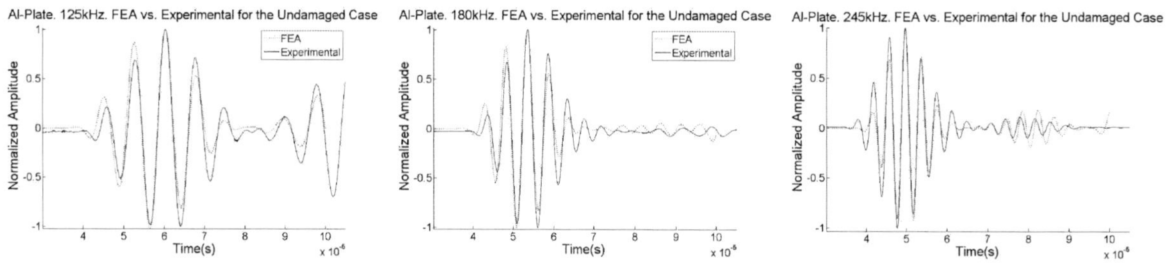

Figure 5. FEA vs. Experimental results for the undamaged aluminum plate. Signal *B* read at transducer P2, as a response to the actuation pulse at central frequencies 125, 180 and 245 kHz, represented from left to right

The presence of the hole in the plate caused an effect on the response signal observed in both FEA and experimental results. Fig. 6 shows an example of this effect at 2 frequencies, 180 and 245 kHz. This effect was higher in the composite case.

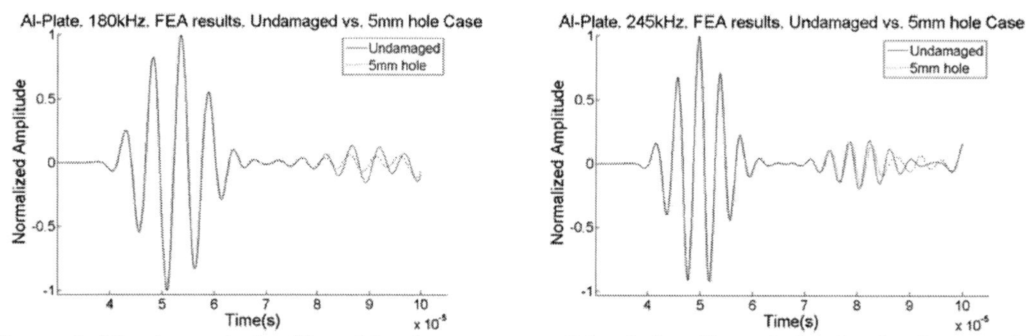

Figure 6. The A_0 mode is affected by the presence of the defect (5mm hole) for the frequencies studied, while the effect on the S_0 wave can be neglected for 180 kHz (left) and 245 kHz (right). Both amplitude and frequency effects are observed on the A_o

After the back-propagation of the time reversal signal, this difference observed in the direct response was also observed in the reconstructed signal, and is given by the drop on the correlation coefficient when correlating with the actuation signal. An example is shown in Fig. 7.

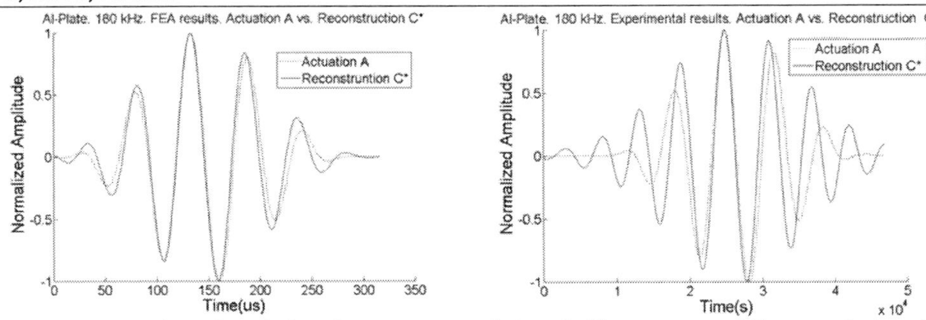

Figure 7. Actuation signal *A* and reconstructed signal *C** represented for correlation. The reconstructed signal is closer to the actuation in the FEA results (left) than in the experimental (right), where the waveform is spread in time

Nevertheless, this drop in correlation is not very significant, as can be seen in Table 2, and the presence of a 5mm hole does not translate into a higher drop in the correlation.

As can be seen in Table 2 the correlation obtained from the middle lobe showed a maximum drop of correlation of 0.15% for any case with respect to the original actuation *A*. The drop of correlation for the full signal was more significant. The maximum difference between correlation calculated as Correlation Coefficient and Maximum Cross-Correlation, was of 1.04%, both showing the same tendencies. Hence either of them can be considered in the analysis. The presence of the 5mm hole increased the correlation coefficient with respect to the undamaged case, opposite to the expected behavior. The influence of the frequency was more predominant in the experimental results, versus the FEA. The same influence was also observed in the CF/PPS plate.

Table 2. FEA and Experimental comparison of the Full signal and Middle Lobe correlation between the original actuation signal *A*, and the reconstructed signal *C** obtained from the time-reversal of the time interval including the S_0 mode

CorrCoeff (%)/ X-Corr (%)					
Frequency (kHz)	(mm) Hole diameter	FEA		Experimental	
		Full Signal	Middle Lobe	Full Signal	Middle Lobe
125	0	96.3/96.5	100.0/100.0	84.0/85.0	99.9/99.9
	5	96.5/96.7	100.0/100.0	85.9/86.3	99.9/99.9
180	0	96.5/96.8	99.8/99.8	74.1/75.0	99.9/99.9
	5	97.0/97.4	99.7/99.6	75.1/76.1	99.9/99.9
245	0	99.0/99.0	98.9/99.9	92.0/92.0	99.8/99.8
	5	98.9/98.9	99.0/99.0	92.6/92.6	99.9/99.9

6. DISCUSSION

The same trends on the influence of actuation frequency on the correlation values observed in the composite plate were observed in the aluminum, in both the experimental results and the numerical simulations. This influence may be caused mainly by the differences between the diameter of the transducers and the wavelength of each mode, at each frequency. In order to capture the whole wave by the transducer, the diameter must be approximately half of the wavelength of the actuation signal [15]. This introduces a limitation on the actuation frequencies for Lamb wave based techniques.

The ability of the TRLW for detection of the type of defect studied here, was demonstrated by the differences in the reconstruction signal between damaged and undamaged results, as seen in Fig. 6. Nevertheless, it was expected to have a drop in the correlation due to the presence of the damage with an increasing hole diameter, and decreasing distance to the line P1-P2 described by the transducers. Unfortunately, these trends were not observed in the correlation coefficients. In some cases the correlation coefficient for the damaged test was higher than the one for the undamaged, but these changes are too small to be taken in consideration.

The drop of correlation for each actuation frequency is not enough to be considered as a criterion for detection of the damage type analyzed in this research (hole through the thickness). The correlation method in time domain as it is shown here is not giving enough information regarding the presence of damage in the composite plate nor in the aluminum. The presence of damage is reflected as a variation in amplitude, time of flight or delay, and a spreading effect. In the TRLW, the amplitude effect is lost first when the response B recorded at P2 (see Fig. 2) is time reversed and then back-propagated towards P1. The received amplitude at P2 has to be amplified in order to be sent back to P1. In addition, the signals A and C^* need to be correlated, and for that their amplitudes need to be normalized. Signal A is normalized with respect to its own maximum. The normalization of C^* can be made with respect to its own maximum, or relative to a second signal, like the reconstruction from the undamaged case, C_{ref}*. In both cases the amplitude of the waves is altered, and so is the information captured in the wave. The time of flight or delay of the wave induced by the hole is lost each time the signal is time-reversed (twice), as a new time reference needs to be set. For the correlation between A and C^* the middle lobe of the signals is forced to coincide. The spreading effect induced by the hole is too small to be revealed by the correlation in the time domain. The analysis in the frequency domain is more suitable in this case.

The limitations of the correlation method as implemented are found for both composite and metallic materials.

7. CONCLUSIONS

The ability of Time Reversal of Lamb Waves for damage detection in aluminum and composite plates has been demonstrated. It can be concluded that the correlation method in the time domain is not accurate enough as to be used as damage detection criterion. The results of correlation coefficient are very dependent on the time interval chosen for both time reversal and correlation. The time reversal interval including modes S_0 and A_0 carries more information of the propagation path. The correlation based on a longer time interval of the reconstructed signal, up to the length of the actuation, gives higher changes in the correlation coefficient.

In order to make this procedure applicable for damage detection, a trustable correlation threshold needs to be defined from the undamaged case as a baseline. Therefore improvements in the reconstruction signal evaluation must be implemented.

The selection of the actuation frequency in combination with the size of the transducers is important in order to ensure a correct interpretation of the results.

Two piezoelectric transducers for actuation and sensing were used. In a real application different transducers configurations might be needed. The development of a numerical tool able to set the capabilities of the one-to-one transducer configuration can be used to guide in the determination of an optimal number and location of the sensors for a specific structure.

References

1. Fink, M., "Time reversal of ultrasonic fields--I: Basic principles", *IEEE transactions on ultrasonics, ferroelectrics, and frequency control*, 1992, Vol. 39, No. 5, pp. 555-566.
2. Ambrozinski, L., Stepinski, T., Packo, P. & Uhl, T., "Self-focusing Lamb waves based on the decomposition of the time-reversal operator using time frequency representation", *Mechanical Systems and Signal Processing,* 2012, Vol. 27, no. 1, pp. 337-349.
3. Chang, F.K., (ed.), "SHM-The Demands and Challenges", *Proc. 3rd Int. Workshop on Structural Health Monitoring*, Publ. Stanford (CA), Sept. 2001, pp. xxi-xxvi.
4. Rose, J.L., Ultrasonic Waves in Solid Media, *Cambridge Univ. Press*, 1999, pp. 101-126.
5. Boller, C., Buderath, M., "Fatigue in aerostructures—where structural health monitoring can contribute to a complex subject", *Phil. Trans. R. Soc. A. 365*, 2007, pp. 561-587.
6. Su, Z., Ye, L., Identification of Damage Using Lamb Waves, *Springer-Verlag Berlin Heidelberg*, 2009, LNACM 48, pp. 15-58.
7. Park, H.W., Kim, S.B. & Sohn, H., "Understanding a time reversal process in Lamb wave propagation", *Wave Motion*, 2009, Vol. 46, No. 7, pp. 451-467.
8. Díaz Valdés, S.H., Soutis, C., "Real-time nondestructive evaluation of fiber composite laminates using low-frequency Lamb waves", *Journal of the Acoustical Society of America*, 2002, Vol. 111, No. 5 I, pp. 2026-2033.
9. Kirikera, G.R., Balogun, O., Krishnaswamy, S., "Adaptive fiber bragg grating sensor network for structural health monitoring: Applications to impact monitoring", *Structural Health Monitoring*, 2011, Vol. 10 (1), No. 1, pp. 5-16.
10. Watkins, R. & Jha, R., "A modified time reversal method for Lamb wave based diagnostics of composite structures", *Mechanical Systems and Signal Processing*, 2012, Vol. 31, pp. 345-354.
11. Poddar, B., Bijudas, C.R., Mitra, M. & Mujumdar, P.M., "Damage detection in a woven-fabric composite laminate using time-reversed Lamb wave", *Structural Health Monitoring,* 2012, Vol. 11, No. 5, pp. 602-612.
12. Barber, J.R., Elasticity, Chapter 30, *Series: Solid Mechanics and Its Applications*, 1992, Vol. 107, 2nd ed., Springer Netherlands, pp. 393-404.
13. Lamb, H., "On Waves in an Elastic plate", *Proceedings of the Royal Society*, A: Mathematical, Physical and Engineering Sciences 93, 1917, pp. 114-128.
14. Sachs, R. G., The Physics of Time Reversal, *University of Chicago Press*, Chicago, 1987 , pp. 11-22.
15. Wilcox, P. D., Lowe, M. J. S. and Cawley, P., "Mode and Transducer Selection for Long Range Lamb Wave Inspection", *Journal of Intelligent Material Systems and Structures*, 2001, Vol. 12 (8), pp. 553 – 565.
16. Documentation Center MathWorks, http://www.mathworks.nl/help/, *MathWorks ©2013*, last accessed Oct. 2013.
17. SensorTech., "Piezoelectric material constants for compliance matrix", http://www.sensortech.ca/site/content/2006-01.pdf , 1990, revision 2006, last accessed Oct. 2013.
18. Chen, L., Dong, Y., Meng, Q., and Liang., W., "FEM simulation for lamb wave evaluate the defects of plates", *Microwave and Millimeter Wave Circuits and System Technology (MMWCST)*, 2012.

ICAST2013: 24th International Conference on
Adaptive Structures and Technologies
October 7-9, 2013, Aruba

ICAST2013 #XXX

Structural Damage Identification by Means of Neural Network (An Inverse Problem Approach Based on Dynamic Characteristics)

Kazuyuki Hanahara[1]* Yukio Tada[1]

[1] Department of Systems Science, Graduate School of System Informatics, Kobe University, Kobe, Japan

Abstract

Damage identification or health monitoring is an important issue especially for a structural system being in operation for a long time. In this study, we deal with an approach to damage identification of structural system by means of a multi-layered neural network. The approach is based on the change in some specific characteristics caused by the damage of structural parts. The relation between the damage and the characteristics is implemented with a multi-layered neural network; the error back-propagation with fixed connection weight is adopted to solve the inverse problem for the damage identification. We discuss the type of characteristics used for the damage identification. Numerical experiments with truss structures are conducted and the feasibility of the proposed approach is examined.

1. INTRODUCTION

Structure is thought to be a kind of silent mechanical system which just withstands its given loading condition as well as its surrounding environment at the place of its installation, in the primitive and traditional sense. In contrast to this traditional stereotype, a structure dealt with in the concept of so-called 'adaptive structure'[1] is not just silent but having various abilities such as actuation, sensing, intelligence and so on.

Structural health monitoring (SHM)[2] is one of such research fields; it is to be implemented to structural systems such as aircraft, spacecraft, automobiles, bridges, buildings, ships and many others. Damage of structural system deteriorates its performance. Unlike natural living systems, such damage can not be cured by itself in the case of engineering structural systems in general. Damaged structural parts or members have to be identified and repaired or replaced in order to regain the original system performance. The damage identification as SHM for this purpose is achieved by means of integrating sensing devices and the information processing capability. Non-destructive testing (NDT)[3] can also be viewed as a related research field of SHM; difference between the terms SHM and NDT can be vague in case.

In this study, we propose an approach to damage identification of structural system by means of a multi-layered neural network[4]. The approach is based on the change in mechanical characteristics of the entire structure caused by the damage on its constituent structural parts. The relation between the local damage and the system characteristics change is implemented as a multi-layered neural network. The error back-propagation technique with fixed connection weights[5] is adopted to solve the inverse

* Corresponding author; email: hanahara@cs.kobe-u.ac.jp

problem of determining the local damage corresponding to the obtained system characteristics change. We discuss some requirements for mechanical characteristics suitable to the damage identification of this type and propose to adopt the plural natural frequencies of the entire structure.

Truss structures are adopted as the concrete example. We use change in axial stiffness of the truss members as the parameter to quantize the local damage degree of the structural parts. The learning data sets consisting of member stiffness change and corresponding change in natural frequencies are computationally obtained; the ordinary error back-propagation learning is performed with the data set to prepare the multi-layered neural networks. A number of numerical experiments are conducted with the prepared neural networks. The feasibility of the proposed damage identification approach is discussed based on the experiment results.

2. STRUCTURAL DAMAGE IDENTIFICATION AS INVERSE PROBLEM

We give a brief overview of the approaches of the structural damage identification. In this study, we adopt an approach of diagnostic type; an inverse problem formulation of the problem is also explained.

2.1 Structural Damage Identification or Health Monitoring: A Brief Overview

There have been a number of researches concerning structural damage identification or structural health monitoring. In addition, such as so-called non-destructive inspection or non-destructive testing can also be included in this field of engineering. The approaches adopted for the damage identification can be categorized as follows:

Sensory nerve system type A number of sensing devices for damage detection are distributed over the target structure system. For example, Giurgiutiu et al.[6] use piezoelectric devices to detect crack or defect; an active technique is adopted. Betz et al.[7] use fiber-optic sensors in an embedded manner in order to detect and localize the damage. The approach given by Takada et al.[8] for detecting bolt loosening can be also categorized as this type.

Scanning type A relatively small number of sensing devices are adopted; scanning and signal processing techniques are combined in order to localize the damage. The eddy current approach[9] and the ultrasonic approach[10] are typical examples of this scanning type. The X-ray computed tomography approach[11] can also be categorized as this type. The approach developed by Takahashi and Hahn[12] is a kind of hybrid between the nerve system type and this scanning type, since it uses a network system attached on the target structure and utilizes a kind of scanning technique for the combinations among them. An approach adopting a kind of ultrasonic radar[13] can also be said as a variation of scanning type in a sense.

Diagnostic type The relation between the sensor-obtained data and the structural damage can be indirect in this case. This damage identification process has two stages; the first stage is the development or the estimation of the relation between the damage and the data to be obtained and the second stage is the identification of damage corresponding to the sensor-obtained data by means of the implemented relation. A damage detection system of this kind is studied by Manning[14]; it is also said as an adaptive structure application since it adopts active truss members. Many of the approaches of this type deal with the dynamic response of the target structure[15-18] such as vibration frequencies and modes. Wang et al.[19] adopt a static response as well and give an approach of damage identification based on both dynamic and static behaviors of the structure. Approaches such as [20] and [21] can also be categorized as this type; these can said to be represented as a combination with the sensory nerve type approach in some sense.

This categorization is an example. Another type of categorization such as based on the type of signal used to detect the damage or defect (ultrasonic, magnetic, electric, etc.) can also be possible.

2.2 Inverse Problem Formulation for Structural Damage Identification

Let $\boldsymbol{x}$ be a state variable vector of a target structure. In our case, $\boldsymbol{x}$ is required to be able to express all the pattern and the magnitude of the possible structural damage. Let $\boldsymbol{y}$ be a structural characteristic vector of the structure that is affected by the assumed structural damage, which is obtained by means of some sensing devices. The relationship between the two variables can be expressed in the following general form:

$$\boldsymbol{y} = \boldsymbol{y}(\boldsymbol{x}). \tag{1}$$

In the case that we have a structural characteristic value $\boldsymbol{y}^*$ which is obtained by means of the sensing devices, the corresponding structural damage pattern $\boldsymbol{x}^*$ is determined by solving the following problem:

$$\text{Find} \quad \boldsymbol{x}^* \quad \text{such that} \quad \boldsymbol{y}(\boldsymbol{x}^*) = \boldsymbol{y}^*. \tag{2}$$

This is a typical inverse problem based on Eq.(1).

3. SOLVE INVERSE PROBLEM BY MEANS OF NEURAL NETWORK

In order to make the article self-contained, we summarize the approach to solve an inverse problem by means of neural network[5].

3.1 Error Back-Propagation Learning for Multi-Layered Neural Network

Figure 1(a) shows a conceptual illustration of a multi-layered neural network. Neurons are depicted as circles and connected in a layered manner. The input and output layers placed at leftmost and rightmost sides are the 0th and Nth layers; the so-called hidden neurons are placed between them and make up the 1st$\cdots$$(N-1)$th layers. The behavior of the ith neuron of the kth layer is modeled as follows:

$$v_i^k = f(u_i^k), \quad u_i^k = \sum_j w_{ij}^k v_j^{k-1} \tag{3}$$

where u_i^k and v_i^k are the internal state and the output value of the neuron and w_{ij}^k is the connection weight between the ith neuron of the kth layer and the jth neuron of the $(k$–$1)$th layer. The characteristic function of a neuron adopted in this study is expressed as

$$f(u) = \frac{2}{1+\exp(-2u)} - 1, \tag{4}$$

which is shown in Fig.1(b). The input-output relation as the entire neural network can be represented in the following form:

$$\boldsymbol{v}^N = \boldsymbol{v}^N(\boldsymbol{v}^0, \boldsymbol{w}) \tag{5}$$

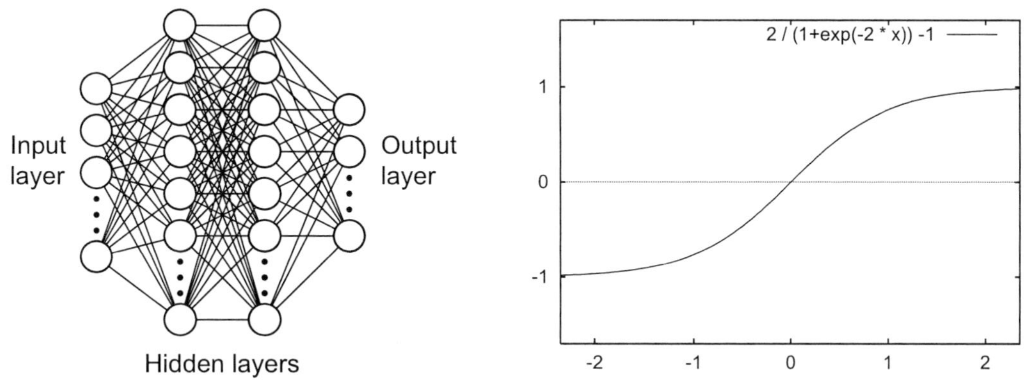

(a) Neural network (b) Characteristic function

Figure 1. Multi-layered neural network

where $\boldsymbol{v}^0 = [v_i^0]$ and $\boldsymbol{v}^N = [v_i^N]$ are the input and output vectors. The connection weights $\boldsymbol{w} = [w_{ij}^k]$ determines the input-output relation.

On the basis of the error function to the desired output value $\underline{\boldsymbol{v}}^N = [\underline{v}_i^N]$ expressed as

$$E = \frac{1}{2}(\boldsymbol{v}^N - \underline{\boldsymbol{v}}^N)^T(\boldsymbol{v}^N - \underline{\boldsymbol{v}}^N) = \frac{1}{2}\sum_i (v_i^N - \underline{v}_i^N)^2, \tag{6}$$

the following iterative modification of the connection weights are conducted for various input values and their corresponding desired output values:

$$w_{ij}^k \leftarrow w_{ij}^k - \Delta w_{ij}^k, \quad \Delta w_{ij}^k = \epsilon \frac{\partial E}{\partial w_{ij}^k} \tag{7}$$

where ϵ is an adequate small coefficient. The modification Δw_{ij}^k is obtained in the following manner:

$$\Delta w_{ij}^k = \Delta v_i^k \cdot f'(u_i^k) \cdot v_j^{k-1}, \tag{8}$$

$$\Delta v_i^k = \epsilon \frac{\partial E}{\partial v_i^k} = \sum_j \Delta v_j^{k+1} \cdot f'(u_j^{k+1}) \cdot w_{ji}^{k+1}, \tag{9}$$

$$\Delta v_i^N = \epsilon \frac{\partial E}{\partial v_i^N} = \epsilon (v_i^N - \underline{v}_i^N). \tag{10}$$

Equation (9) denotes the error back-propagation process from the $(k+1)$th layer to the kth layer.

3.2 Solve Inverse Problem with Neural Network of Fixed Connection Weights

We deal with the inverse problem (2) in terms of the following error minimization problem:

$$\text{Minimize} \quad E = \frac{1}{2}(\boldsymbol{y}(\boldsymbol{x}) - \boldsymbol{y}^*)^T(\boldsymbol{y}(\boldsymbol{x}) - \boldsymbol{y}^*) \quad \text{with respect to} \quad \boldsymbol{x}. \tag{11}$$

The solution can be obtained by means of an iterative gradient procedure as

$$\boldsymbol{x} \leftarrow \boldsymbol{x} - \Delta\boldsymbol{x}, \quad \Delta\boldsymbol{x} = \epsilon\frac{\partial E}{\partial \boldsymbol{x}} \tag{12}$$

based on an adequate initial value of $\boldsymbol{x}$. After the input-output relation (5) of the neural network has been trained to represent the relation (1), the gradient $\Delta\boldsymbol{x} = \epsilon(\partial E/\partial \boldsymbol{x})$ can be obtained as the modification to the input layer at the neural network, $\Delta\boldsymbol{v}^0 = \epsilon(\partial E/\partial \boldsymbol{v}^0)$, in terms of Eq.(9) as the result of the error back-propagation process, since the error function (6) for the learning and the error in the minimization problem (11) to solve the inverse problem are essentially the same at this point. This enables us to solve the inverse problem (2) in terms of the minimization problem (11) by means of the neural network. Note that the connection weights of the neural network are fixed in the case of the inverse problem solving process, so that the implemented input-output relation should not be changed.

4. STRUCTURAL CHARACTERISTICS FOR DAMAGE IDENTIFICATION

We discuss the structural characteristics that is considered to be suitable to the proposed damage identification approach based on the inverse problem formulation and the error back-propagation process of multi-layered neural network.

4.1 Requirements for Structural Characteristics

Conceptually, any structural characteristics that are affected by the possible structural damage can be adopted for the inverse problem approach. On the basis of the practicality consideration, however, the following requirements should be taken into account for the characteristics:

(a) exhaustiveness – damage on any part of the structure affects the characteristics, **(b) discriminability** – change in the characteristics should be different in accordance with the part and magnitude of the damage, **(c) pattern generability** – learning patterns for the neural network has to be generated in a practical manner, **(d) in situ damage detection** – the process should be applicable to the target structure as it is, possibly under the operating condition, **(e) robustness** – influence of the aging of the structure and various environmental change should be compensated and **(f) less affectiveness** – damage detection process itself should not affect the performance of the structure.

The requirements (a) and (b) are both essential in the case of the inverse problem approach. The requirement (c) is because our approach is based on the neural network that acquires the input-output relation by learning. Especially in the case of a large-scale structure, it is not practical to remove and transport the structure to the laboratory for inspection; the requirement (d) is specified for this reason. General requirements for the damage identification or the health monitoring of structure are also given as (e) and (f).

4.2 A Promising Candidate Adopted: Natural Frequencies

On the basis of the above consideration, we adopt the change in plural natural frequencies of the entire structure to be caused by the damage as the structural characteristics used for the proposed damage identification.

The most important requirement for the proposed neural network approach is (c) in the above list.

It is not practical in general to carry out enough number of vibration experiments in order to prepare the learning pattern table, but some numerical approach to calculate the natural frequencies is generally available. There can be an issue of accuracy in the case of such a numerical approach for the learning pattern generation; however, it is not the absolute value of the natural frequencies but their relative change due to the damage that matters in our case.

The requirements (d) and (f) are also expected to be satisfied. There are a number of ways to measure the vibration behavior of a structure in a low affective manner, such as a small accelerometer. The vibration of the structure need not be caused by means of an artificial manner such as hammering; the vibration caused by natural disturbances such as wind can also be expected to be available, in the premise of some adequate filtering or signal processing technique.

The learning capability of the multi-layered neural network can manage the requirement (e). This is also applicable to the case that there are a number of individual structures manufactured based on a single common specification. One neural network that learned the relation between the damage and the corresponding change in natural frequencies of an individual structure can be duplicated to be applicable for all other structures, after the adjustment by means of this learning process is performed.

The two basic requirements (a) and (b) may not be completely satisfied in some case, since accurate natural frequencies of higher vibration modes of an actual structure are not always easy to obtain in general. In addition, the influence of damage at the tip-end neighborhood of a structure on its entire vibration characteristics is generally not significant. These conditions have to be taken into consideration.

5. DEMONSTRATION WITH TRUSS STRUCTURES

The proposed damage identification approach is demonstrated with a number of numerical experiments. The feasibility as well as the limitation at this stage is discussed.

5.1 Five-unit Trusses

Two five-unit trusses of different topologies are examined; the feasibility as well as the fundamental limitation of the proposed approach is demonstrated.

5.1.1 Problem Specification

Figures 2(a) and 3(a) show the truss structures adopted in the numerical experiments; Fig.2(a) has asymmetric topology while Fig.3(a) has symmetric topology. All the square units are assumed to be $1\text{m}\times 1\text{m}$; all the truss members are steel rod of 10mm diameter. A 1kg mass is assumed to be attached to the top node of the trusses. Damage is given in the form of degradation of stiffness of truss members; the maximum magnitude of the damage is assumed to be 50% of the stiffness and the number of truss members to be damaged is assumed to be two to the maximum.

5.1.2 Preparation of Neural Network

The specifications of the multi-layered neural networks adopted in this study are given in Table 1. In the first two examples, five natural frequencies are adopted to examine the proposed damage identification approach; that is, the number of output neurons are five in these cases. The numbers of input

Table 1. Neural network specification

Target truss	Result	Number of	Number of	neurons		Total number of	
units	type*	Fig.	modes	Input	Output	Hidden	BP-learning
5	A	2(c), 4(c)	5	22	5	100, 100	300,000,000
5	S	3(c)	5	27	5	100, 100	300,000,000
5	A	4(d)	10	22	10	200, 200	250,000,000
10	A	5(a)-(d)	10	42	10	90, 60, 60	200,000,000

∗ 'A' for asymmetry truss, 'S' for symmetry truss

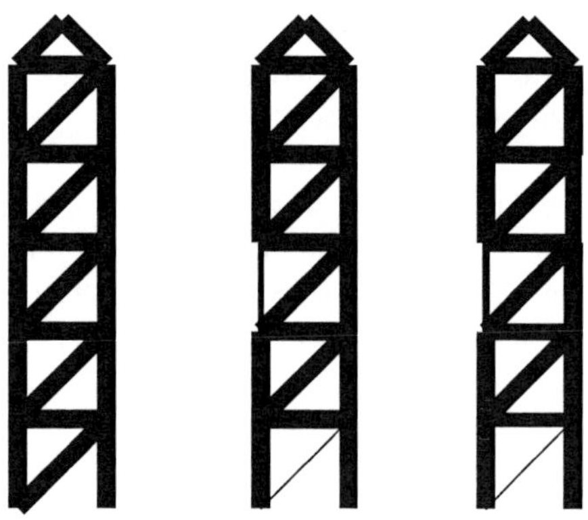

(a) undamaged (b) damaged (c) identified

Figure 2. Asymmetry truss of 5 units (I)

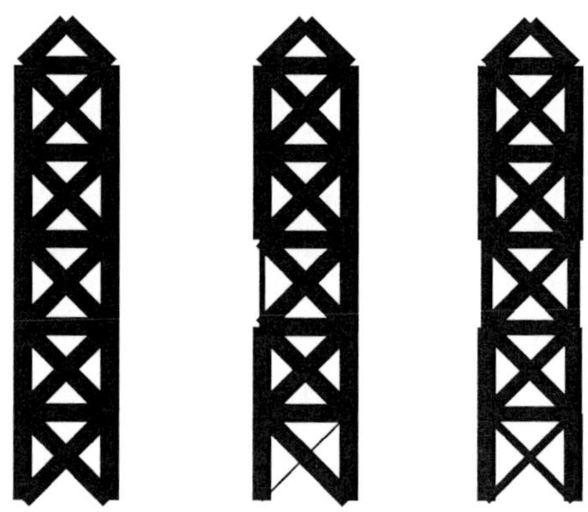

(a) undamaged (b) damaged (c) identified

Figure 3. Symmetry truss of 5 units

neurons are the same as the numbers of members of corresponding trusses, that is, 22 for the asymmetric truss and 27 for the symmetric truss, respectively. Both neural networks have two hidden layers each of which consists of 100 neurons.

It should be noted that the input and output values of the neural networks are confined between -1.0 and $+1.0$. We apply the following normalization to the values of member stiffness and the natural frequencies: 0.0 for the non-damaged member stiffness and -0.95 for the 50% damaged stiffness, and 0.0 for the natural frequencies of non-damaged case and -0.95 or $+0.95$ for the maximum altered corresponding natural frequencies due to the damage.

As indicated in Table 1, the error back-propagation learning is conducted 300 million times for both neural networks. The number of neurons of hidden layers and the number of iterations of error back-propagation learning are considered to be sufficient for both neural networks; the average error norm at the output is less than 0.004.

5.1.3 Numerical Experiments

Figures 2(b) and 3(b) show the trusses with damage. The damages given to the two truss members are respectively 30% and 20%. It should be noted that the magnitude of the damage is exaggerated in the figures. Figures 2(c) and 3(c) are the corresponding damage identification results. It is clearly confirmed

Table 2. Comparison of natural frequencies (5-unit truss)

Truss		Natural frequencies (Hz)				
Type	Fig.	1st	2nd	3rd	4th	5th
Asymmetry	2(a)	29.468	119.599	170.227	260.962	401.506
	2(b)	29.052	112.416	167.562	248.654	390.971
	2(c)	29.043	112.427	167.548	248.669	390.898
	4(b)	29.103	116.165	165.122	257.134	401.225
	4(c)	29.103	116.164	165.124	257.139	400.923
Symmetry	3(a)	28.887	143.349	159.988	331.743	484.984
	3(b)	28.632	138.060	157.509	322.500	479.959
	3(c)	28.623	138.060	157.520	322.502	479.937

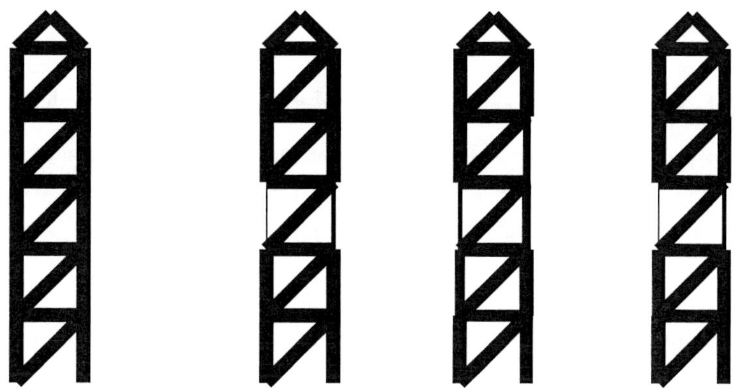

(a) undamaged (b) damaged (c) identified (by 5 frequencies) (d) identified (by 10 frequencies)

Figure 4. Asymmetry truss of 5 units (II)

that the damaged members are identified in the case of asymmetric truss shown in Fig.2.

The damage identification result shown in Fig.3(c) does not fully coincide with the damaged truss shown in Fig.3(b). On the basis of the natural frequencies corresponding to these trusses shown in Table 2, we can say, however, that the inverse problem (2) based on the natural frequencies and the member stiffness values is solved from the viewpoint of the natural frequencies of the damaged truss and the damage-identified truss. This is confirmed in terms of the values indicated as 3(a), 3(b) and 3(c) in Table 2; the difference of the values between the cases 3(b) and 3(c) are insignificant compared to the cases 3(a) and 3(b) or 3(a) and 3(c). This is a fundamental limitation of the current approach based only on the natural frequencies of the entire structure, since the influence of a left-side damage and the corresponding right-side damage cannot be distinguished theoretically.

Figure 4 shows a similar case. In this case, we can see a difference between the damaged truss and the damage-identified truss shown in Figs.4(b) and 4(c). The reason of this insufficient identification is similar to the previous symmetric case; this is also confirmed in terms of the values of natural frequencies indicated as 4(b) and 4(c) in Table 2. This insufficiency of damage identification is, however, not funda-

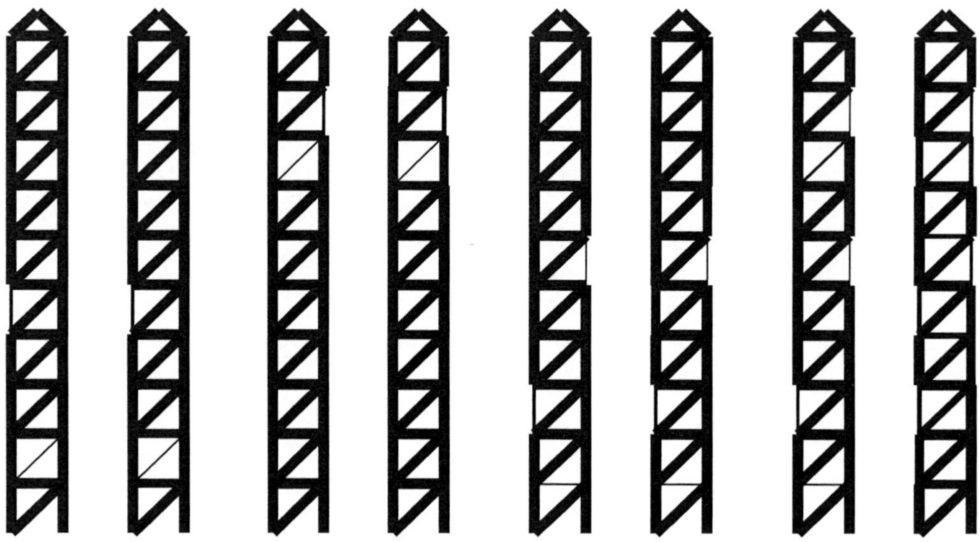

(a) 2 members (b) 2 members (c) 3 members (d) 5 members

Figure 5. Ten-unit asymmetry truss examples of various damage patterns

mental and can be avoidable. As shown in Fig.4(d), we can correctly identify the damage in the case that ten natural frequencies are adopted for the damage identification instead of five. The specification of the neural network in this case is also shown in Table 1.

5.2 Ten-unit Truss

Figure 5 shows the damage identification results in the case of an asymmetric ten-unit truss. This is based on ten natural frequencies; the specification of the neural network adopted in this numerical experiment is shown at the bottom line in Table 1. The error back-propagation learning for the neural network is also performed in terms of the learning pattern which is calculated based on the combinations of two damaged members and ten natural frequencies.

Figures 5(a) and 5(b) show the damage identification results in the case of two damaged members. It is confirmed that the damage at the relatively tip-end side members is successfully identified as shown in Fig.5(b).

Figures 5(c) and 5(d) show the results of three-member and five-member damage cases. It should be noted that the neural network adopted for the damage identification is trained based on the learning pattern consisting only of the two-member damage cases. However, the three-member damage case is sufficiently dealt with as shown in Fig.5(c) and the damage identification of the five-member case can said to be performed fairly well as shown in Fig.5(d). This is considered to be because of the generalization capability of the neural network to some extent; however, we have not yet clearly confirmed the limitation of the identification capability for such damage patterns that are not prepared in the learning pattern of the neural network.

6. CONCLUDING REMARKS

A structural damage identification technique by means of multi-layered neural network is proposed from the viewpoint of inverse problem. The approach is based on the error minimization capability of the back-propagation process. The basic idea and its implementation by means of the error back-propagation with fixed connection weights have been explained. We have also discussed the structural characteristics for the proposed inverse problem approach. The natural frequencies of the target structure are adopted in this study.

A number of numerical experiments have been conducted with truss structures. The basic feasibility of the proposed approach is demonstrated; however, it is also confirmed that there is a fundamental limitation with the proposed approach in the case that the damage identification is performed based only on the natural frequencies of the entire structure. An applicability of the approach to the case that is not preliminary considered at the time of learning process of the neural network is also demonstrated.

Specific capability of the proposed damage identification approach is still under investigation. We are currently examining some additional structural characteristics in order to cope with the limitation confirmed in this study. Sensing devices and signal processing methodologies for the practical application are also considered as the future works.

References

1. e.g. Wada, B. K., Fason, J. L. and Crawley, E. F., "Adaptive Structures", *Journal of Intelligent Material Systems and Structures*, Vol. 1, No. 2, 1990, pp. 157-174.
2. e.g. *Structural Health Monitoring*, an international journal.
3. e.g. *NDT & E International* (formerly known as *Non-Destructive Testing* and *NDT International*).
4. Rumelhart, D. E., McClelland, J. L. and PDP Research Group, *Parallel Distributed Processing (Explorations in the Microstructure of Cognition)*, 1986, The MIT Press.
5. Tanaka, M., Hanahara, K. and Seguchi, Y., "Configuration Control of the Truss-Type Parallel Manipulator by the Modular Neural Network Model", *JSME International Journal Series III*, Vol. 35, No. 1, 1992, pp. 89-95.
6. Giurgiutiu, V., Zagrai, A. and Bao, J. J., "Piezoelectric Wafer Embedded Active Sensors for Aging Aircraft Structural Health Monitoring", *Structural Health Monitoring*, Vol. 1, No. 1, 2002, pp. 41-61.
7. Betz, D. C. et al., "Structural Monitoring Using Fiber-optic Bragg Grating Sensors", *Structural Health Monitoring*, Vol. 2, No. 2, 2003, pp. 145-152.
8. Takada, Y., Okugawa, M. and Atsumi, T., "Supervisor with Adaptive Observer for Bolt Loosening Diagnosis by Using Smart Washer", *Proceedings of 22nd International Conference on Adaptive Structures and Technologies*, 2011, Paper No. 041.
9. Merckle, J. et al., "Intelligent Sensing for Non-Destructive Testing Using Eddy Currents", *NDT International* Vol. 23, No. 6, 1990, pp. 335-344.
10. Saka, M. and Fukuda, Y., "NDT of Closed Cracks by Ultrasonic Propagation along the Crack Surface", *NDT & E International*, Vol. 24, No. 4, 1991, pp. 191-194.
11. Lambrineas, P., "X-Ray Computed Tomography Examination of Inshore Mine-Hunter Hull Composite Material", *NDT & E International*, Vol. 24, No. 4, 1991, pp. 207-213.

12. Takahashi, K. and Hahn, H. T., "Towards Practical Application of Electrical Resistance Change Measurement for Damage Monitoring Using an Addressable Conducting Network", *Structural Health Monitoring*, Vol. 11, No. 3, 2012, pp. 367-377.
13. Giurgiutiu, V. and Bao, J. J., "Embedded-Ultrasonics Structural Radar for In Situ Structural Health Monitoring of Thin-wall Structures", *Structural Health Monitoring*, Vol. 3, No. 2, 2004, pp. 121-140.
14. Manning, R. A., "Structural Damage Detection Using Active Members and Neural Networks", *AIAA Journal*, Vol. 32, No. 6, 1994, pp. 1331-1333.
15. Sundaresan, M. J. et al., "Experimental Damage Detection on a Wing Panel Using Vibration Deflection Shapes", *Structural Health Monitoring*, Vol. 2, No. 3, 2003, pp. 243-256.
16. Haritos, N. and Owen, J. S., "The Use of Vibration Data for Damage Detection in Bridges: A Comparison of System Identification and Pattern Recognition Approaches", *Structural Health Monitoring*, Vol. 3, No. 2, 2004, pp. 141-163.
17. Chortis, D. I., Chrysochoidis, N. A. and Saravanos, D. A., "On the Detection of Buckling on Composite Strips Based on Frequency and Damping Monitoring", *Proceedings of 22nd International Conference on Adaptive Structures and Technologies*, 2011, Paper No. 106.
18. Xu, B., Song, G. and Masri, S. F., "Damage Detection for a Frame Structure Model Using Vibration Displacement Measurement", *Structural Health Monitoring*, Vol. 11, No. 3, 2012, pp. 281-292.
19. Wang, X. et al., "Membership-Set Identification Method for Structural Damage Based on Measured Natural Frequencies and Static Displacements", *Structural Health Monitoring*, Vol. 12, No. 1, 2013, pp. 23-34.
20. Habib, F., et al., "Structural Health Monitoring (SHM) of Bonded Composite Repairs", *Proceedings of 22nd International Conference on Adaptive Structures and Technologies*, 2011, Paper No. 032.
21. Hufenbach, W., et al., "In-Situ Integrity Assessment of a Smart Structure Based on the Local Material Damping", *Proceedings of 22nd International Conference on Adaptive Structures and Technologies*, 2011, Paper No. 062.
22. Muromaki, T. et al., "Estimation of Loading Conditions of Failed Crane-hook (An Image-based Approach with Knowledge and Simulation)", *International Journal of Reliability and Safety*, Vol. 6, No. 1/2/3, 2012, pp. 130-147.

Study on optimum connecting points of boom-membrane structure as a basic model for larger lightweight space structure

Ayako Torisaka[1], Yoshitaka Sato[2], M.C. Natori[3], Takeshi Akita[4], Hiroshi Yamakawa[5], Nobuyuki Kobayashi[6]

[1] 1st Author, Dept. of Mechanical Engineering, Aoyama Gakuin University, Sagamihara, Japan
[2] 2nd Author, Dept. of Modern Mechanical Engineering, Waseda University, Shinjyuku
[3] 3rd Author, Dept. of Modern Mechanical Engineering, Waseda University, Shinjyuku
[4] 4th Author,Dept. of Mechanical Science of Engineering, Chiba Institute of Technology, Narashino
[5] 5th Author, Dept. of Modern Mechanical Engineering, Waseda University, Shinjyuku
[6] 6th Author, Dept. of Mechanical Engineering, Aoyama Gakuin University, Sagamihara

Abstract

This paper presents a first step of suggesting optimization designing on a composite of lightweight membrane and boom structure. In this structure, booms keep rigidity of the whole structure. We believe such structure can be used for deep space exploration in the future as a following and developed version of IKAROS which has already in full success of its self-deployment, self-power generation by thin solar films, and space navigation missions. Or in short-sighted developing vision, this type of structure is going to be applied to the deorbit hardware of low-orbit satellites to earn much drag for atmosphere entry within smaller years. So here we focused on the basic model like a former mentioned type of boom-membrane structure and studied about optimization strategy, and we'd like to suggest to add our research points on the currently recognized as common standard models of deorbit hardware which has just two terminal connecting points on each boom between membrane. Here we studied the efficiency of considering connecting points between boom and membrane from the viewpoint of the reaction force of the booms and wrinkle amount on membrane those are having trade off relations. These parameters are going to be the most critical factor in the future to architect large lightweight membrane structure. Moreover, we formulated this problem as a discrete parameter optimization problem and discussed its efficiency of the optimized model. Furthermore, need of the hoop parts are mentioned as a basic unit of a large repetitive architecture such like solar power plant concept in space.

1. INTRODUCTION

Recent success of membrane structure represented by solar sail (IKAROS[1]) broaden the possibilities of space exploration and its potential has been developed actively since then. Since it is a tension element, for future large scale space structure systems centrifugal force due to spinning is useful in order to stabilize it, and there already appeared a spinning solar sail spacecraft to demonstrate such basic structure concept.

But for those in rather near future a corresponding compression element such as a boom should be used from the viewpoint of easy maneuverability of the whole spacecraft system, and the mechanical characteristics of compound structure systems consisted of membrane elements and boom ones must be clarified. So one of the prevailing structure aiming for deep space exploration or constructing modular structure is four deployable booms with a square membrane type[2]-[5]. Detail design study of a square solar sail architecture was carried out by Greschik etc. [6] and one of his summarize is that optimal design efficiency is not degraded dramatically by slight to moderate boom cross section variations. Then he also proposing the stripped square architecture with diagonally boom and continuously suspension between membranes, but the study about the optimum configuration of suspension layout (connection layout) has not been carried out. So the optimization study from the view point of design configuration should be carried out.

Focusing on the use of cable, Sakamoto etc. proposed surrounding membrane structure with weblike cables along the membrane perimeter to realize alleviating wrinkle in the membrane and reduction in corner cable tensions [7]. When viewed in light of membrane suspended by four diagonal boom structure, the tension members are already exist that is a boom, but its deformation does not effectively absorb the disturbance and directly transmit to membrane which will cause wrinkle. So another different role should be given to cable that is mainly working on structural stiffness for in-plane modes those would be appear in higher frequencies.

From above, the problems on this research are summarized below.

(1) Proposal of optimum suspension layout between membrane and booms to realize high stiffness, less wrinkle, less load on the booms.
(2) Proposal of cable use to improve dynamic performance, which is mainly focusing on in-plane modes appearing in higher modes.

We studied about the above points with the ORIGAMI-based membrane model suspended by CFRP booms that authors and other members has been developing[5].

2. BOOM-MEMBRANE STRUCTURE

This section describes the analytical model of boom-membrane solar sail as shown in Figure 1 which has been developed by authors and other members aiming for the CFRP booms and a membrane wrapped together on the same hub by using membrane-folding pattern [5]. The basic assemble concept is that the four radial rib booms are connected to the main satellite and supporting membrane. Here, how to decide the connecting points are of our interest in this research, and efficiency of applying cable is also discussed in following chapters.

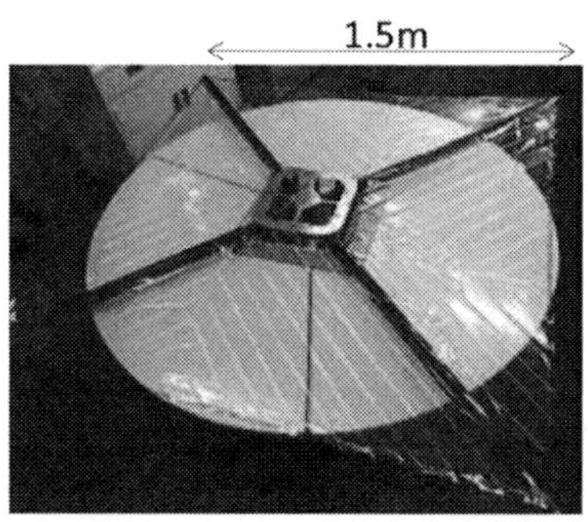

Figure 1. ORIGAMI-based membrane model suspended by four diagonal CFRP booms

2.1. General outlook and material property

Generally, however we can take many variety of solar sail's shape as a squared one or hexagonal one etc.., here in this paper we are focusing on the squared shape and treating the 1/4 model as shown in Figure 2.
Since asymmetrical vibration modes are not considered with this model, booms are assumed to move only in the out-of-plane direction. It is also assumed that the satellite on the center is large and heavy, so the satellite can be treated as fixed to the original point in the coordinate.

Material properties are decided as Table 1 according to our previous study, the reason of taking these values is that to correspond the future comparative study with laboratory model experiments. Here, the material of the membrane is poly-imide film (25μm) and the boom material is SMA (Shape Memory Alloy). We assume the length of the boom is 1 m, 0.0001m in height, 0.015m width.

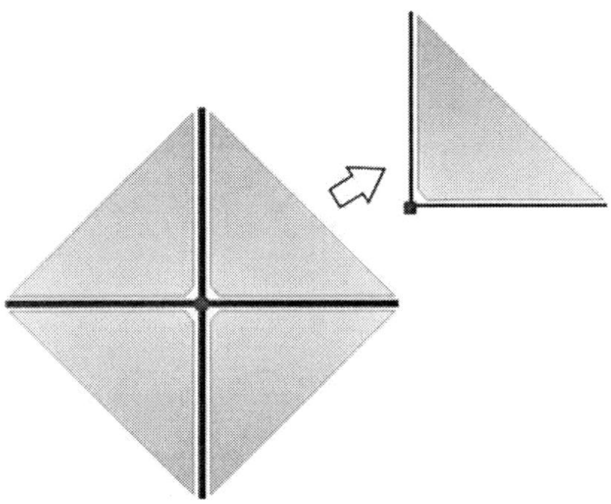

Figure 2. Square model and 1/4 model

Table 1 Material properties

	Young's modulus [GPa]	Poisson's ratio	Shear modulus [GPa]	Density [kg/m3]
Membrane (poly-imide)	3.19	0.34	-	1420
Boom (SMA)	60.0	-	20.6	6500

2.2. Analysis model

The initial shape of membrane is an acute-angled triangle truncated near the origin as shown in Figure 3. The membrane is connected to booms through penalty springs, and it is tensioned through the change of three parameters, d and δ. d is the basic distance between membrane and boom elements, δ is the adjustment angle of the additional distance between the elements. Near the origin a small right-angled isosceles triangle with the edge length of s is cut from the original membrane.
Here, δ is determined to make strain of membrane becomes uniform, which means δ becomes to be a function of s and d.
Regarding with connecting points, both tips of one boom are always connected as default, and we are focusing on the region between these two tips, then here we express the value of design variable as 1 or 0

(connected or not connected), and we decided the expression rule of the model as Figure 4 comes to be [0100] for example.

Further, the total spring stiffness varies due to the total number of connecting points when using the uniform value as a penalty spring, we modified it to be variable to keep spring stiffness uniformly.

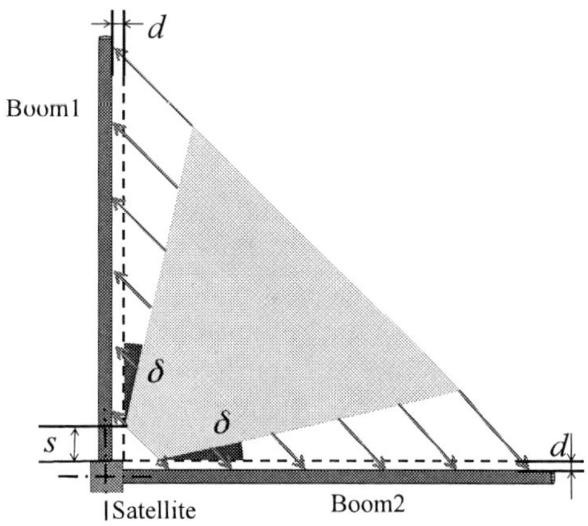

Figure 3. Analysis model

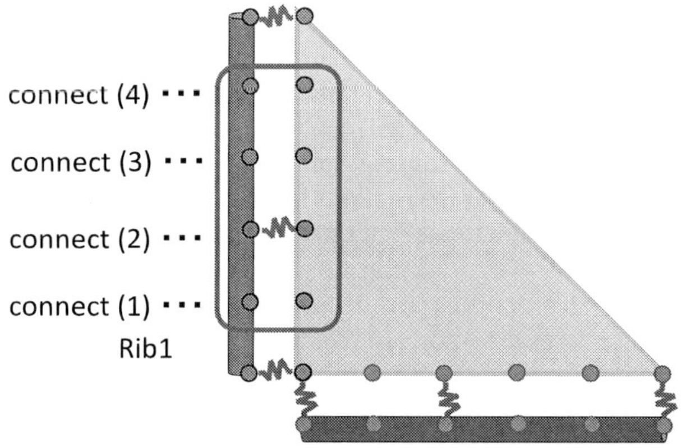

Figure 4. Connecting points of [0100] model

As for wrinkle evaluation which is the most interest when talking about sail structure, however, here we omit the detail formulation of wrinkle intensity due to the space of paper, we use its value as an evaluation of wrinkle and we call that as wrinkle intensity which is expressed in Eq.(1)

$$\phi_w = \frac{1}{2}\boldsymbol{\varepsilon}_w^T \mathbf{C} \boldsymbol{\varepsilon}_w \tag{1}$$

Where, $\boldsymbol{\varepsilon}$ is wrinkle mode deformation, $\mathbf{C}$ is elasticity matrix.

$$\phi = \phi_e + \phi_w \tag{2}$$

This wrinkle intensity is derived from the Eq (2) which shows the total strain energy can be efficiently decomposed into elastic energy part and apparent energy part (=zero energy) which is the wrinkle intensity. Wrinkle intensity means the strain energy density function due to geometrical contraction corresponded to wrinkle [9], which is based on the tension-field theory.

3. PROPOSAL OF A BASIC STRUCTURE FOR LARGER LIGHTWEIGHT SPACE STRUCTURE

Here explains the optimization of a solar sail structure and propose it as a basic model of a larger, light weight and square shaped solar sail.

3.1. Optimization problem

3.1.1. Relation between 1st mode frequency and wrinkle intensity

Basically boom-membrane structure come to need trade off designing on the point of wrinkle intensity and reaction force on the boom. Generally there is a relation such like no wrinkle model can be realized when it is highly tensioned, in other words reaction force become larger, and on the other hand, small reaction force means the boom don't support the membrane and it has much wrinkle.

Although this relation is true in one axis direction structure, the planar structure can't always adopt this way of thinking especially when the structural rigidity is not uniform due to the connecting points between boom and membrane, so the optimization technique should be applied to find out the structural designing strategy. Table 2 shows an example of the relation between reaction force and wrinkle intensity of all suspension pattern with 5 x 5 meshed model which has four design parameter as a connection point. It's obvious that the tradeoff relation of reaction force and wrinkle intensity is disappeared due to the layout of connection points.

Table 2 Material properties

Connect points	0000	0001	0010	0011	0100	0101	0110	0111	1000	1001	1010	1011	1100	1101	1110	1111
Reaction Force x 10^8[N]	2.75	3.20	2.90	3.21	2.60	2.97	2.75	3.02	2.30	2.75	2.53	2.84	2.30	2.66	2.48	2.75
Wrinkle Intensity x 10^8	2.29	2.27	2.14	2.16	1.59	1.65	1.55	1.59	0.377	0.315	0.313	0.246	0.110	0.179	0.171	8.09 x10^{-6}

3.1.2. Aspiration Level Method

First, here explains the confliction of taking linear summation as an evaluation function for optimization problems, and proposing the introduction of aspiration level method.

Generally in the MOP (Multi Objective Problem), it has two or more evaluation element and a set of feasible designs make Pareto Frontier between these evaluation elements as shown in Figure 5. Regarding with the models on Pareto Frontier, one of the evaluation element can be improved only under the sacrifice of another evaluation element. So each model on the Pareto Frontier is known as an optimized model respectively. Hence the designer need to have a certain criterion to pick up a best model for his own problem because all of the models on Pareto Frontier is optimum as mentioned in previous.

One of the popular and simple solution is taking a linear summation as Figure 6, but we need to be careful to take this method. For example, suppose the Pareto models are making convex set like Figure 5, it doesn't always have positive correlation between weight ω_i and obtained model as shown in Figure 6.

So we need to take care that the possibility of easily coming down to the carpet bombing like designing when we treat a MOP due to the lack of enough searching ability which coming from the simple linear evaluation function.

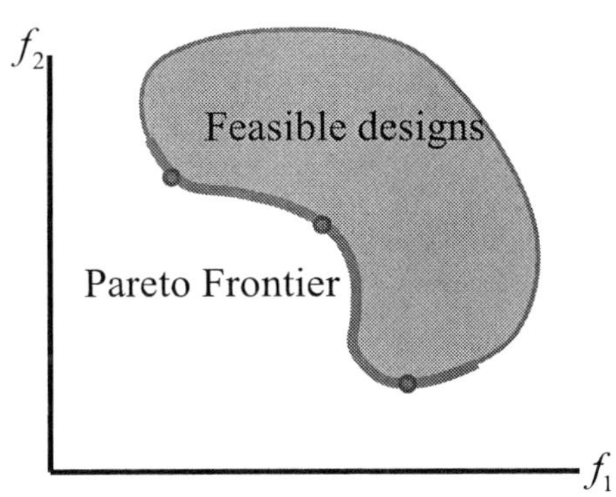

Figure 5. Pareto Frontier

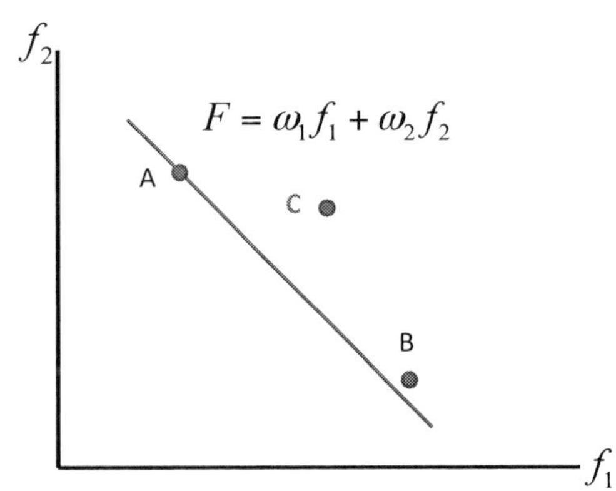

Figure 6. Linear summation

Then we propose to introduce Aspiration Level method as an evaluation function. The process of this method needs in advance to decide aspiration level $\overline{f_i}$ on each evaluation element that is equal to the lowest performance we want to keep. Also ideal point f_i^* should be set in advance and be better to be chosen among higher performance values which the system can't realize. This method searches the optimum design according to equation (3)~(5) which has an ability to keep much better design even after attaining the aspiration level performance because the equation accepts negative value.

$$\min\left[\max_{1\le i\le r}\omega_i(f_i(\mathbf{x})+\bar{f}_i)+\alpha\sum_{i=1}^{r}\omega_i f_i(\mathbf{x})\right] \tag{3}$$

$$\omega_i=\frac{1}{\bar{f}_i-f_i^*} \tag{4}$$

$$f_i^*\le\min\left\{f_i(\mathbf{x})\mid\mathbf{x}\in X\right\} \tag{5}$$

Here, f_i^* is ideal point, $\bar{f}_i$ is aspiration point respectively. Usually, α is 10^{-6}.

3.1.3. Formulation for structural optimization of solar sail

Here summarize the optimization of connecting position between boom and membrane from the view point of MOP with 1st mode, reaction force and wrinkle intensity.

Design variable	: $[p_1, p_2, .., p_i, .., p_{15}]$, ($p_i$ =1 or 0)
Evaluation function	: $\min\left[\max_{1\le i\le r}\omega_i(f_i(\mathbf{x})+\bar{f}_i)+\alpha\sum_{i=1}^{r}\omega_i f_i(\mathbf{x})\right]$
Optimization algorithm	: GA (Population: 50, Calc. generation : 100 gen.)

Aspiration level are chosen as the performance of [000000000000000] model because this model is the currently popular model among most researches in worldwide, and we are challenging to find much better design than this aspiration level model. Further, design variables we are focusing on is discrete, so the heuristic search algorithm which is suitable for strong nonlinear problem should be used and we decided to use GA (Genetic Algorithm) in this research.

3.2. Suggested structure

3.2.1. Optimized design

Figure 7 shows the optimum design from the view point of wrinkle, 1st mode freq. and reaction force on booms. This suggests two design characteristics.

1. 1/3 length from the root of a boom should be continuously connected
2. About 1/5 length point from the tip of a boom should be connected.

Dynamic characteristic are shown in Figure 8, and it is obvious that the dominant factor to the behavior of a solar sail is not membrane but booms. Figure 9 shows wrinkle intensity of the optimum model and Figure 10 shows the principal stress. Regarding with Figure 10, red color means the principal stiffness is positive on the element, green color means the minimum principal stress is zero, which means there appears wrinkle (wrinkle region).

Then focusing on the edge part of a membrane along the boom, optimum model has longer red colored region that acquires more areas of tension-fields on a membrane, which means the model has less wrinkle intensity. Also the dynamic characteristics in higher order can be also improved or adaptable from the view point that it regulate the structural stiffness.

Combination to the design characteristics 1 and design characteristics 2 also increase not only the 1st mode freq. but also 2nd mode freq. This is because the connecting point on the 1/5 length distance from the tip reduces the deformation of the boom around its tip which means the structure becomes locally rigid.

This can be seen in Figure 10 as the red colored region is appeared around the tip. This design characteristics 2 also makes wrinkle on the oblique line of a membrane, in other words, it avoid the stress concentration on that region, so the possibility of tearing the membrane can be decreased.

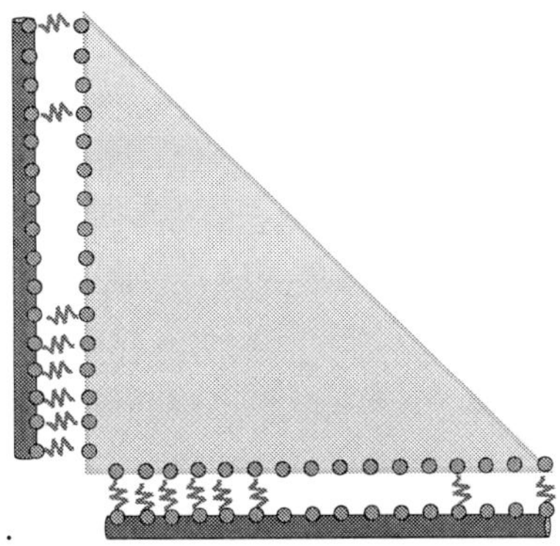

Figure 7. Optimum connecting points

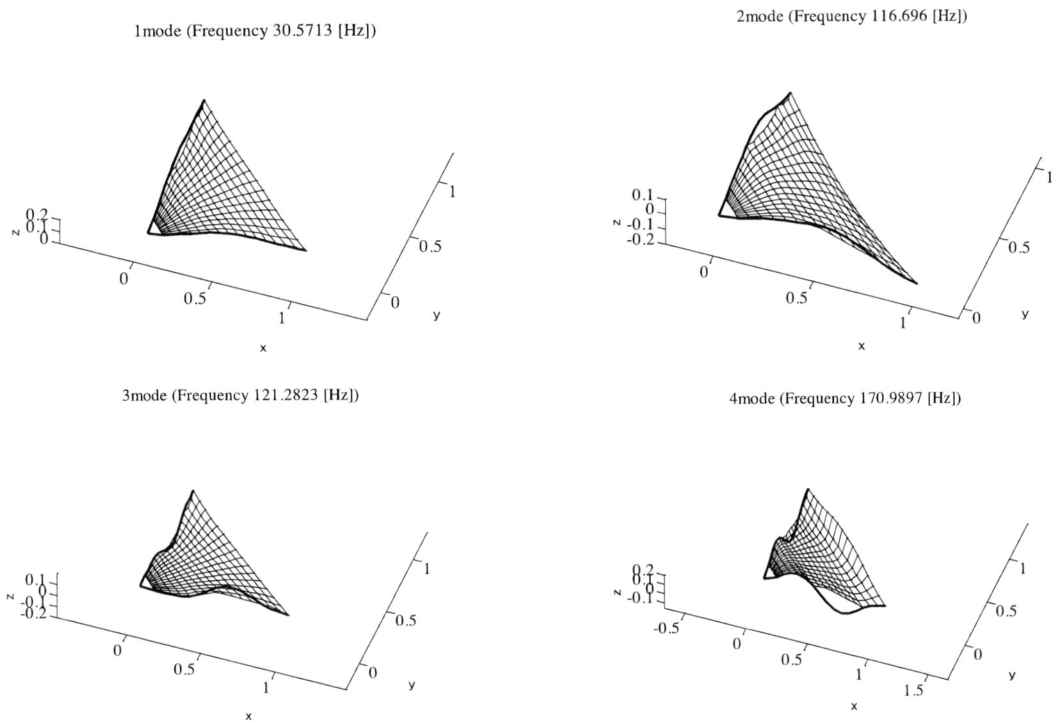

5mode (Frequency 174.1048 [Hz])

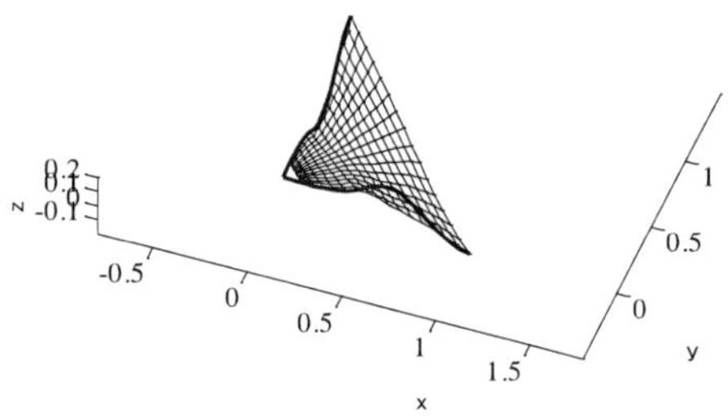

Figure 8. Modes of the optimum model

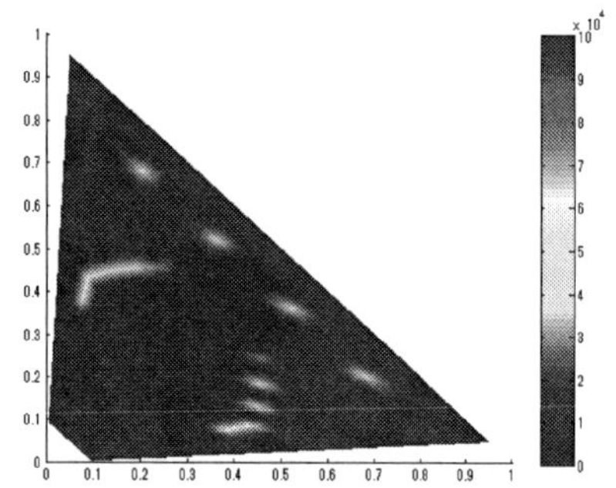

Figure 9. Wrinkle intensity of the optimum model

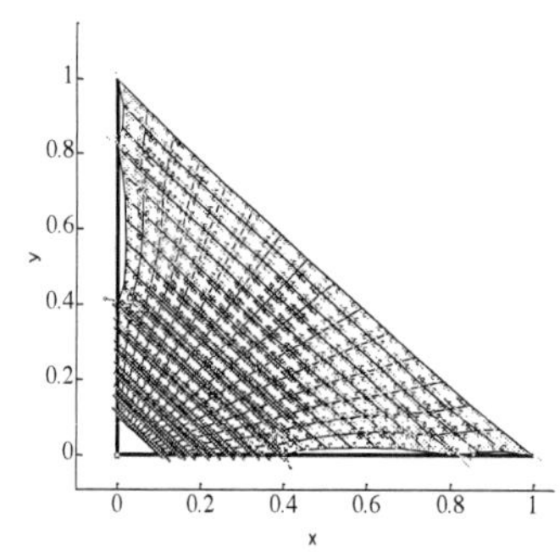

Figure 10. Principal stresses of the optimum model

3.2.2. Discussion

Table 3 shows the comparison of Structural characteristics of optimum model and aspiration level model. Column in the middle (②) is the optimization result with only a reaction force and wrinkle evaluation using Aspiration Level Method. Although its reaction force and wrinkle shows better characteristics than the optimum model (①) which additional 1st mode freq. is took in account as an evaluation function, the difference in the evaluation value of reaction force between model ① and ② are judged as small because generally the value of reaction force and wrinkle can be varying different order of magnitude. So the increase ratio of the 1st mode freq. is considered as more important in the aspiration level method. Further, the model ② is eliminated from the optimum candidate in the optimization process which include1st mode freq. evaluation because it doesn't satisfy the aspiration level.

In summarize, although model ② is the best design in statistic evaluation, model ① should be took as an optimum design in the view point of actual operation which needs evaluation of dynamic characteristics.

And the importance of connecting point on the 1/5 length from the boom tip is also shown in the view point of evaluation function which contains dynamic characteristics evaluation.

Table 3 Structural characteristics of optimum model and other model

	□Optimum (1st mode, Reaction force, Wrinkle)	②Optimum (Reaction force, Wrinkle)	③Aspiration level model

	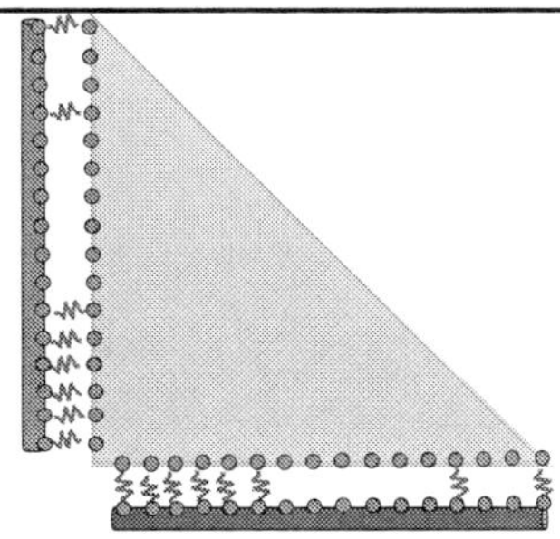	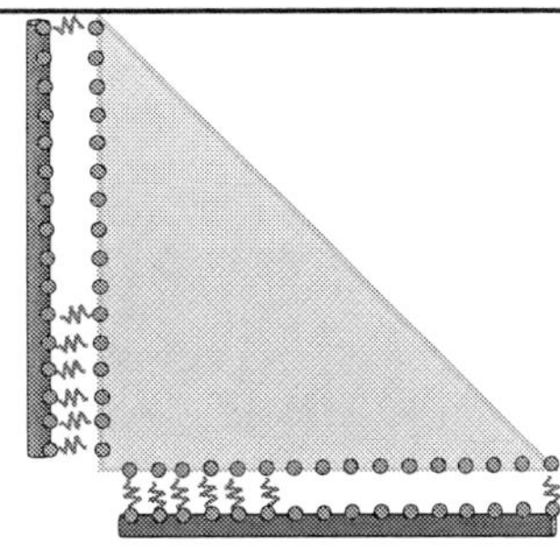	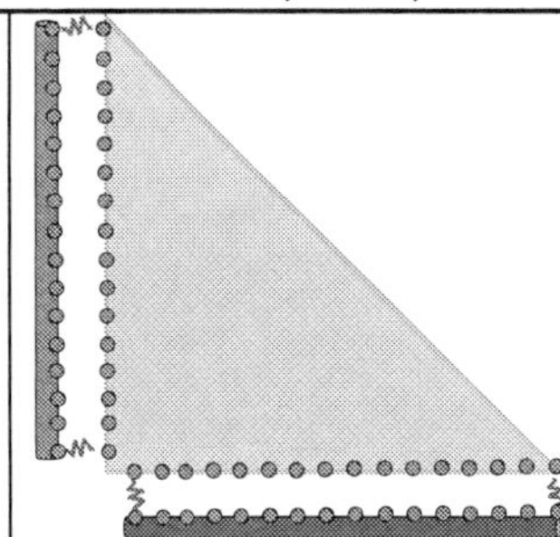
model	**[11111000000100]**	**[11111000000000]**	**[00000000000000]**
Reaction Force $\times 10^8$ [N]	2.075	1.786	2.750
Max of wrinkle Intensity $\times 10^7$	3.695	3.485	45.38
1st mode [Hz]	30.57	26.28	28.36
Wrinkle Intensity	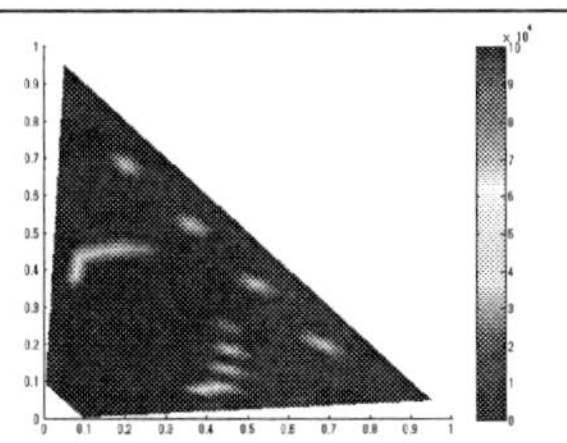	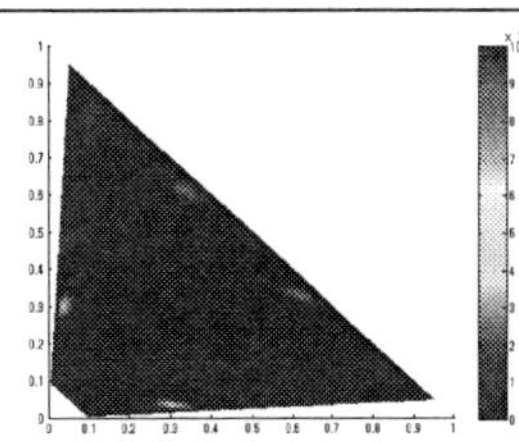	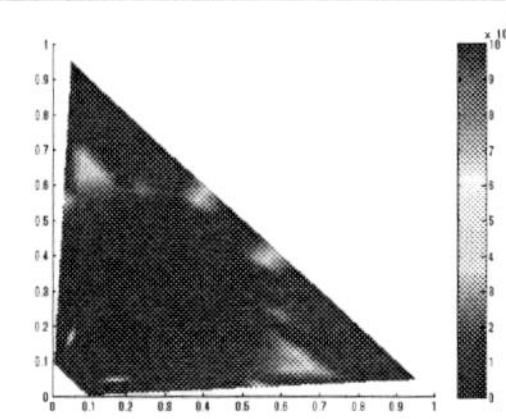
Principal Stresses	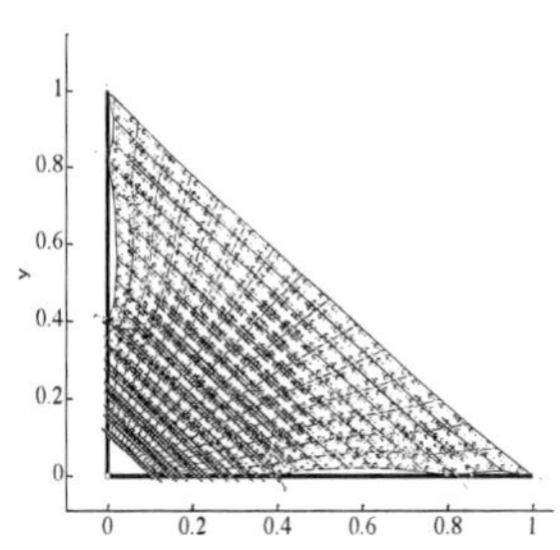	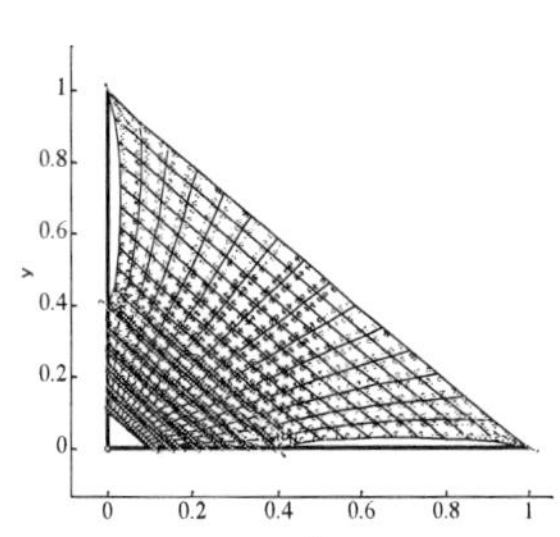	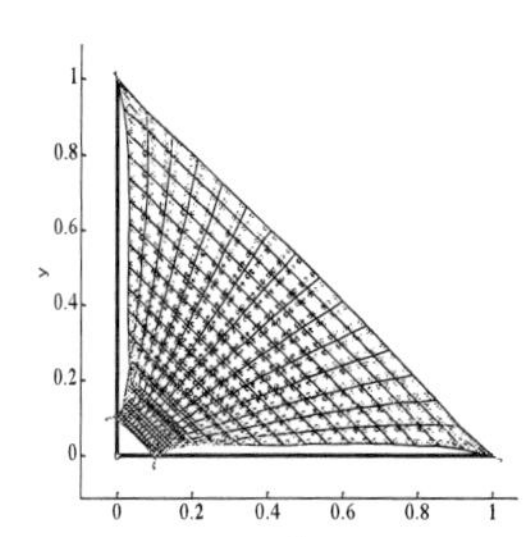

4. EFFICIENT USE OF CABLES AND DISCUSSIONS

This section shows the efficiency of using a cable with the higher order dynamic characteristics. We suggest to put cables on an optimum [11111000000100] model (①) in previous chapter with which we expand our following arguments. Wrinkle intensity and principal stress are same with any of Case1 to Case3 as shown in Figure 12 and Figure 13 here. Additionally, [11111000000000] with cable connected model has stress concentration on the oblique line of a membrane, so we considered that the connecting point on the 1/5 length position is always needed and cable doesn't work as a substitute of the physical connection on that point even if the cable is connected to the point.

Here the three comparing cases of cable applying position are shown in Figure 11 to investigate the effect of cable positioning to higher order dynamic characteristics. Case2 shows that cables are connected to the point of union at the 1/5 length displacement from the tip of a boom, and Case1 and 3 shows cables are put to the inside or outside of the 1/5 length point.

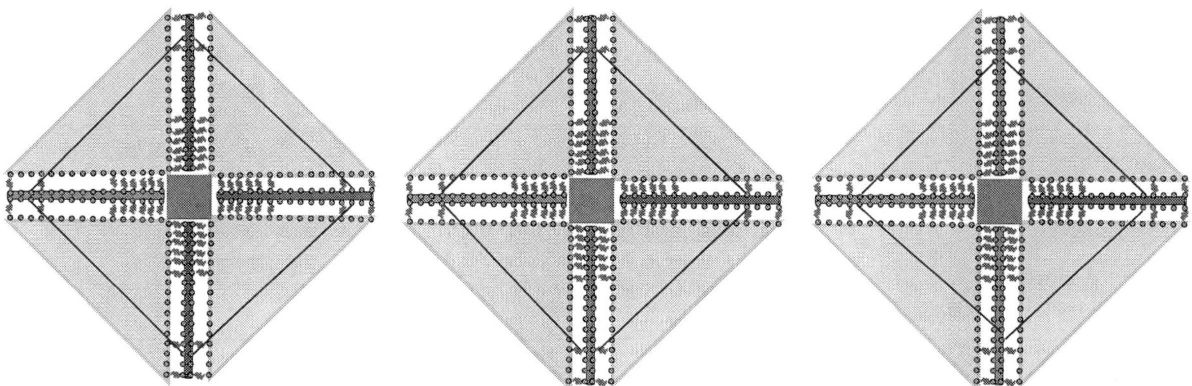

Case1 Case2 Case3

Figure 11. Full model with cables

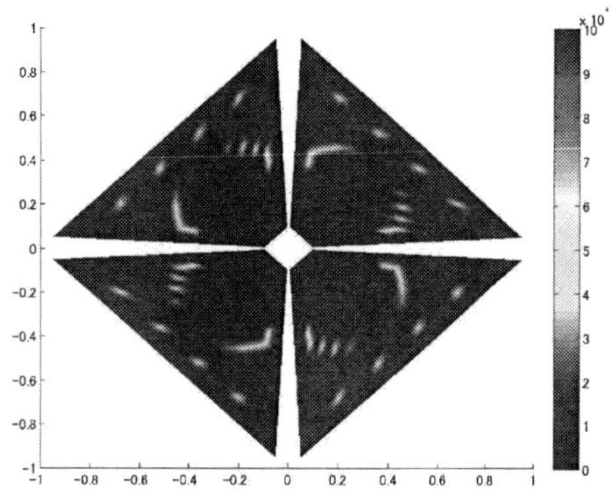

Figure 12. Wrinkle intensities (full model, Case3)

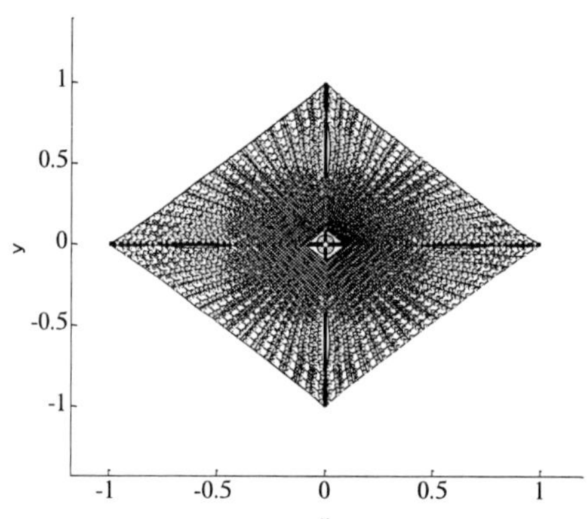

Figure 13. Principal stresses (full model, Case3)

Then Table 4 shows the dynamic characteristics of each model, and we can find the reverse of mode order such like the 2nd and 3rd (symmetrical) modes of no cable model are moved to 3rd and 4th modes in cable using model. Further, it's obvious that the use of cables enables to increase the eigen frequency of these symmetrical mode. On the other hand, cables don't work on the entire out-plane model such like bowl-shaped mode (same with 4th mode of no cable model or 2nd mode of cable use model.) Further, more inside cable position than Case3 doesn't work to increase eigen frequencies on 3rd and 4th mode any more. Incidentally, these characteristic of cable use are same in [1111100000000] model.

Table 4 Eigen frequencies

Base structure	11111000000100			
Cable position	No cable	Case1	Case2	Case3
1st mode [Hz]	30.56	30.56	30.56	30.56
2nd mode [Hz]	86.79	116.64	116.64	116.64
3rd mode [Hz]	86.79	120.05	137.10	135.92

4th mode [Hz]	116.64	120.05	137.10	135.92
5th mode [Hz]	121.24	134.82	150.19	170.95

Then Figure 14 shows modal displacement in z (out-plane) direction of a boom. Cable position like Case3 is the most effective in reducing the displacement amplitude of boom. So the best cable position can be summarized as applying a little bit inside of 1/5 length distance from the tip of a boom to attain improvement of dynamic performance in hirer order mode also.

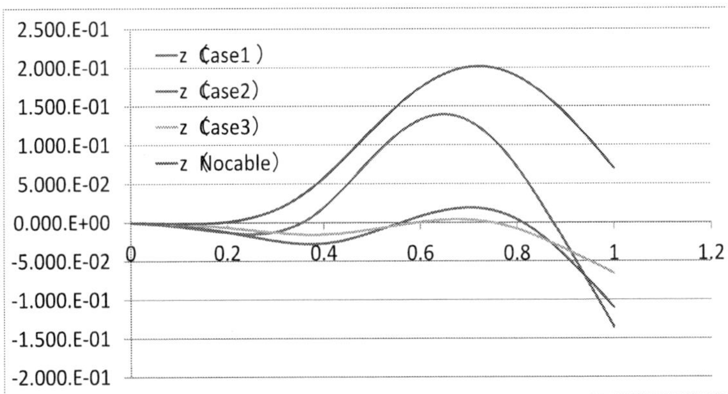

Figure 14. Modal displacement of a boom (z-axis)

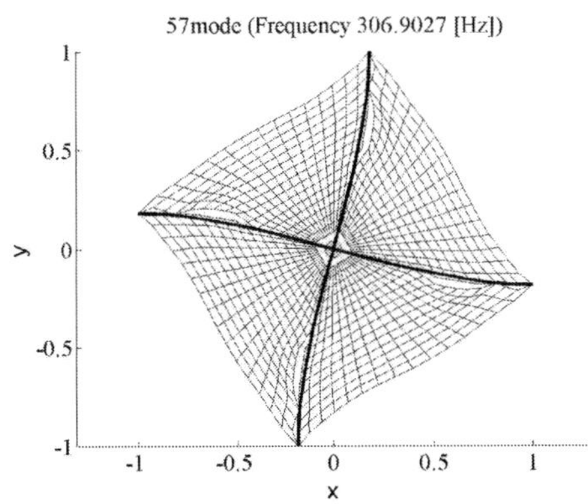

Figure 15. In-plane mode (No cable model)

Focusing on in-plane mode which will cause much wrinkle on the membrane, one more efficiency of cable use can be seen from the fact that any of the cable use model doesn't have in-plane mode like Figure 15 of no cable model. So this model can be said as especially suitable for the structure such like hating in-plane deformation, solar array use on the membrane for example.

5. Conclusions

Basic structural concept of boom- membrane structure for larger light weight space structure are proposed by using GA+ aspiration level method with 1/4 model. Two structural design strategy are obtained and then we proposed the cable use on the full model and investigated its efficiency. They are summarized as below.

- ✓ 1/3 length from the root of a boom should be continuously connected to ensure much tensioned area on the membrane.
- ✓ About 1/5 length point from the tip of a boom should be connected to the membrane. This connection enables the structure satisfy the aspiration level on 1st mode frequency, reaction force and wrinkle intensity, and as realize better performance as possible. Also this connection realize to avoid the stress concentration on the oblique line of the membrane.
- ✓ Cable should be put on a little bit inside of 1/5 length point from the tip of a boom to avoid in-plane mode of the full model and also to increase the higher order dynamic performance.

These above points would be a designing strategy for large boom- membrane structure which considers dynamic characteristics in any of the low to higher orders.

Acknowledgements

This study has been supported by Japan Society for the Promotion of Science (JSPS), Grant-in-Aid for Young Scientists (B), Grant Number 25820413.

References

1. Sawada, H., Mori, O., Okuizumi, N., Shirasawa, Y., Miyazaki, Y., Natori, M., Matunaga, S., Furuya, H., and Sakamoto, H., 2011, "Mission Report on The Solar Power Sail Deployment Demonstration of IKAROS," *Proc. of 52nd AIAA/ASME/ASCE/AHS/ASC Structures, Structural Dynamics and Materials Conference, Denver, Colorado.*, 2011, AIAA 2011-1887,
2. O.R. Stohlman, V. Lappas, "Deorbitsail: a Deployable Sail for De-orbiting," *Proc. of 54th AIAA/ASME/ASCE/AHS/ASC Structures, Structural Dynamics, and Materials Conference*, Boston, Massachusetts, USA, Apr. 2013, AIAA 2013-1806.
3. O.R. Stohlman, J. M. Fernandez, V. J. Lappas, M. Hillebrandt, M. Straubel, C. Hühne, "Testing of the Deorbitsail Drag Sail Subsystem," *Proc. of 54th AIAA/ASME/ASCE/AHS/ASC Structures, Structural Dynamics, and Materials Conference*, Boston, Massachusetts, USA, Apr. 2013 , AIAA 2013-1807.
4. Juan M. Fernandez and Vaios J. Lappas, "The Completely Stripped Solar Sail Concept," *Proc. of 53rd AIAA/ASME/ASCE/AHS/ASC Structures, Structural Dynamics and Materials Conference*, April 2012, AIAA 2012-1745
5. H.Sakamoto, H.Furuya, Y.Satou, M.C.Natori, A.Watanabe, N.Kawabata, R.Sakai,N.Pkuizumi, O.Mori, Y.Shirasawa, M.Takai, N.Katsumata, A.Torisaka, R.Funase, "ORIGAMI-BASED MEMBRANE STORAGE AND DEPLOYMENT TECHNOLOGY FOR DE-ORBITING SATELLITES," *Proc. of 64th International Astronautical Congress,Beijing, China*, Sept 2013, IAC-13-B4.6A.4
6. G. Greschik, M. M. Mikulas, "Design Study of a Square Solar Sail Architecture", *J, of spacecraft and rockets.*,Vol. 39, No. 5, September-October 2002
7. H.Sakamoto, K.C.Park, "Dynamic Wrinkle Reduction Strategies for Cable-Suspended Membrane Structures, " *J, of spacecraft and rockets.*,Vol. 42, No. 5, September-October 2005
8. Akita, T., Nakashino, K., Natori, M.C., and Park, K.C., "A simple computer implementation of membrane wrinkle behavior via a projection technique," *International Journal for Numerical Methods in Engineering*, 72(10), 2007, 1231-1259.
9. T. Akita and M. C. Natori.: Sensitivity Analysis Method for Membrane Wrinkling Based on the Tension-Field Theory, AIAA Journal, Vol. 46, pp.1516-1527, 2008.
10. N. Kogiso, K. Toyama, and T. Akita, "Wrinkle Intensity Minimization Design of Space Membrane Structure, "*Proc. of 12th AIAA/ISSMO Multidisciplinary Analysis and Optimization Conference*, Sept. 2008, AIAA 2008-6030.

Development of Novel Seat Cushions for Helicopter Aircrew Vibration Mitigation

Yong Chen*, Viresh Wickramasinghe, and David Zimcik

[1] National Research Council Canada – Aerospace, Ottawa, Canada

Abstract

This paper investigates the use of novel energy absorbing materials to reduce vibration transmitted through helicopter pilot seat cushions. A type of engineered material known as the Hybrid Air Cushioning System (HACS) was integrated into the design of helicopter seat cushions in addition to conventional foam materials. The prototype seat cushions were preliminarily evaluated as the bottom seat cushion on a Bell-412 non-armoured pilot seat through mechanical shaker tests. The promising cushion designs were further evaluated on a Bell-412 helicopter through extensive flight tests and aircrew configurations. The pilot whole-body vibration levels were assessed in accordance with ISO 2613-1:1997 standard. The flight test results demonstrated that the prototype cushions were able to reduce significantly the pilot whole-body vibration. Therefore, the integration of HACS with Bell-412 seat cushion can serve as a low cost solution to mitigate the vibration levels transmitted to helicopter pilots.

1. INTRODUCTION

The helicopter has become an important aerial transportation platform due to the unique capability of vertical taking-off, landing and its ability to hover. However, the ride quality in a helicopter cabin is generally unpleasant due to high levels of vibration and noise. Helicopter cabin vibration is primarily excited by the aerodynamic and inertial loads at the N/rev harmonic frequencies of the main rotor speed. The helicopter floor vibration is transmitted through the aircrew seat cushions and exposes all aircrew to a whole-body vibration environment. Short-term exposure to mechanical vibration transmitted to human body increases fatigue, degrades comfort, interferes with aircrew performance and affects operational safety [1]. Long-term exposure of helicopter aircrew to whole-body vibration is known to contribute to occupational health issues including aircrew neck pain and back pain injuries among other human-interface related factors in helicopter operations [2, 3].

These health issues become more severe as more equipment is integrated onto the military helmet and flown for long hours during military operations. Recent experience of the Canadian RCAF aircrew in Afghanistan reported mission times routinely over six hours while wearing helmets fitted with NVGs. In these cases, the seat cushion performance directly impacts the aircrew health and comfort. To mitigate the health risks and improve the ride comfort, helicopter seat cushions should be designed to dissipate the vibration energy to reduce the transfer of dynamic forces to the aircrew.

However, helicopter seat structures are designed primarily to meet crashworthiness requirements while vibration isolation is achieved passively through the seat cushion materials. The impedance properties of helicopter seat cushions are difficult to optimize without compromising pilot handling qualities and seat comfort. Most helicopter seat cushions are designed to be relatively firm to avoid

* Corresponding author, eric.chen@nrc-cnrc.gc.ca

excessive displacement during flight manoeuvres and also to meet the crashworthiness safety requirements, while the performance for aircrew vibration protection typically becomes a secondary requirement. Moreover, varying aircrew size, posture and flight configurations also impose compromises in the design of helicopter seat cushions.

This paper presents the development of novel pilot seat cushions for the Bell-412 helicopter to mitigate adverse health issues caused by pilot exposure to the high whole-body vibration helicopter environment. The performance of the prototype pilot seat cushions were evaluated on the NRC Bell-412 helicopter through extensive flight tests that involved representative flight conditions and pilot configurations.

2. PROTOTYPE SEAT CUSHION DEVELOPMENT

2.1 Helicopter Seat Cushion Requirements

The vibration level experienced by helicopter aircrew is influenced by several factors including flight condition, aircrew size, posture, and dynamics of the seat frame among many other factors. Modification to the seat cushion is considered as a low-cost solution that could be implemented in the near-term to mitigate the vibration induced health risks to the helicopter aircrew.

To provide effective vibration protection to aircrew, passive vibration isolation techniques require the resonance frequency of a seated aircrew and associated seat frame system to be significantly lower than the vibration input frequency [4]. Since helicopter vibration is generated from a combination of harmonics of the main rotor speed typically between 3 to 5Hz, such a design would require the stiffness of a helicopter seat cushion to be impractically low to isolate the low frequency harmonics of the vibration excitation, especially the 1/rev. However, MIL-STD-81771A states that the helicopter cushion design must avoid introducing dynamic overshoot of the system to guarantee stability while ensuring ride comfort and durability [5]. The helicopter seat design standard also requires that the cushion is not displaced or separated from the seat during aircraft maneuvers, crashes, or occupant motion. For vibration isolation, the MIL-STD-85510 stipulates that the "occupied and unoccupied seat shall be free of resonance within the frequency range of the aircraft in which it will be used and no amplification shall occur" [6]. These design requirements must be considered in the design of novel seat cushions for any helicopter seat applications.

In previous investigations, the Bell-412 pilot seat system has been found to increase the transmission of vertical vibration at low frequencies in the vicinity of the primary human whole-body resonance (4-8 Hz) while attenuating the transmission of vibration at the higher frequency range [7]. Therefore, an approach is proposed to employ the properties of the existing Bell-412 seat cushion for high frequency vibration reduction while integrating novel energy absorbing cushion materials to reduce the vibration amplification at the 1/rev harmonics to mitigate the vibration exposure of the pilot.

2.2 Energy Absorption Materials

Based on literature review, a type of engineered cushion material known as the Hybrid Air Cushioning System (HACS) has been selected to dampen the 1/rev vibration peak observed on the Bell-412 helicopter seat. This material was designed to mitigate vertical impact energy through a series of internal airflow ports and passages to dissipate the vibratory energy [4, 5]. As shown in Figure 1, the HACS consists of a series of hexagonal cells with interconnecting vents which force the airflow inside the cushion to choke when exposed to dynamically compressed load. It is this choked airflow that dissipates the energy and redirects the energy laterally instead of through the thickness of the cushion.

Although the HACS has not been investigated previously for helicopter seat vibration suppression applications, it has been used by the US Navy on the Cobra AH-1W helicopter seat to improve its crash safety. Therefore, this material has been proven to comply with the flammability requirement in accordance with the FAR 25.853 standard, as required for aircraft seat cushion applications.

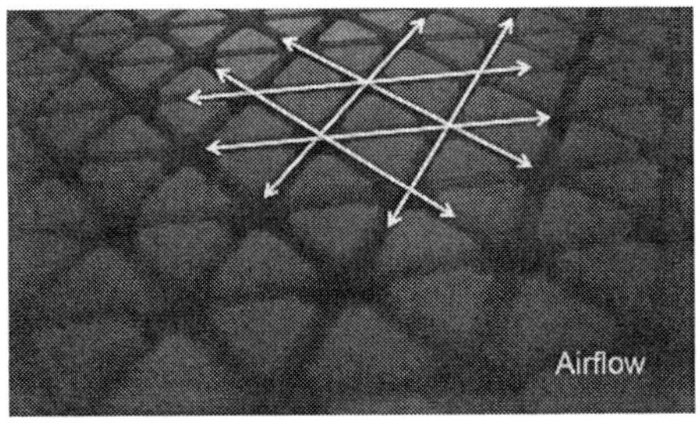

Figure 1. HACS cushioning structure with interconnected vents

2.3 Prototype Seat Cushion Design

The prototype pilot seat cushion design included one layer of HACS for low frequency energy absorption and one layer of cushion foam material used in conventional aircraft seats for high frequency vibration absorption. The two layers of materials are wrapped in a durable seat fabric for packaging and endurance.

One prototype cushion design is shown in Figure 2. The cushion foam material was of a similar type as used in the original Bell-412 pilot seat cushions. The prototype cushion fabric was designed with an opening at the bottom side to accommodate the HACS pads. This arrangement enabled the ability to conduct performance refinement by permitting easy modification of the cushioning system to accommodate the evaluation of other types of HACS pads or their combinations.

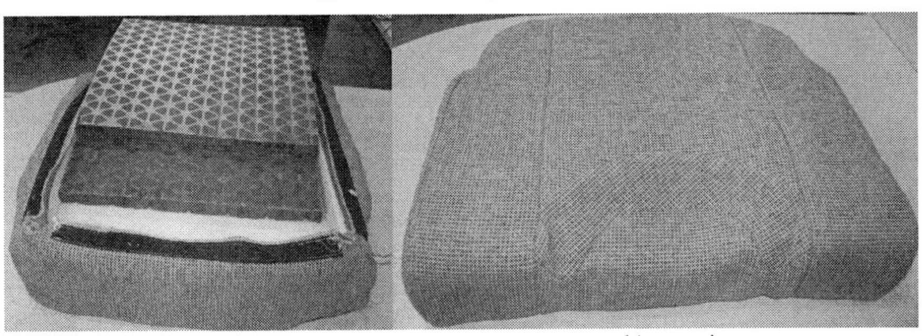

a) opened bottom view b) top view

Figure 2. Prototype seat cushion design

3. ASSESSMENT OF SEAT CUSHION PERFORMANCE

In this investigation, the prototype seat cushion was designed to retrofit a Bell-412 non-armored pilot seat. The international standard ISO 2631-1:1997 titled "Mechanical Vibration and Shock-Evaluation of Human Exposure to Whole-Body Vibration Part 1: General Requirement" was adopted to assess the test subjects exposure levels to whole-body vibration [6]. The ISO 2631-1:1997 is a general metric used for the assessment of whole-body vibration of people in vehicles (air, land and water), machinery and industrial activities that expose them to periodic, random and transient mechanical vibration which may affect their comfort, duties and health. In addition to the definition of measurement and evaluation methods, the ISO 2631-1:1997 also provides current information and guidance on the possible effects of vibration on human health in Annex B, comfort and perception in Annex C, which enables qualitative assessment of the performance of the original and prototype seat cushions.

3.1 Vibration Evaluation Method

The weighted root-mean-square (r.m.s) acceleration defined in the ISO 2631-1:1997 is a commonly used method to evaluate the overall magnitude of vibration with high frequency content and low crest factor applications. Crest factor is a term defined as the modulus of the ratio of the maximum instantaneous peak value of the frequency-weighted acceleration signal to its r.m.s value. Since the crest factor [6] of Bell-412 vibration is below 9 based on previous flight test results, this method is sufficient to describe the severity of vibration. In accordance with this method, the recorded vibration data is processed to calculate the weighted r.m.s acceleration within the frequency range between 0.5 and 80Hz using the appropriate frequency weighting factors and multiplying factors depending on the directions of the vibration data.

The total value of the weighted r.m.s. acceleration, determined from vibration measured in orthogonal coordinates is calculated as the following:

$$a_{total} = \sqrt{k_x^2 a_{wx}^2 + k_y^2 a_{wy}^2 + k_z^2 a_{wz}^2} \tag{1}$$

where kx, ky and kz are the multiplying factors, and awx, awy and awz represent the weighted r.m.s accelerations in respective directions.

The weighted r.m.s acceleration in each of the orthogonal directions is derived as follows:

$$a_w = [\sum_i (w_i a_i)^2]^{1/2} \tag{2}$$

where aw is the frequency-weighted acceleration, wi is the weighting factor for ith one-third octave band and the ai is the r.m.s acceleration for the ith one-third octave band. All these parameters are defined in the ISO2631-1:1997 standard.

3.2 Evaluation of Vibration Effects on Aircrew Health Risk

As stated in ISO 2631-1:1997, biodynamic and epidemiological studies have given evidence for an elevated risk of health impairment due to long-term exposure with high-intensity whole-body vibration. Mainly the lumbar spine and the connected nervous system may be affected due to vibration. It is also assumed that the environmental factors such as body posture, temperature, and humidity levels can also contribute to muscle pain. However, it is unknown if these factors can contribute to the degeneration of discs and vertebrae. Guidance to these effects of vibration on human health risk is listed in Annex B of the ISO 2631-1:1997 standard.

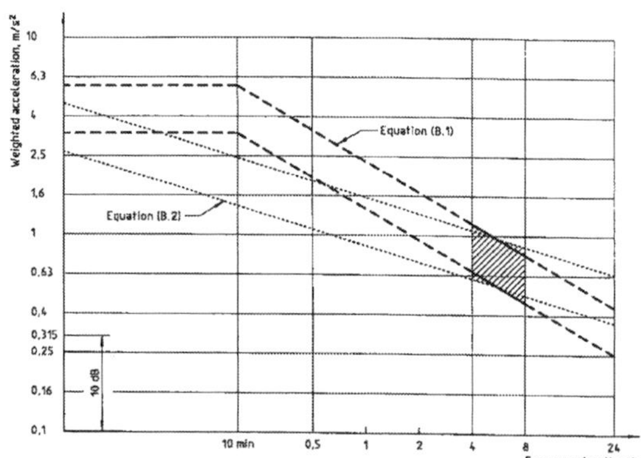

Figure 3. Health guidance caution zone (ISO 2631-1:1997)

Increased duration and increased vibration intensity is also known to introduce increased vibration dose which increases the health risk. However, there are insufficient data to show a quantitative relationship between vibration exposure and risk of health effects. Hence, it is not possible to assess whole-body vibration in terms of the probability of health risk at various exposure magnitudes and durations. Exposure time dependant health guidance is provided as Annex B of ISO2631-1997 as shown in

Figure **3** which can be used to qualitatively assess the impact on aircrew health.

3.3 Evaluation of Vibration Effects on Aircrew Comfort

Annex C of ISO 2631-1:1997 provides a uniform and convenient method of indicating the subjective severity of the vibration. It should also be noted that many factors combine to determine the degree to which discomfort may be noted or tolerated. An accurate assessment of the acceptability of the vibration levels and the formulation of vibration limits can only be based on the knowledge of a group of factors.

Since acceptable values of vibration magnitude for comfort depend on many factors which may vary with each application, a limit is not defined in ISO2631-1:1997. However, Table 1 presents approximate indications of likely reactions to various magnitudes of vibration total values in public transportation platforms, which can be used to assess qualitatively aircrew ride comfort.

Table 1 Comfort and perception rating according to vibration levels (ISO2631-1:1997)

Vibration Level (weighted r.m.s, m/s^2)	Comfort Rating
<0.315	Not uncomfortable
0.315-0.63	A little uncomfortable
0.5-1	Fairly uncomfortable
0.8-1.6	Uncomfortable
1.25-2.5	Very uncomfortable
>2	Extremely uncomfortable

4. EVALUATION OF SEAT CUSHION PERFORMANCE ON MECHANICAL SHAKER

It is understood that the aircrew whole-body vibration levels depend on many factors including the impedance properties of the seat cushion, aircrew size and posture as well as the vibration input. Therefore, the prototype pilot seat cushions were designed to allow adjustment of the impedance properties by changing HACS material type and combination of inserts. Mechanical shaker tests were conducted to evaluate the vibration mitigation performance of each cushion combination by using consistent test subjects and representative vibration input profiles.

The Hybrid III mannequins were used as the test subjects in the preliminary evaluation of seat cushions using mechanical shaker tests. The Hybrid III mannequins are full-scale anthropomorphic test devices (ATD) that simulate the dimensions, weight proportions and articulation of the human body with acceptable bio-fidelity. Compared to previous generations of mannequins, the Hybrid III mannequin is designed with improved neck response in forward flexion and head rotation that better simulates the dynamics of humans.

In the preliminary investigation, the prototype bottom seat cushions were installed to the seat pan of the Bell-412 non-armoured pilot seat by replacing the original bottom seat cushion. The retrofitted Bell-412 seat was fastened on a vertically oriented 10,000lb(f) mechanical shaker. The mechanical shaker was controlled to provide simulated helicopter vibration profiles in the Z axis of the seat frame. A typical test configuration is shown in Figure 4.

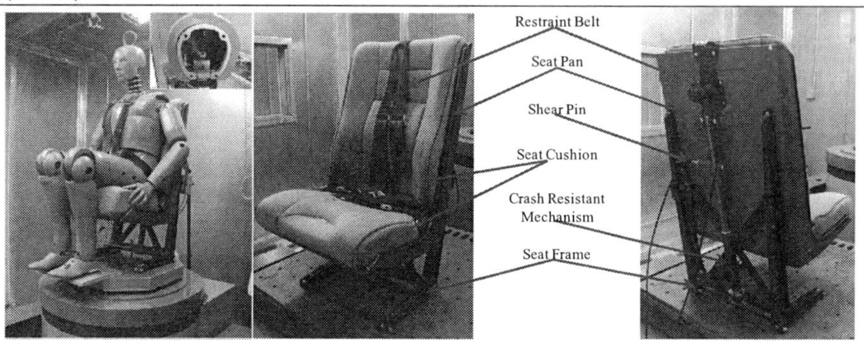

Figure 4. Mannequin test configuration on the mechanical shaker table

In accordance with the ISO 2631-1:1997, formed rubber pads fitted with tri-accelerometer were installed at the interfaces of the bottom seat cushion and the mannequin buttock as well as the backrest cushion interface near the mannequin torso location. Additional tri-axial accelerometers were also installed inside the mannequin to measure the vertical vibration response of the mannequin head, chest and torso under simulated vibration profiles.

The vibration input profile was a combination of four harmonic sine tones at 5.4, 10.8, 21.6 and 43.2Hz and broadband random vibration spectrum between 3 and 50Hz, and the overall vibration level was 0.2 g r.m.s. This profile was to simulate the vibration spectrum measured on the Bell-412 floor near the pilot seat location.

An original Bell-412 bottom seat cushion was also tested on the shaker table to base-line the performance of the original bottom cushion design. The mannequin whole-body vibration levels were measured and processed in accordance with the ISO 2631-1:1997. The obtained vibration levels were used to establish a set of baseline parameters to enable quantitative evaluation of the vibration mitigation performance of the prototype seat cushions.

Among the 22 prototype seat cushions evaluated, one prototype seat cushion noted as DART-Red-Red demonstrated favourable vibration mitigation performance on all of the three test subjects which including a 5th percentile female, a 50th percentile male and a 95th percentile male mannequin. This cushion design included two urethane (Red) HACS pads and one layer of conventional cushion foam wrapped with durable seat fabric. The shaker test results demonstrated reduction of weighted r.m.s vibration levels at the bottom cushion interface as well as the mannequin head location simultaneously when compared to the original Bell-412 bottom cushion. Moreover, the vibration levels measured at the mannequin chest and torso locations were also decreased significantly. The vibration mitigation performance of the DART-Red-Red prototype seat cushion is listed in Table 2.

Table 2 Prototype seat cushion performance in shaker test

Mannequin Subject*	Reduction - Bottom Cushion ISO-weighted r.m.s	Reduction-Head ISO-weighted r.m.s	Reduction-Chest un-weighted r.m.s	Reduction-Torso un-weighted r.m.s
95th percentile male	-9.0%	-8.1%	-7.5%	-19.5%
50th percentile male	-13.5%	-11.4%	-6.7%	-11.3%
5th percentile female	-13.3%	-1.3%	-3.6%	-4.2%

* Based on USA anthropometry data

The shaker test results verified that the DART-Red-Red prototype seat cushion not only reduced the vibration energy at the bottom seat cushion interface, but also decreased the whole-body vibration response of the mannequins effectively. Therefore, this prototype seat cushion showed potential to mitigate the whole-body vibration exposure of the Bell-412 pilots by improving occupant health and comfort in flight conditions.

However, despite the improved bio-fidelity of Hybrid III family compared to previous mannequin designs, it is important to note that mannequin subjects are passive devices that simulate dynamic features of the human body, while the human body is a complicated active system that adjustments adaptively in response to external stimulations. Therefore, the vibration reduction performance of the prototype seat cushion was further validated through flight tests using representative pilot configurations.

5. EVALUATION OF BELL-412 PILOT WHOLE-BODY VIBRATION EXPOSURE

The pilot whole-body vibration exposure on the original Bell-412 seat cushion was measured and analyzed according to the ISO 2631-1:1997 to establish baseline parameters for the performance evaluation of the prototype seat cushion. The pilot vibration levels were also needed to assess the vibration effects on Bell-412 pilot's health risk and comfort qualitatively.

5.1 Measurement of Pilot Whole-Body Vibration

The NRC four-bladed Bell-412HP helicopter shown in Figure 5 was used in this investigation [7]. It is equipped with the same type of non-armoured crash-worthy pilot seat as used in the previous shaker tests. Two identical pilot seats were used on the Bell-412, one for the pilot on the right side and the other for the co-pilot on the left side of the aircraft. In the flight test, the pilot represented a 50th percentile male and the co-pilot represented an 85th percentile male, and the test configuration is shown in Figure 6.

Figure 5. NRC Bell-412HP helicopter

Figure 6. Pilot test configuration

The flight test plan included representative flight conditions of the Bell-412 helicopter to quantify the pilot vibration exposure. During the flight test, the aircraft was maintained at each specified flight conditions to provide statistically stationary data sequence. The vibration data was recorded using the LMS Test.Lab data acquisition system. In the most representative flight conditions such as 60kt and 120kt level flight, the vibration data was recorded for 120 seconds. In other flight conditions the recording was limited to 60 seconds. Moreover, the flight profile was designed to provide two data points for each flight conditions to establish repeatability. Details of the flight profile are listed in Table 3.

Table 3: Details of the flight profile

Sequence	Flight Condition	Duration (s)	Sequence	Flight Condition	Duration (s)
1.	Ground	60	8.	60kt Climb	60
2.	Hover 5feet	120	9.	120kt Level	120
3.	Translational	60	10.	60kt Level	120
4.	60kt Climb	60	11.	140kt Descend	60
5.	60kt Level	120	12.	Translation Lift	60
6.	120kt Level	120	13.	Hover 50feet	120
7.	140kt Descend	60	14.	Ground	60

Figure 7. Formed accelerometer mount configuration on the aircrew seat

In accordance with the ISO 2631-1:1997, two formed rubber mounts packed with PCB tri-axial accelerometers were used to measure the vibration level of each pilot, one at the bottom cushion interface and the other at the backrest cushion interface. Both rubber mounts were taped to the cushion surfaces firmly to avoid movement due to pilot operations. The rubber mount configuration is shown in Figure 7. The helmet vibration of both pilots was also measured as a reference index during the flight.

5.2 Pilot Whole-Body Vibration Response

The whole-body vibration levels of the 50th percentile pilot in representative flight conditions were measured and evaluated. Typical vibration spectra of the 50th percentile pilot in 120kt forward flight condition are shown in Figure 8 and Figure 9 respectively.

It was clearly shown that at the pilot seat rail location, the dominant vibration peak was generated from the 4/rev harmonics of the main rotor speed, which was approximately 22Hz. Other harmonic peaks included the 8/rev, 2/rev and 1/rev. However, at the bottom cushion interface, the dominant peaks included the 1/rev harmonics at 5.5Hz and the 4/rev harmonics at 22Hz. The increased amplitude of the 1/rev harmonics was primarily resulted from the aircrew body reaction force.

Comparatively, the dominant vibration peak at the pilot helmet location was the 1/rev harmonics while other peaks including the 4/rev harmonics have been isolated by the seat cushion as well as the human body. The movement of the pilot head/helmet related to operation of the vehicle was also identified especially in the X direction, but usually shown as very low frequency segment in the spectral diagrams. Similar features were also identified on the 85th percentile pilot on the left seat in the same flight conditions.

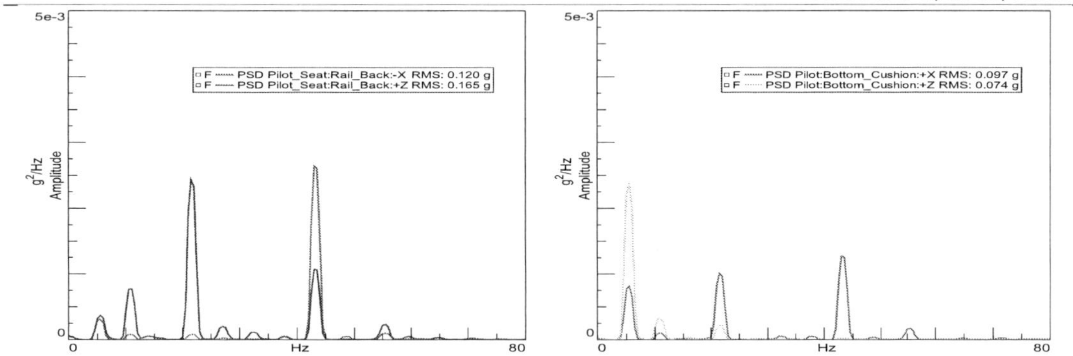

Figure 8. Vibration spectra at the seat rail and bottom cushion locations of the pilot

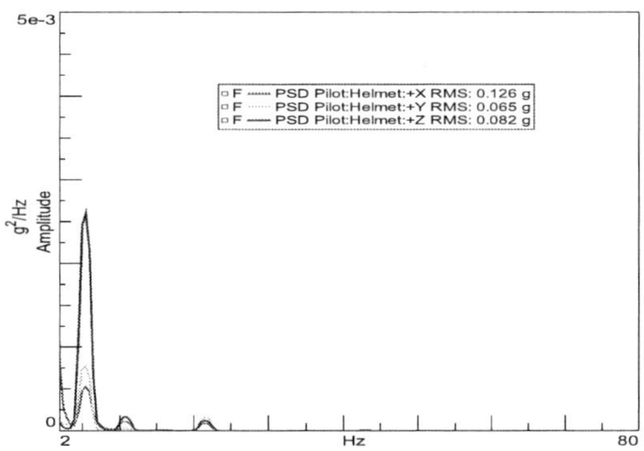

Figure 9. Vibration spectra at the helmet location of the 50th percentile pilot

Based on the vibration spectra measured on the 50th percentile pilot and 85th percentile pilot, it can be concluded that it is necessary to reduce the amplitudes of the 1/rev and 4/rev harmonic peaks simultaneously to mitigate the whole-body vibration levels of the Bell-412 pilots at the bottom cushion interface. Moreover, at the pilot helmet location, it is more important to reduce the amplitude of 1/rev harmonic peak to relieve the dynamic load applied to the pilot neck and back while the 4/rev and 2/rev harmonic peaks were insignificant compared to the amplitude of the 1/rev harmonics.

5.3 Pilot Health and Comfort Assessment

5.3.1 50th Percentile Male Pilot

The whole-body vibration levels of the 50th percentile pilot were found to vary significantly depending on flight conditions using the original Bell 412 pilot seat cushion. The worst vibration levels were observed during the high speed forward flight conditions in terms of both health risk and comfort. For example, in terms of health risk, the weighted r.m.s vibration levels at 120kt and 140kt were evaluated as 0.087g and 0.075g respectively. In terms of comfort and perception, the weighted r.m.s vibration levels at 120kt and 140kt were evaluated at 0.083 g and 0.069g respectively. The increased pilot whole-body vibration levels at high speed flight conditions were mainly due to the higher vibration input from the helicopter floor, which may be related to the change in main rotor state of the vehicle.

Since there is a lack of statistical information on flight conditions and durations of the Bell-412 fleet in typical missions, the vibration levels in the seven selected representative flight conditions were averaged linearly to provide a preliminary estimation of the pilot whole-body vibration level exposure. The averaged vibration level was derived as 0.055g in terms of health risk. This was marginally beyond the recommended threshold of 0.05g according to Annex B of ISO 2631-1:1997; therefore, the vibration effect on the 50th percentile pilot was qualified as "Caution" in terms of health risk. As recommended by the ISO 2631-1:1997, actions should be taken to reduce the aircrew vibration levels under the assumption of exposure to 4-8 flight hours per day.

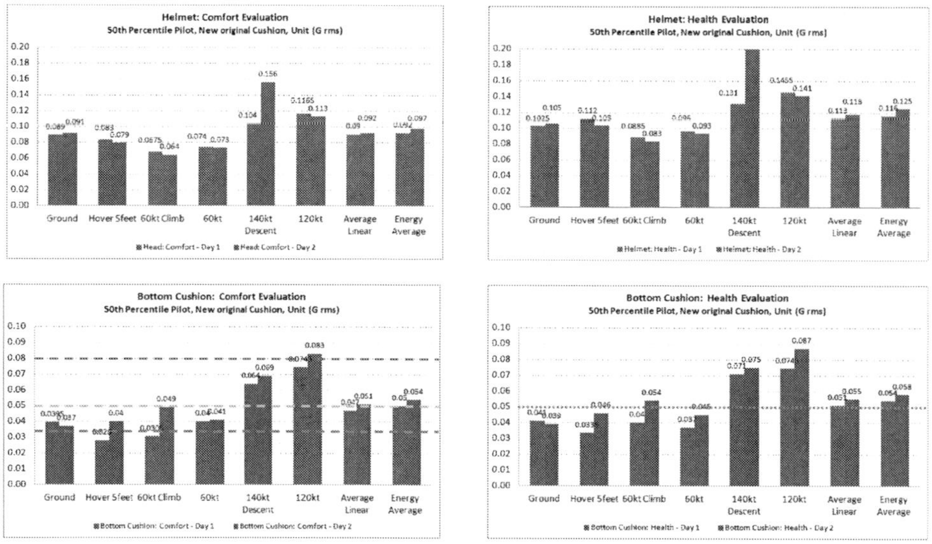

Figure 10. The 50th percentile pilot vibration levels on the original seat cushion

Similarly, in terms of pilot comfort and perception, the averaged vibration level of the 50th percentile pilot was derived as 0.051g, which was rated as "Fairly Uncomfortable" according to Annex C of ISO 2631-1:1997. In particular, the vibration level at 120kt forward flight condition was 0.083g, which was rated as "Uncomfortable". The whole-body vibration exposure levels of the 50th percentile male pilot in representative flight conditions are shown in Figure 10. To enable qualitative assessment, the green guideline in comfort evaluation chart indicates "A Little Uncomfortable"; the yellow guideline indicates "Fairly Uncomfortable" and the red guideline indicates "Uncomfortable". In the health risk evaluation chart, the green guideline indicates "Caution Zone".

However, it is also important to note that the human whole-body vibration exposure threshold is time-dependent. Reduced flight mission duration can increase the threshold, and vise versa. This relationship is expressed below.

$$a_{w1} \times T_1^{1/2} = a_{w2} \times T_2^{1/2} \tag{4}$$

Where a_{w1} and a_{w2} are the weighted r.m.s acceleration for the first and second exposure respectively, and T_1 and T_2 are the corresponding exposure durations for the first and second exposure. Based upon this relationship it is worthwhile indicating that the use of the "linear equal rating" of the flight conditions is a conservative representation of the vibration exposure for pilots flying military missions.

5.3.2 85th Percentile Male Pilot

The whole-body vibration levels of the 85th percentile male pilot were also evaluated based on the vibration data measured at the bottom cushion interface in each flight conditions. The weighted r.m.s vibration levels are shown in Figure 11.

The vibration levels of the 85th percentile male pilot showed similar trends as the 50th percentile pilot on the original Bell-412 bottom cushion. The measured vibration levels also varied with flight conditions, and the worst vibration exposure level was observed at high speed flight conditions. The average of weighted r.m.s. levels was 0.039g in terms of health risk based on vibration measured at the bottom cushion interface. Therefore, it was below the threshold of 0.05g, and qualified as "Safe" in accordance with the ISO 2631-1:1997. The 140kt descent represented the worst scenario for vibration exposure, which generated a weighted r.m.s vibration level of 0.060g to the pilot. This was the only flight condition inside the "Caution" zone.

In terms of aircrew comfort, the average of weighted r.m.s. vibration levels in 7 representative flight conditions was 0.035g, which was qualitatively rated as "A Little Uncomfortable" according to the ISO2631-1:1997.

Comparing the vibration levels of the 50th percentile pilot and 85th percentile pilot, it was clearly shown that the 50th percentile pilot was exposed to higher vibration levels in comparable flight conditions. The averaged vibration level of the 50th percentile pilot was approximately 32.8% higher in terms of health risk, and 35.2% higher in terms of comfort when compared to the 85th pilot. The difference was mainly related to the pilot size, posture and non-linear impedance properties of the bottom seat cushion. Therefore, the helicopter seat cushion properties should be customized according to the size of the pilot to provide enhanced vibration mitigation to the helicopter pilot.

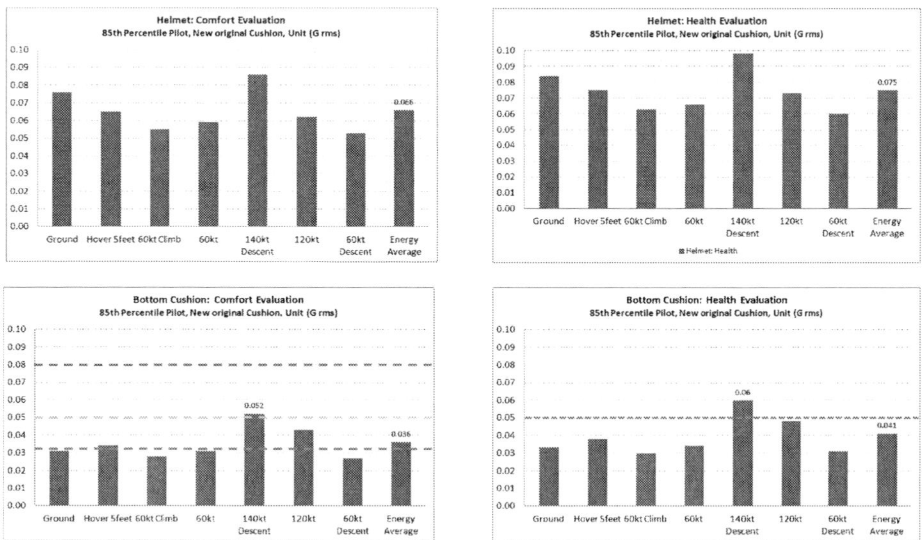

Figure 11. The 85th percentile pilot vibration levels on the original seat cushion

5.4 Assessment of Pilot Helmet Vibration

It is understood that the military helicopter aircrew neck strain and back pain injuries may also be related to the use of helmet mounted equipment. The introduced dynamic force can be estimated from acceleration levels measured at the helmet location. Therefore, the vibration levels at the pilot helmet location were evaluated as a reference index in this investigation. The pilot helmet vibration levels were processed using the same technical method as the bottom cushion vibration, and the results of the 50th percentile pilot and 85th pilot are also shown in Figure 10 and Figure 11 respectively.

Since the vibration spectra at the helmet location were dominated by the 1/rev peak at 5.5Hz, the weighted r.m.s vibration value was generally similar to the r.m.s values of the un-filtered vibration. However, compared to the bottom cushion, the helmet vibration levels were significantly higher. For example, the average of weighted r.m.s vibration levels of the 50th percentile aircrew over the 7 representative flight conditions was 0.118g which in terms of health risk was even beyond the "Immediate Stop" threshold of 0.115g as recommended by Annex B of ISO 2631-1:1997. Similarly, the averaged vibration level was 0.092g which in terms of comfort was also rated as "Uncomfortable" according to Annex C of ISO 2631:1-1997.

For the 85th percentile co-pilot, the average of weighted r.m.s vibration levels at the helmet location was 0.075g and 0.066g in terms of health risk and comfort, respectively, which were also significantly higher than the levels of 0.039g and 0.035g at the bottom seat cushion interface.

However, these values were less severe compared to the 50th percentile pilot. Unfortunately, there is no appropriate standard for reference when evaluating the severity of vibration at the helmet location.

6. PILOT WHOLE-BODY VIBRATION ON THE PROTOTYPE SEAT CUSHION

6.1 50th Percentile Pilot

The DART-Red-Red prototype cushion demonstrated favourable performance on both the 50th percentile pilot and 85th percentile pilot in the flight test. The weighted r.m.s levels of the 50th percentile pilot in the flight test are shown in Figure 12.

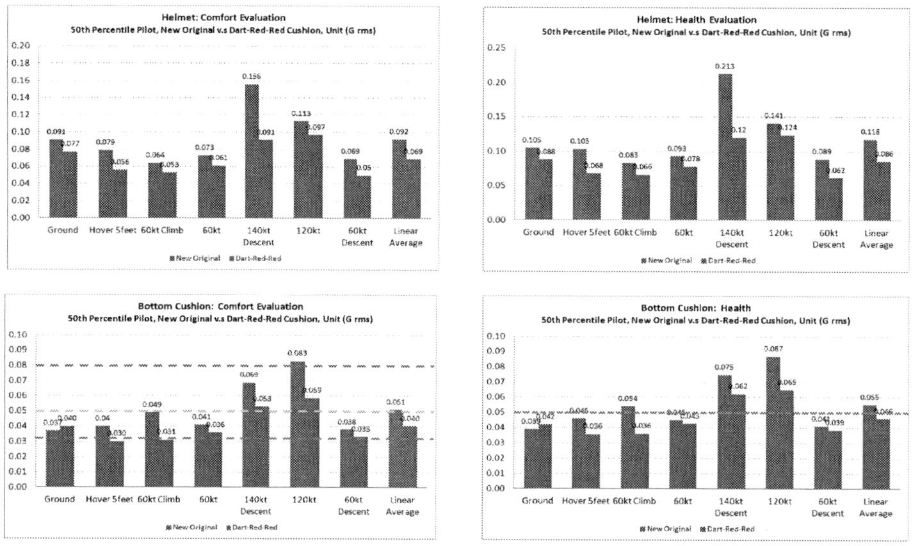

Figure 12. The 50th percentile pilot vibration levels on the DART-Red-Red cushion

As shown in Figure 12, the highest vibration levels of the 50th percentile pilot were observed in the high speed flight conditions such as the 120kt forward and 140kt descent, which was consistent to the case of the original Bell-412 seat cushion. The linear average of weighted r.m.s vibration levels in the seven flight conditions were reduced by 16.6% in terms of pilot health risk and 20.9% in terms of pilots ride comfort when compared to the original Bell-412 bottom cushion. Qualitatively, the health risk was reduced from "Caution" to "Safe" based on the assumption of exposure to 4-8 hours flight mission daily. More importantly, the weighted r.m.s vibration levels measured at the helmet location was also reduced significantly by 26.7% in terms of health risk and 24.9% in terms of comfort. These results confirmed that the prototype DART-Red-Red cushion was effective at reducing the vibration at the bottom cushion interface while simultaneously mitigating the vibration levels at the pilot helmet location.

The DART-Red-Red cushion also demonstrated consistent improvement of the ride quality rated at the seat bottom interface in a majority of flight conditions with the exception of the ground flight condition. It was believed that the slight increase in the weighted r.m.s vibration level during the ground run condition was due to the body and posture movement of the pilot while handling the vehicle.

The frequency weighted vibration spectra (ISO2631-1:1997) of the 50th percentile pilot at 120kt forward flight condition are shown in **Error! Reference source not found.** where the red line represents the original cushion and the blue line represents the prototype Dart-Red-Red cushion. The vibration level at the bottom cushion interface was reduced by 28.9% in terms of health risk and 25.3% in terms of comfort. These were achieved mainly by the reduction of the 1/rev peak at 5.4Hz. In the Z direction, the amplitude of 1/rev peak was reduced by 65.8%, and the overall vibration level was reduced by 28.8%. Similarly, at the helmet location, the 1/rev peak was reduced by 46.0%, and overall vibration level was reduced by 22.7% in the Z direction when compared to the original Bell-412 bottom seat cushion. Detailed data are shown in Table 4.

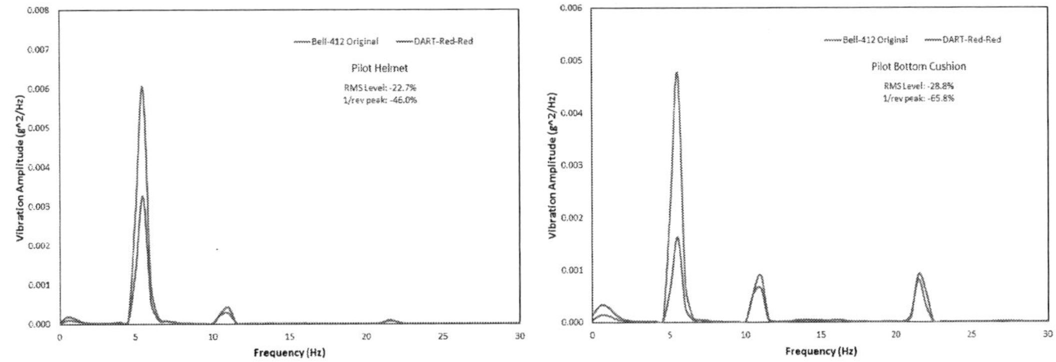

a) Vibration in the Z direction of bottom cushion b) Vibration spectra in the Z direction of helmet

Figure 13. Plot vibration spectra on the original and prototype seat cushions

Table 4 Performance of the prototype DART-Red-Red seat cushion

Aircrew	Reduction at Bottom Cushion Interface				Reduction at Pilot Helmet			
Parameters	1/rev (Z)	1/rev (X)	weighted r.m.s level		1/rev (Z)	1/rev (X)	weighted r.m.s level	
			Comfort	Health			Comfort	Health
120 kt-50th aircrew	-65.8%	-41.4%	-28.9%	-25.3%	-46.0%	-13.0%	-13.0%	-12.1%
140 kt-50th aircrew	-69.7%	-21.6%	-23.2%	-17.0%	-22.7%	-44.0%	-41.7%	-43.7%
weighted r.m.s level			-13.3%	-6.0%			-24.9%	-26.7%
120 kt-85th aircrew	-13.4%	18.1%	-2.3%	0.0%	-13.8%	-63.1%	-6.5%	-9.6%
140 kt-85th aircrew	-62.5%	-44.0%	-19.2%	-13.3%	-28.5%	-30.5%	-7.0%	-3.1%
weighted r.m.s level in			-8.5%	-6.5%			-2.3%	-0.2%

6.2 85th Percentile Pilot

The prototype DART-Red-Red cushion also demonstrated effective vibration reduction to the 85th percentile pilot in comparison to the use of the original cushion. The weighted r.m.s vibration level was reduced by 6.5% in terms of health risk and 8.5% in terms of comfort at the bottom seat cushion interface. It is important to note the overall whole-body vibration levels of the 85th percentile pilot were relatively low, and qualified as "Safe" in terms of health risk and "A Little Uncomfortable" in terms of comfort according to ISO2631-1:1997. The reductions of vibration in each flight conditions are shown in Figure 14.

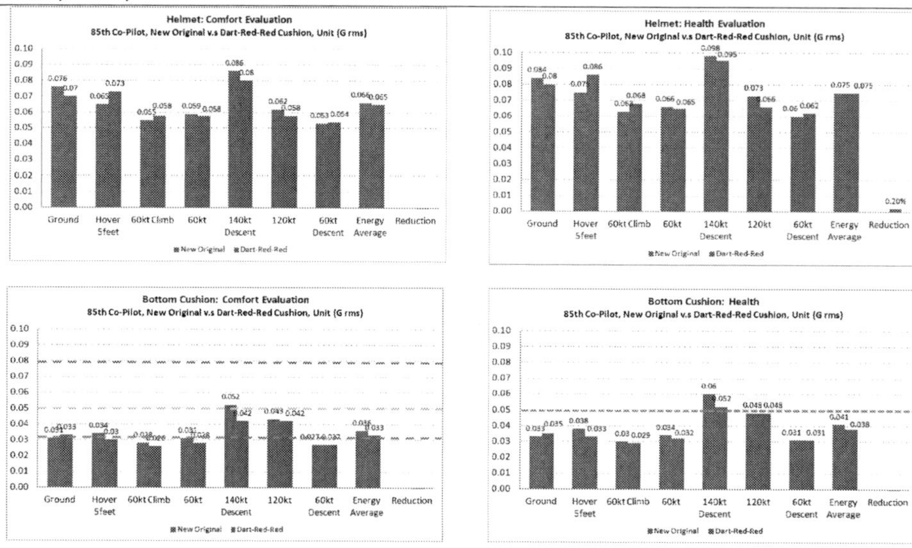

Figure 14. 85th percentile pilot vibration levels on the DART-Red-Red cushion

7. BACKREST CUSHION VIBRATION EFFECTS

The prototype DART –Red-Red seat cushion was tested as the bottom cushion while the original Bell-412 backrest cushion was maintained in all test conditions. Despite the effective mitigation to pilot whole-body vibration levels, the impact to vibration levels at the backrest cushion interface was also evaluated.

The vibration levels at the backrest cushion interface were found to be relatively lower than the bottom cushion interface, as shown in Figure 15. The 140kt descent and 120kt forward flight conditions introduced the highest vibration levels to the pilots as measured at the backrest cushion interface. For the 50th percentile pilot, the average of weighted r.m.s levels in seven flight conditions was 0.040g at the backrest cushion interface while the level was 0.054g at the bottom seat cushion interface. For the 85th percentile pilot, the vibration at the backrest cushion interface was 0.032g as compared to the 0.036g at the bottom seat cushion interface.

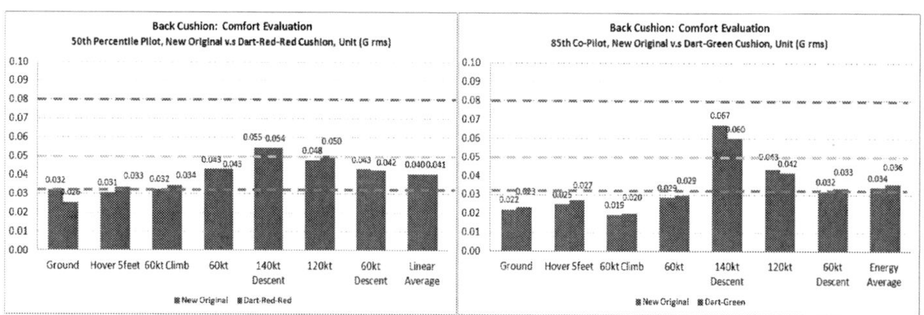

Figure 15. Vibration levels at the back cushion location of 50th percentile Pilot

As shown in Figure 15, the change of bottom cushion did not affect the vibration levels at the backrest cushion interface significantly. In the case of the 50th percentile pilot, the averaged vibration level was 0.042g as compared to 0.040g on the original Bell-412 bottom cushion. Similarly, the vibration level of the 85th percentile pilot at the backrest cushion interface was 0.036g using the Dart-Red-Red bottom cushion as compared to 0.034g using the original bottom cushion. Therefore, despite the changes in the bottom seat cushion, consistent vibration levels were observed at the backrest cushion interface in all tested flight conditions for the same pilot configuration.

Further examining the vibration in three directions, the X direction showed dominantly higher vibration levels than the Y and Z directions at the backrest cushion interface. According to ISO2631-1:1997, there is a lack of evidence showing the effect of backrest vibration on health. However, it is important to note that vibration levels at the backrest interface were similar to the bottom seat cushion interface, and also exceeded the "A little Uncomfortable" threshold. Therefore, there may be benefit to retrofit the Bell-412 back cushion with the HACS to further reduce the pilot whole-body vibration exposure.

8. CONCLUSION

This paper investigated the development of novel helicopter seat cushions to mitigate pilot whole-body vibration exposure. The prototype seat cushion included a novel energy absorbing material layer and a conventional foam material layer wrapped in durable cushion fabric for packaging. Mechanical shaker tests and flight tests on a Bell-412 helicopter have demonstrated improved vibration mitigation performance to the pilots. Based on the investigation, the following statements have been concluded.

1. The helicopter pilots are exposed to a whole-body vibration environment, and the vibration levels varied with pilot size as well as flight conditions.
2. The vibration levels of the 50th percentile pilot measured at the bottom seat cushion interface and the pilot helmet location were significantly higher than the 85th percentile pilot.
3. The prototype DART-Red-Red seat cushion demonstrated significant vibration reduction to all of the test subjects.
4. Due to the impact of pilot size on the whole-body vibration levels, helicopter seat cushions should be customized for each pilot to provide enhanced vibration mitigation performance.
5. Considering impact of varied flight conditions and pilot configurations, an adaptive seat/cushion system would be most suitable to minimize the vibration effects on pilot health and improve comfort by actively adjusting seat cushion impedance properties.

Acknowledgements

This is a collaborative project with Department of National Defence (DND) – Directorate Technical Airworthiness Engineering Support (DTAES) Human System Integration/Human Factors Section which provided partial funding as well as guidance.

References

1. Smith, D., "Characterizing the Effects of Airborne Vibration on Human Body Vibration Response," *Journal of Aviation, Space, and Environmental Medicine*, 2002, Vol. 73, No. 3, pp. 36-45.
2. Castelo-Branco, N. A., and Rodriguez, E., "The Vibroacoustic Disease – An Emerging Pathology," *Journal of Aviation, Space, and Environmental Medicine,* 1999, Vol. 70, No. 1, pp. 1-6.
3. Adams J., "Results of NVG-Induced Neck Strain Questionnaire Study in CH-146 Griffon Aircrew," Report No. TR2004-153, Defence Research and Development Canada, 2004.
4. Colvin, D.P. and Potter, C. S., "Cushioning Structure," US Patent 5,030,501, 1991.
5. Colvin, D.P. and Potter, C. S., "Cushioning Structure," US Patent 5,518,802, 1996.
6. International Standard ISO2631-1:1997, "Mechanical Vibration and Shock – Evaluation of Human Exposure to Whole-Body Vibration – Part1: General Requirements", Second Edition, May 1st, 1997, corrected and reprinted July 1st, 1997.
7. Bell 412EP Product Specifications, Accessed 24 January 2012, Bell Helicopter, http://www.bellhelicopter.com/en_US/Commercial/Bell412/1291148331974.html.

ICAST2013 #paper number

Static and dynamic thermo mechanical characterization of a bio-compatible shape-memory polymer

Pauline Butaud[1*], Anne Maynadier[2], Nicolas Brault, Vincent Placet[1], Morvan Ouisse[1], Emmanuel Foltête[1], Cécile Rogueda-Berriet[1]

[1] FEMTO-ST Applied Mechanics Department, Besançon, FRANCE
[2] LaMCoS Laboratoire de Mécanique des Contacts et des Structures, Lyon, FRANCE

Abstract

Shape memory polymers encounter a growing interest over the past ten years particularly because their eventual bio-compatibility leads to many bio-medical applications. They also present many benefits for the design of micro-adaptive systems for deployment or controlled damping materials. Indeed, the SMPs are polymeric smart materials which have the remarkable ability to recover their primary shape from a temporary one when submitted to an external stimulus.
The present study deals with the synthesis and the thermo-mechanical characterization of a thermally-actuated SMP. The polymer considered hereafter is a chemically cross-linked thermoset. It is synthesized via photo polymerization (UV curing) of the monomer *tert*-butyl acrylate (tBA) with the crosslinking agent poly(ethylene glycol) dimethacrylate (PEGDMA) and the photoinitiator 2,2-dimethoxy-2-phenylacetophenone (DMPA).
A mechanical characterization has been performed using three kinds of tests: quasi-static tensile tests, tensile dynamic mechanical analysis (DMA) and modal tests. The Young's modulus and the Poisson ratio are determined at ambient temperature using the first technique. The DMA is used to determine the evolution of viscoelastic properties as a function of the temperature and the frequency under harmonic loading. The modal analysis is employed to identify the viscoelastic properties of the material at higher frequency. A comparison of the results obtained by these three experimental methods highlights their complementarity.

1 INTRODUCTION

Shape memory polymers (SMPs) belong to the class of smart materials. These materials have found growing interest throughout the last few years. Indeed, in some aspects, SMPs are more efficient than the well-known metallic shape memory alloy (SMA's). SMPs particularly offer recoverable strain that can reach 400% (to be compared to 8% for SMAs), light weight, easy manufacturing and cheapness compared to SMA [1]. They can be used in a large variety of applications, e.g. actuators, electromechanical systems, clothing manufacturing, morphing and deployable space applications, control of structures, self-healing, biomedical devices and so on [2, 3]. The SMPs have the ability of changing their shape in response to an

* pauline.butaud@femto-st.fr

external stimulus [4], most typically thermal activation. When the SMP is heated above the glass transition temperature T_g, it is soft and rubbery and it is easy to change its shape. If the SMP is subsequently cooled below T_g, it retains the given shape (shape fixing characteristic). When heated again above T_g, the material autonomously returns to its original permanent shape [5]. The shape-memory effect is in principle a behavior inherent to all polymers. However, polymers that exhibit truly exploitable shape-memory effect must demonstrate a sharp transition temperature and a rubbery plateau, along with relatively large strain capacity without local material damage [6]. The memory effect of SMP is quite depending on the level of external constraint. For example if the stiffness of the constraining material is equal to the stiffness of the shape memory polymer in the rubbery state, the polymer will only recover half of its total strain [7].
The increasing use of SMP for quasi-static and dynamic applications, under various temperature ranges, has made necessary the characterization of these materials over wide frequency bands [8]. Three kinds of tests have been performed in this study. First, quasi-static tensile test are used at ambient temperature and at a controlled loading rate permit to identify the Young's modulus. Then, a dynamic mechanical analysis (DMA) is used to determine the evolution of viscoelastic properties as a function of the temperature and loading frequency. Finally, a modal analysis has been performed to reach higher loading frequencies. In this last technique, a finite element model is employed to identify the viscoelastic properties of the material from the dynamic measurements. The main purpose of this work is to check the validity of the time-temperature equivalence [9] obtained from the DMA measurements, and more exactly to check the capacity of DMA on a restricted frequency range and at various temperature levels to predict accurately the viscoelastic properties on a wider frequency range. The predicted viscoelastic properties are compared to the measurements collected using quasi-static and modal test. This is an important query considering that measurement combining several experimental methods is particularly time-consuming.

2 MATERIALS AND EXPERIMENTAL METHODS

2.1 Material and specimen preparation

As a representative thermally-actuated shape-memory polymer, a chemically-crosslinked thermoset polymer recently studied by Yakacki et al. [10] and Srivastava et al [11] is chosen. Following the procedure described by these authors, the SMP is synthesized by manually mixing 95 wt% of the monomer *tert*-butyl acrylate (tBA) with 5 wt% of the crosslinking agent poly(ethylene glycol) dimethacrylate (PEGDMA) (with typical molecular weight $M_n = 550$g/mol). The photoinitiator, 2,2-dimethoxy-2-phenylacetophenone (DMPA), is added to the solution at a concentration of 0.5 wt% of the total weight. The liquid mixture is then injected between two glass slides separated by a 3 mm spacer. The polymerization is initiated by exposing the solution to UV light during 10 minutes and achieved by heating the polymer at 90°C for 1 hour.
One of the obtained plates, measuring 200×150 mm, was machined with a CNC machine. Dog-bone shaped specimens were produced from the polymer plate for quasi-static tensile tests. The gauge length is 35 mm, and the rectangular cross section is 2.9×6.8 mm. In the same plate, samples for DMA testing were cut to $110 \times 4 \times 2.9$ mm. Another plate was produced with the exact same protocol for the modal analysis measurements.

2.2 Mechanical tests

2.2.1 Quasi-static tests

The quasi-static tensile test have been performed using a universal commercial testing machine (Instron 6025) with a capacity of ± 100kN. The specimen, clamped using wedge action grips, was subjected to a displacement control feedback at a rate of 1 mm/min. The test was performed at ambient temperature 21°C (± 1°C). The engineering longitudinal strain was measured by a laser extensometer (EIR LE05) with a displacement accuracy of approximately 1 µm. The gauge part of the specimen for the strain measurement is 33 mm in length. The longitudinal Young's modulus was determined by the slope of the stress-strain curve at 0.1% of engineering strain (tangent modulus technique).

2.2.2 Dynamic mechanical analysis

Viscoelastic properties (storage modulus, loss modulus and loss factor) are measured using a Bose Electroforce 3200 apparatus every 5°C in isothermal conditions. Temperature varies between -28°C and 30°C, the frequency of the solicitation from 0.1 Hz to 10 Hz. The temperature is stabilised during 2 minutes before measurements to ensure a homogeneous temperature inside the specimen. The heating rate is about 0.1°C/sec between each plateau temperature. The temperature is measured using a thermocouple positioned in a reference sample put in the heating chamber close to the tested specimen.
The compliance of the test set-up was taken into account for the determination of the viscoelastic properties.
A sinusoidal tensile load was applied on the sample with a mean amplitude of 2 N and a peak-to-peak amplitude of 3 N, in order to test the specimen in the linear viscoelastic range. Loading-unloading tests were also made on this equipment, at temperature below and above the glass transition. A tensile load from 0.1 N to 5 N was applied on the sample with a constant rate of 0.2 N/s, between -15 °C and 30 °C every 15 °C. After the glass transition, between 70 °C and 90 °C (every 10°C), a tensile ramp of 0.05 N to 0.15 N was applied on the sample at a rate of 0.005 N/s. At high temperature, a more sensitive load cell was used considering the high decrease in rigidity of the tBA/PEGDMA and to ensure accurate measurements. It should be emphasized that at this temperature level the specimen stiffness is only of 0.1 $N.mm^{-1}$.

2.2.3 Modal analysis

The studied structure is a tBA/PEGDMA rectangular plate with dimensions $157.7 \times 45.5 \times 3.1$ mm (± 0.1 mm). The plate is suspended in order to realize free-free conditions. Contactless actuators and sensors are used. The external force is applied thanks to a voice-coil actuator with a permanent magnet glued in a corner of the plate. This assembly is placed in a thermal chamber with a temperature control between 22 and 36 °C every 7°C (± 2°C). A broadband random excitation is applied between 150 and 4000 Hz. The plate response is measured by using a laser vibrometer (Polytec OFV-505) successively focused on 28 reflecting stickers spread on the plate. The modal frequencies, the modal damping and the mode shapes are then identified by using Modan© (homemade modal analysis software). At the same time, a finite element model is established on Patran™2012 under the hypothesis that the material is homogeneous and isotropic, with a Poisson's ratio $v = 0.37$ (determined from quasi-static tests), a mass density $\rho = 1004\ kg/m^3$ (determined by measuring the plate's mass and volume) and a presumed Young's modulus $E = 2000\ MPa$. The magnet is included in the numerical model ($E = 210\,000\ MPa, v = 0.33$, $\rho = 7460\ kg/m^3$). Finally a model-test correlation is performed on AESOP software and a model updating procedure is conducted to obtain the mechanicals parameters of the tBA/PEGDMA at the modal frequencies.

2.2.4 Infrared thermography

Thermal field was monitored by infrared thermography using a CEdip Jade III MWIR camera. It provides 320×240 pixels images. The acquisition frequency is set to 25 Hz and the integration time is 430 µs. The measuring range was 5-90°C. The measurements were performed during quasi-static tests and DMA to visualize the possible overheating in the sample.

2.2.5 Kinematic fields measurement

An optical camera (Ueye 1Mpixel 8bit) was used to assess the displacements fields in the gauge part of the specimen using digital image correlation (DIC). The camera was equipped of a telecentric lens (0.20x Techspec Silver series telecentric lens - 63073 - Edmund Optics) that prevents measurements from optical distortions. It also ensures a large depth of field (of approximately 6 mm) which guarantees the sharpness of images even in case of necking. Image acquisition was triggered using a Labview application at a frequency of 1 frame per sec. DIC was performed with a free home-made global correlation code, implemented in Octave and described by Maynadier et al. [12]. Quadratic elements were chosen for meshing the area of interest. The strain fields were computed from the displacement fields by pixel-wise difference gradient.

3 RESULTS

3.1 Quasi-static characterization

Figure 1 shows the stress-strain curve obtained from the quasi-static test. The apparent Young's modulus is approximately 1500 MPa, the ultimate stress (32 MPa) is reached at a strain of 1.8% and the elongation at rupture is about 60 %.

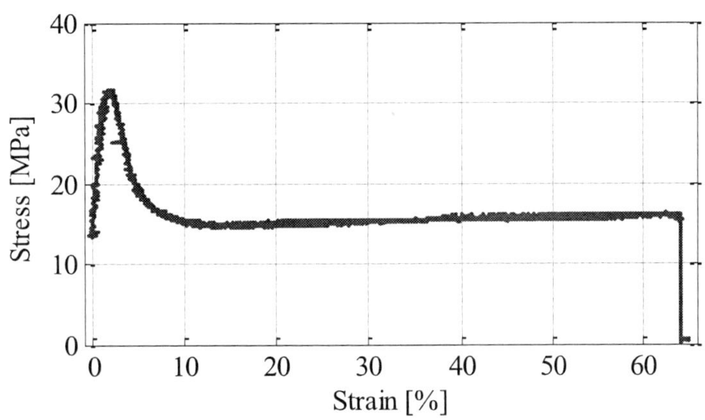

Figure 1. Stress – strain curves in tensile test at 1 mm/s, for tBA/PEGDMA, at ambient temperature

Infrared thermography was used for visualization of the temperature field at the surface of the specimen during this test. As shown in figure 2, the thermo-mechanical behavior of the tBA/PEGDMA in quasi-static is quite similar to the classical polymer's behavior: there is a cooling during the elastic phase, then a warming during the viscoelastic phase. Heat release is localized first in the necking zone and then heat propagates along the sample. Finally, rupture occurs with high heat release localized in a very sharp area around the crack tip.

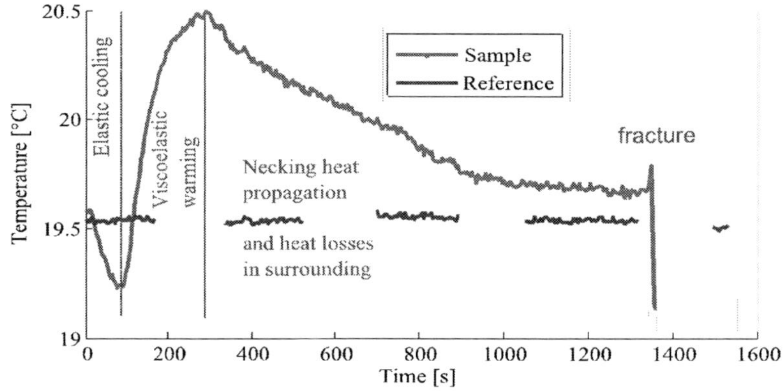

Figure 2. Temperature variation on the centerline of the sample compare to the ambient temperature at 1 mm/s

The Poisson's ratio was determined using the strain fields determined using DIC. The value is approximately 0.37 (Figure 3).

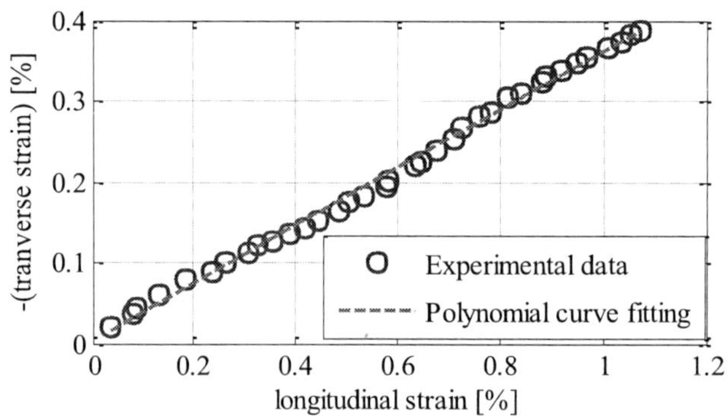

Figure 3. Identification of the Poisson's ratio around 0.37 at 1 mm/s

Figure 4 shows the high influence of temperature on the Young's modulus. Its value is almost constant, around 2060 MPa from -20°C to 40°C and also constant but 2500 times lower (0.8 MPa) above 60°C. This abrupt change in the elastic modulus reveals a phase transition between 40°C and 60°C.

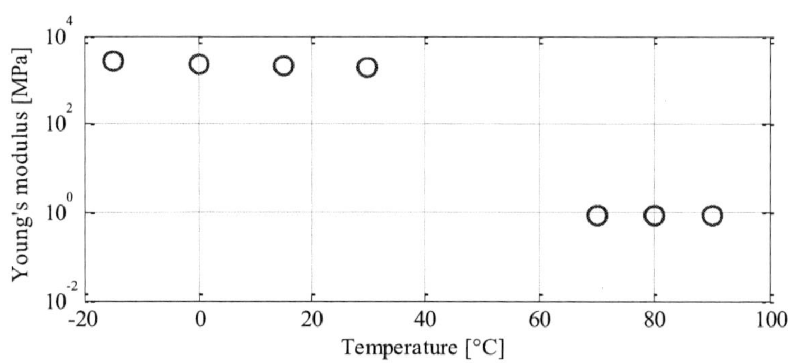

Figure 4. Stress – strain curves in tensile test at several temperature

The temperature variation registered by infrared thermography was not significant during this kind of tests; the variation level is lower than the noise level. The hypothesis of isothermal test is verified, there is not self-heating of the sample under mechanical solicitation.

3.2 Dynamic mechanical analysis

The results of the DMA tests are shown in Figure 5. The storage modulus decreases with the temperature and increases with the frequency. On the contrary, the loss factor increases with the temperature and decreases with the frequency. These trends are coherent with the classical behavior of a polymer without measurements on the transition area. Moreover the values of E' and $\tan\delta$ at 1 Hz are consistent with the results on the tBA/PEGDMA of Ortega et al. [13].

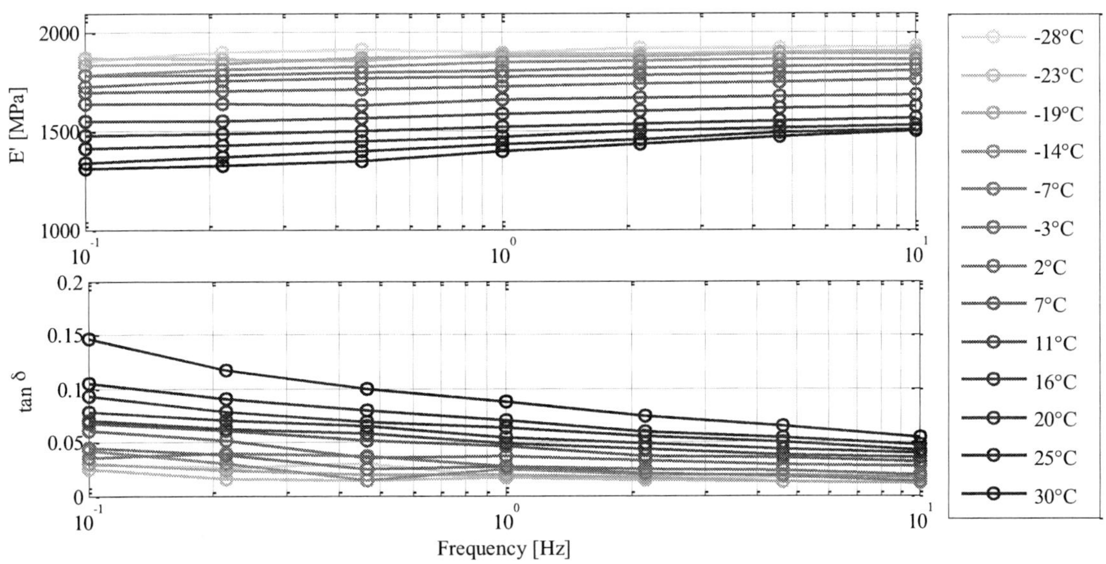

Figure 5. E' and tan delta according to the frequency and for several temperatures

Curves of E' vs. frequency at one temperature can be shifted to overlap with adjacent curves (Figure 5). The time-temperature superposition shift factors a used for E' must be the same for $\tan\delta$ to get time-temperature equivalence [14]. In our case, the shift factors a are obtained through an optimization procedure (classical least square method) and are equivalent for both E' and $\tan\delta$, therefore the time-temperature equivalence is validated in a specified temperature and frequency ranges for the tBA/PEGDMA. The master curves of the storage modulus and $\tan\delta$ are given in Figure 6.

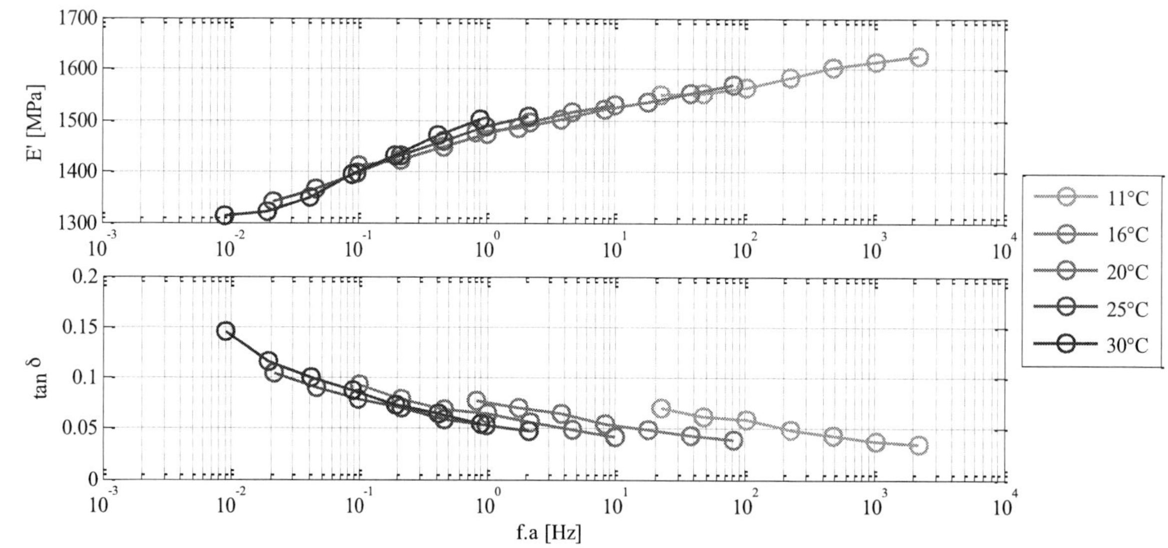

Figure 6. Master curves of E' and $\tan\delta$ according to the reduced frequency with a reference temperature of 20°C

3.3 Modal analysis

Depending on the temperature, 11 to 13 modes are identified after the model-test correlation (Figure 7).

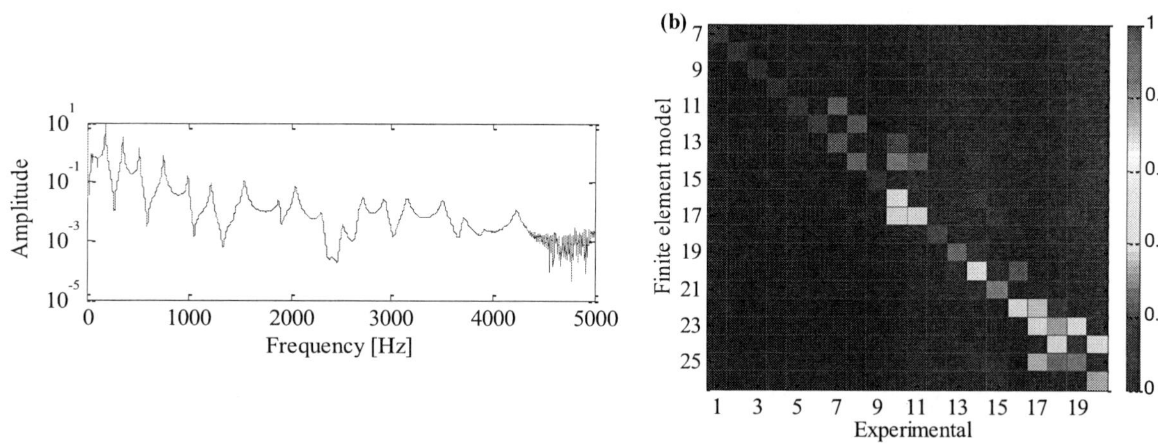

Figure 7. (a) FRF and **(b)** MAC Matrix for 22°C experimental data

Just as for DMA, the obtained Young's modulus and $\tan\delta$ of the tBA/PEGDMA depend on the temperature and the frequency as shown in Figure 8.

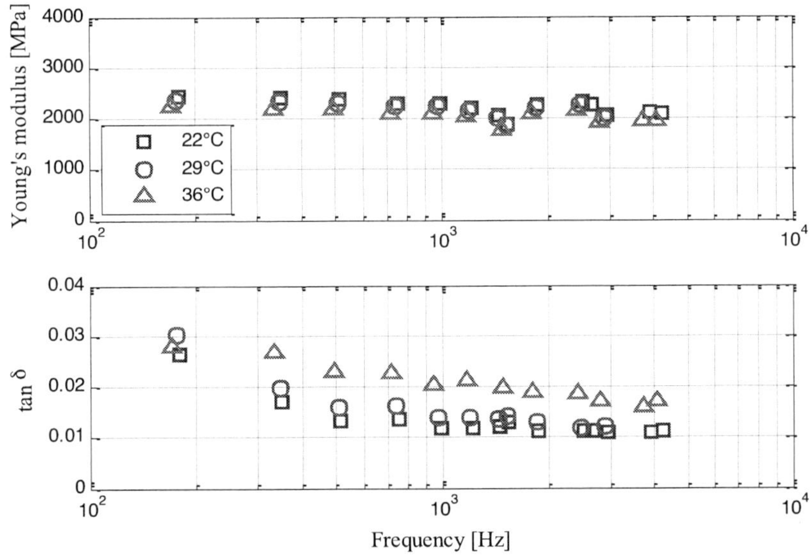

Figure 8. E' and $\tan\delta$ according to the frequency and for several temperatures

As expected, the loss factor increases with the temperature and decreases with the frequency. Concerning the storage modulus, it decreases as planned with the temperature but its variation with the frequency is unusual, as it decreases too. This trend is not well understood yet, even if some hypotheses are considered. For example, the Poisson's ratio used in the finite element model was determined in quasi-static regime whereas it could fluctuate with the frequency like the storage modulus. Other hypotheses like the isotropic character of the material are currently under investigation.

4 DISCUSSION

This study intends to compare the master curves obtained with the DMA and the results collected by the modal analysis and the quasi-static tests. Based on the time-temperature equivalence determined by DMA, the storage modulus predicted by this equivalence should be equal with the quasi-static and modal modulus, likewise for the loss factor. The DMA master curves and the results of quasi-static tests and modal analysis in temperature beforehand translated thanks to the shift factor are plotted on a same graphic (Figure 9). To define the frequency of the quasi-static test, the loading-unloading test period was taken in account, thus the frequency of this test at low temperature is 0.02 Hz.

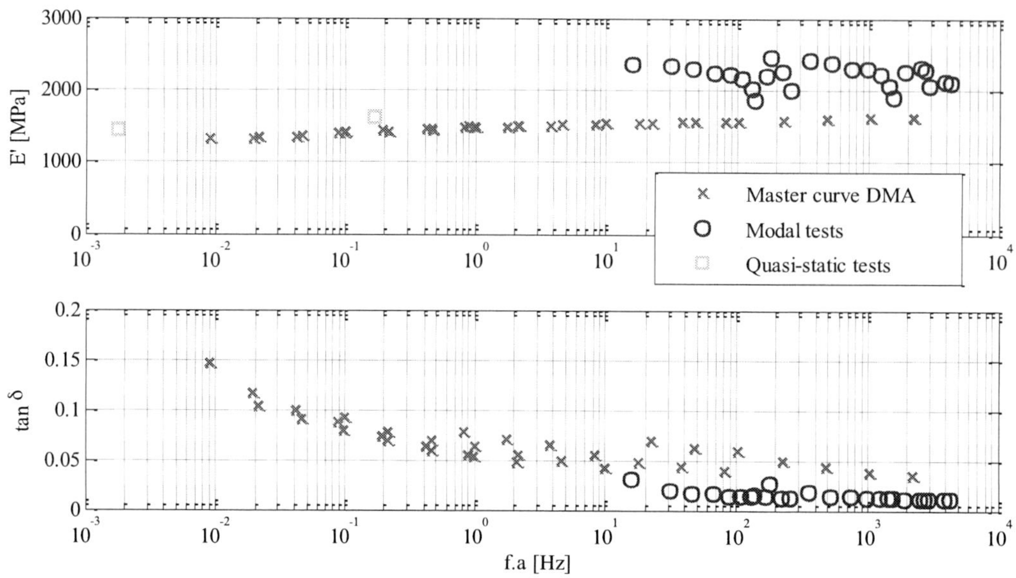

Figure 9. Comparison of experimental methods on E' and $\tan\delta$

The order of magnitude of E' at low frequency is respected. The time-temperature equivalence data are consistent with the quasi-static results. At high frequency, there is a gap of about 50 % between the storage modulus prediction and the modal results. This is not observed for the loss factor, for which the comparison is quite conclusive.

There are some options to explain this gap between the modal tests and the master curve. First of all, the hypothesis of isothermal test could be unverified i.e. the sample could undergoes self-heating under mechanical solicitation. This possibility was eliminated for DMA tests with infrared thermography but not for the modal analysis. Secondly, as it was mentioned previously, the Poisson's ratio could be not correctly defined in the finite element model. Finally, the most important distinction between modal analysis and the two others experimental methods is the kind of solicitations. Indeed, whether for quasi-static or DMA, the test is in tensile on a beam sample, contrary to the modal analysis for which the test is in bending on a plate and for smaller strain amplitude. It therefore appears necessary to study the behavior of the tBA/PEGDMA in tensile and compression, and check whether a distinction. Lastly, a post-treatment is required for the modal data thanks to a finite element model, contrary to DMA and quasi-static data which are used unaltered.

5 CONCLUSION

The objective of this study was to highlight the time-temperature equivalence by comparing three experimental methods allowing the identification of the storage modulus over frequency and temperature. This equivalence has been checked on a sample tested with the DMA. Quasi-static tests and modal analysis have been carried out to explore a wider frequency band. Even though the results are rather consistent, the time-temperature equivalence is not yet established. Further investigations are needed to clarify the tBA/PEGDMA behavior. The correlation between quasi-static, DMA and modal analysis remains to be carried out.

References

1. Pretsch Thorsten, "Review on the functional determinants and durability of shape memory polymers", *Polymers*, Molecular Diversity Preservation International, 2010, Vol. 2, pp. 120-158
2. Behl, M. and Lendlein, A., "Shape-memory polymers", *Materials Today*, 2007, Vol. 10, pp. 20-28
3. Dietsch, B. and Tong, T., "A review-: Features and benefits of shape memory polymers (SMPs)", *Journal of advanced materials*, 2007, Vol.39, pp. 3-12
4. Ratna D. and Karger-Kocsis J., "Recent advances in shape memory polymers and composites: a review", *Journal of Materials Science*, 2008, Vol. 43, pp. 254-269
5. S.A. Wilson, R.P.J. Jourdain, Q. Zhang, R.A. Dorey, C.R. Bowen, M. Willander, Q.U. Wahab, S.M. Al-hilli, O. Nur, E. Quandt, et al, "New materials for micro-scale sensors and actuators: An engineering review", *Materials Science and Engineering: R: Reports*, 2007, Vol. 56, pp. 1-129.
6. B. Sillion, "Shape memory polymers", *Acta Chim*, 2002, Vol. 3, pp. 182-188.
7. Liu, Yiping and Gall, Ken and Dunn, Martin L and Greenberg, Alan R and Diani, Julie, "Thermomechanics of shape memory polymers: uniaxial experiments and constitutive modeling", *International Journal of Plasticity*, 2006, Vol. 22, pp. 279-313.
8. A. Lendlein and S. Kelch, "Shape-memory polymers", *Angewandte Chemie International Edition*, 2002, Vol. 41, pp. 2034–2057.
9. N. Okubo, "Preparation of Master Curves by Dynamic Viscoelastic Measurements", *SII NanoTechnology Inc.*, 1990, Vol. 6.
10. Yakacki, Christopher Michael and Shandas, Robin and Lanning, Craig and Rech, Bryan and Eckstein, Alex and Gall, Ken, "Unconstrained recovery characterization of shape-memory polymer networks for cardiovascular applications, *Biomaterials*, 2007, Vol. 28, pp. 2255-2263.
11. Srivastava, Vikas and Chester, Shawn A and Anand, Lallit, "Thermally actuated shape-memory polymers: Experiments, theory, and numerical simulations, *Journal of the Mechanics and Physics of Solids*, 2010, Vol. 58, pp. 1100-1124.
12. A. Maynadier, S. Carbillet, V. Placet, F. Amiot, A. Seror, A. Longuet, C. Mary, "Fatigue behaviour of D.A. INCONEL 718 alloy under tension-torsion loading". *In PhotoMechanics*, Montpellier, 2013
13. Ortega, Alicia M and Kasprzak, Scott E and Yakacki, Christopher M and Diani, Julie and Greenberg, Alan R and Gall, Ken, "Structure--property relationships in photopolymerizable polymer networks: Effect of composition on the crosslinked structure and resulting thermomechanical properties of a (meth) acrylate-based system", *Journal of Applied Polymer Science*, 2008, Vol. 110, pp. 1559-1572.
14. Dealy, John and Plazek, Don, "Time-temperature superposition—a user's guide", *Rheol. Bull*, 2009, Vol. 78, pp. 16-31

Control Technology of Transonic Aerodynamic Characteristic by Partial Bump

Da Wei Li, Jin Hao Qiu *, Rui Nie, Hong Li Ji

State Key Laboratory of Mechanics and Control of Mechanical Structures
(College of Aerospace Engineering, Nanjing University of Aeronautics&Astronautics, Nanjing, 210016, China.)

ABSTRACT

This paper studied the technology of flow passive control that can increase the wing lift and weaken the wing drag by methods of fluid mechanics. The researches about shock control bump mainly focused on the two dimensional airfoil in the past. However, the two dimensional airfoil has limitations because of wingtip effect. In this article, the computational methods are expanded to a three dimensional wing model to verify the validity of the passive control technology. The computational results showed that the surface contour bump can spread and weaken the shock wave. Based on the two dimensional airfoil, the transonic drag reduction can achieve 2.0%-12.0%, while the maximum drag reduction of three dimensional model can reach 6.4%. When the span of the local bump is one tenth of the three dimensional wing model, the lift and drag coefficients are different from two dimensional airfoil results. The best control effect of limited wingspan model with local bump is achieved when the bump control brought to the position where is near to the 80% of the wing tip.

1. INTRODUCTION

Drag of civil aircraft does not only influence the operating costs of the airlines but it also has a direct impact on fuel consumption and environmental pollution. Schneider [1] pointed out that one percent drag reduction will save about 400,000 liters per year, per aircraft for the typical current far-distance airline. That means about 5000 kg of noxious would not be discharged. Because CO_2 is a long-lived greenhouse gas, reducing its emission is a key long-term goal. Drag and weight reduction are the two most potent technologies, as reported by Green [2]. Since the beginning of the jet–age there has already been a large increase in aerodynamic efficiency of civil plane. Further improvements regarding aircraft aerodynamics performance are expected from flow control technology. Nowadays, the most long–range civil plane cruise to transonic speed which can maximize the operational range as well as cruise speed. This speed region is characterized by occurrence of shock waves, associated with increase of drag.

Shock Control Bumps (SCBs) are a promising way to reduce wave drag on transonic wings. Dargel, Rodde and Archambaud [3] once reported that shock control bumps applied in wing-body configuration. Qiu and Li [4] have proposed a new active control method that based on piezoelectric elements. The piezoelectric elements buried on upper surface of wing to induce periodic disturbance in boundary layer. How the periodic disturbance affects the lift and shock was studied by computational fluid dynamics (CFD). Ashill and Fulker [5-6] utilized local displacements of the aerofoil surface geometry to control the shock as well as interaction of shock and boundary layer. Zhu Y [7] has performed an extensive parametric study of ramps and bumps by using solutions of the Navier-Stokes equations. For near-sonic airplanes, wind-tunnel tests of several airfoils in the near-sonic region were conducted in the 1950s by NACA in the America [8]. In 2011, Li PF and Zhang BQ [9] studied the way of decreasing the shock wave drag by bump. Tao [10] attempted to control the shock wave drag by careful design bump on the wing.

Many works have been done to improve the aerodynamic performance of design point through passive bump control devices. However, the passive bump control device would destroy the structure of the wing when it covers the whole wing. In this paper, how bump affects aerodynamic performance has been researched when the bump span is one-tenth the wing. In 2001, NACA's Aircraft Morphing Project began a series of examinations of surface contour of available bumps to reduce aircraft wing drag. This project was given two applied goals: short term and long term. The short term goal is

Jin Hao Qiu, qiu@nuaa.edu.cn

focused on the design of a surface contour that limited to 20% of the airfoil chord. The long term goal is focused on the development of a surface contour bump that would cover the last 40% of airfoil chord. The application of short term would be suitable for current transonic aircraft wings. The long term application is new and requires an integration of bumps and other technologies from the beginning of the aircraft wing design process. In this paper, the upper surface contour bump is studied from 20% to 40% of RAE 2822 airfoil chord, especially when it is 30%. The objectives of the current research included the length, height and position of the contour bump on a cruise wing upper surface. As a control method, the bump can spread and weaken the shock wave, thereby reduce transonic drag.

2. Model and Bump Design

As show in Figure 1, a two dimensional RAE2822 airfoil [11] model with bump and structured grids have be showed. Figure 2 is the chord wise section profiles of the RAE2822 and the bump. The chord of wing c is 1m, chord wise length of bump L is $0.3c$, height of the bump H is $0.001c$, the bump range is from left border (A_1) $0.4c$ to right border (A_2) $0.74c$, and the X-coordinate of bump crest point (A_3) x_3 is $0.57c$. When the $x_3 = 0.57c$, a point on the top surface of the RAE2822 was shift up $0.001c$ which is A_3. A_1 and A_2 are located on the airfoil. The chord wise section profile of bump is small arc on the upper surface of the airfoil, as show in Figure2. The small arc is identified by A_1, A_2 and A_3. The join points of bump and wing did not require smooth connection and the bump can be easily installed on the wing, therefore, the bump is free to use in the engineering.

Figure 1 RAE2822 grid.

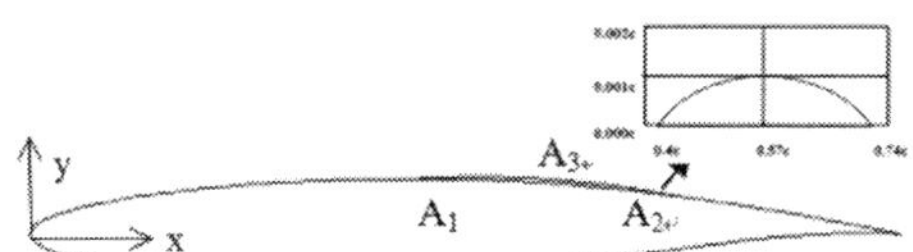

Figure 2 Chordwise section profiles of airfoil and bump.

If the flow direction is parallel to the airfoil, the quadrangular structured mesh is more practical, therefore the quadrangular structured mesh is used to discretize the two dimensional model. In this two dimensional model, the mesh contains 2.6×10^5 grid points and is stretched in the wall-normal direction with more clustering in the airfoil. In wall units, the closest grid point to the wall is at $y^+=2$, the superscript '+' indicates usual normalization by the friction velocity u_τ and /or the kinematic viscosity of fluid ν, in the absence of control, unless otherwise noted [12]. On the basis of two dimensional model, a three dimensional model of RAE2822 is tested and verified the technology of control. The chord length of 3D wing model c_2 is 1m, and the wingspan of the 3D wing model S is $3c_2$. The three dimensional finite wingspan model and finite span bump can be found in Figure 3, the bump span is one-tenth of the finite wing in Figure 3(a), and the bump span in the Figure 3(b) is equal to the span of finite wing model. The overall span bump has been divided into ten parts along the span wise (z). These bumps from left to right are first bump, second bump, third bump,..., ninth bump and tenth bump. The chord wise section of the bump is small arc which can be found in Figure 2. When the bump has been stretched along the wing span, the top view of the bump is square, as show in Figure 3(a).

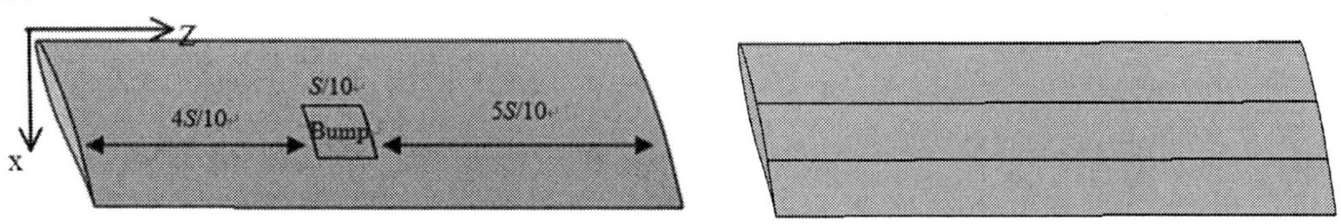

(a) A finite wing with a bump at five tenths of the span. (b) Finite wing with all span bump.

Figure 3 Finite wing with bump.

The span of flow field is $6c_2$, therefore the wing tip of model is in the flow field. The chord wise section in the wing root of computational domain is symmetrical boundary condition and it located the left to 3D wing model (z=0). The chord wise section in the maximum span of the computational domain is pressure far field boundary condition and it located the right to the wing model ($z=6c_2$), which use characteristic information to determine the flow variables at the boundaries. The symmetrical boundary condition and the structure grids can be found from the Figure4. The pressure far field boundary conditions are used to model a free stream condition at infinity and should be placed far enough from the object of interest in order to effectively approximate the true infinite extent condition. In this case the far field boundaries are located approximately 10 chord lengths from the airfoil surface.

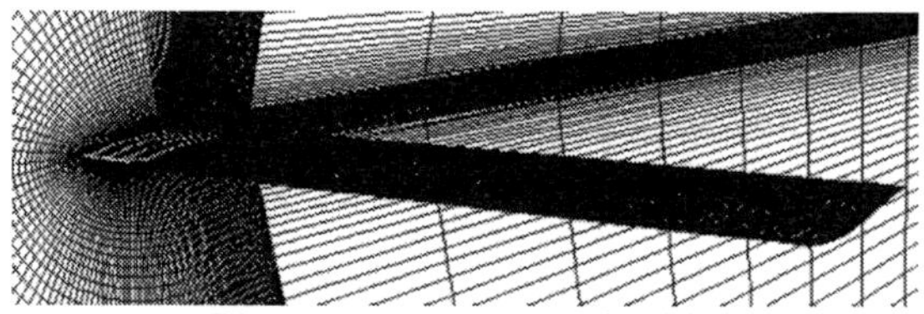

Figure 4 3D Structural grids.

3. Computational Methods

A node-based finite volume formulation is used to discretize the governing equations. The governing equations are the time-dependent Reynolds-averaged Navier-Stokes equations. The equations are expressed as a system of conservation laws relating the rate of change of mass, momentum, and energy in a control volume of area to the fluxes of these quantities through the volume. The non-dimensionalized equations are given as

$$A\frac{\partial Q}{\partial t}+\oint \vec{F}_i\cdot\hat{n}dl-\oint \vec{F}_v\cdot\hat{n}dl=0$$

where $\hat{n}$ is the outward-pointing unit normal of the control volume. The vector of dependent variables Q, $\vec{F}_i$ and $\vec{F}_v$ are the inviscid and viscous flux vectors respectively. Ref. 13 provides an in depth discussion of the conservation equation and flow solver.

The effects of turbulence are simulated by the Spalart-Allmaras turbulence model. The Spalart-Allmaras model solves directly for the eddy viscosity form a transport equation, and it was designed specifically for aeronautics and aerospace applications involving wall-bounded flows and has been shown to give good results for boundary layers subjected to adverse pressure gradients. The transported variable in the Spalart-Allmaras model, $\tilde{\nu}$, is identical to the turbulent kinematic viscosity except in the near-wall region. The transport equation[14] for $\tilde{\nu}$ is

$$\frac{\partial}{\partial t}(\rho\tilde{\nu})+\frac{\partial}{\partial x_i}(\rho\tilde{\nu}u_i)=G_v+\frac{1}{\sigma_{\tilde{\nu}}}\left[\frac{\partial}{\partial x_j}\left\{(\mu+\rho\tilde{\nu})\frac{\partial\tilde{\nu}}{\partial x_j}\right\}+C_{b2}\rho\left(\frac{\partial\tilde{\nu}}{\partial x_j}\right)^2\right]-Y_v+S_{\tilde{\nu}}$$

where ρ is density of the air, G_v is the production of turbulent viscosity, and Y_v is the destruction of turbulent viscosity that occurs in the near-wall region due to wall blocking and viscous damping, $\sigma_{\tilde{\nu}}$ and C_{b2} are the constants and ν is the molecular kinematic viscosity, $S_{\tilde{\nu}}$ is a user-defined source term. In this paper, 2D RAE2822 airfoil is studied when the free-stream Ma is 0.73 and angle of attack α is 3.19° , which corresponds to case 9 in the experimental data of Cook et al[11]. The Reynolds Number based on chord length of the airfoil and free-stream flow condition is 6.5×10^6 which is realistic flight Reynolds numbers, therefore, the simulated flow fields are assumed to be fully turbulent and compressible. Because the calculations are done in free air, the angle of attack has been modified to offset the wind-tunnel effects. In the calculation, the angle of attack α is 2.45° . It should also be noted that all airfoil computations and discussion are under appropriate boundary conditions to simulate the free stream flow fields, and do not include the effects of wind tunnel wall.

4. Results Analysis

All data presentation and discussion below are based on the constant chord (c=1m), the Reynolds Number is 6.5×10^6 when the model is the two dimensional model of RAE2822 [15]. The effect of bump contour is studied on the design point which Ma is 0.73, Temperature T is 223.3K, Operating Pressure p is 26500 Pa and Dynamic viscosity μ is 1.3947×10^{-5}. The drag coefficient reduction is presently defined by

$$DC_d = \frac{(C_d)_{on} - (C_d)_{off}}{(C_d)_{off}} \times 100\%$$

where $(C_d)_{on}$ and $(C_d)_{off}$ denote the drag coefficient with and without control respectively.

4.1 Validation of Method

Figure 5 shows the computational and the experimental [11] pressure distributions (C_p) of the 2D RAE2822 aerofoil, the Figure 5 presents that the computation results are in good agreement with experiment. But there is something to point out that the pressure distributions are slightly different on the leading edge, as well as on the trailing edge. The paper attempts to increase this accuracy on the both edge by reducing the convergence criterion, increasing the density of the mesh and choosing other turbulence model. Unfortunately, these methods can not bring an increase of accuracy of the pressure distributions on the leading edge and the trailing edge.

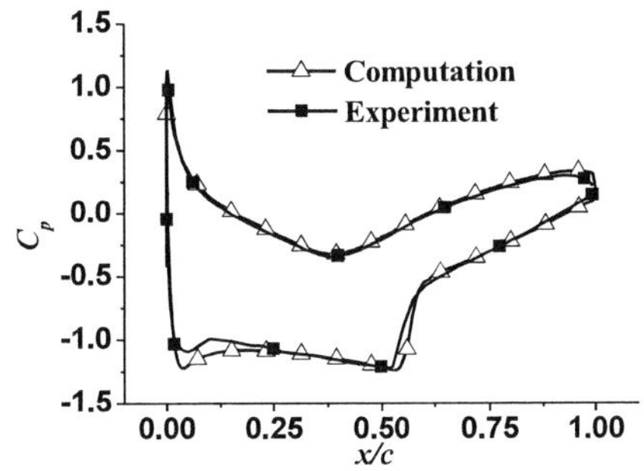

Figure 5 Pressure distributions.

Table 1 shows other kinds of aerodynamic characteristics in the experiment and the computation. Through the comparison of simulation and experiment, it is can be found that the errors of C_l and C_d are less than 2%, the errors of the simulating calculation are proved acceptable. The comparison mentioned above indicates that the methods of simulation are correct and effective.

Table 1 Other aerodynamic characteristic

state	C_l	C_d	Shock(x/c)
Experiment	0.803	0.0168	0.52
Computation	0.7912	0.0189	0.54

4.2 **Effect of Bump Position**

Figures 6 and 7 demonstrate how the bump position affects the drag coefficient and lift-drag ratio, when L^+ is 2000.0 and H^+ is 20.0. The both figures indicated that the drag coefficient and lift-drag ratio have strong sensitivity to the position of the bump. The passive bump control has significant drag reduction in a wide range of lift coefficients, not only just at the design point. At the highest lift coefficient of 0.7991, the drag coefficient reduction is achieved 10.6% for the bump control, when the x-coordinate of the bump contour crest x^+ is 3666.6, at the same time, the lift-drag ratio increased by 12.6%. When crest of the bump is close to the shock, the bump can serve the purpose of reducing shock strength and wave drag. The bump do not weaken shock, when the position is from the back of the wing, moreover, it will accelerate the flow, which do not help to reduce the shock wave drag. In the condition of pre-biased position, the front shock is produced before the supersonic flow was depressed in the leading edge of the bump, it will accelerate, even may cause a new stronger shock. The new shock will increase the wave drag, increase the boundary layer thickness in the behind the shock wave, even induce the interaction between shock and boundary layer, all these results will do damage drag reducing. It is an optimal position for the reduce drag when x^+ is closed to 3666.6, L^+ =

2000.0 and $H^+ = 20.0$.

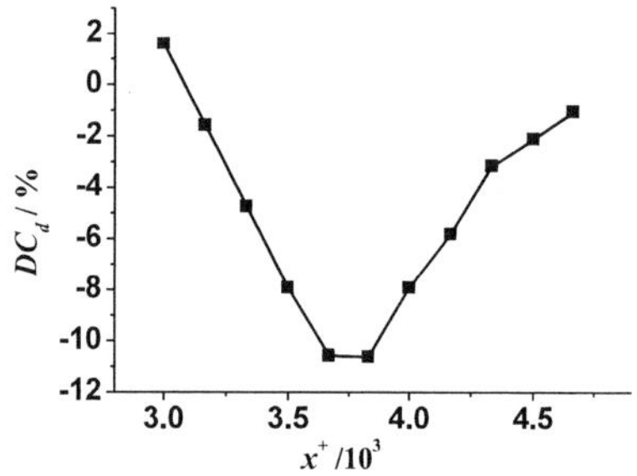

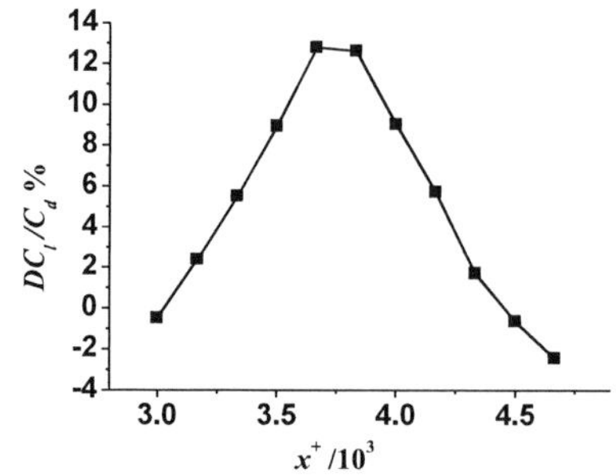

Figure 6 Effect of drag coefficient of position. **Figure 7** Effect of lift-drag ratio of bump position.

4.3 Effect of Parameters of Contour Bump

The effects of bump contour parameters are complicated. A single parameter cannot describe the aerodynamic effect. Both parameters will be discussed together.

(1) Effect of height and position

The bump is fixed at different positions to learn the effect of the bump height follow from the prior simulated results. As show in Figure 8, bump height was examined when L^+=2000.0, x^+=3500.0 or 3666.6 or 3833.3. When the H^+ = 6.6, the drag reductions are no more than 8% for all positions. However, DC_d drops rapidly in all cases when H^+ is from 6.6 to 20.0, especially, when x^+=3666.6 and 3833.3. It concluded that the greatest amount of drag reduction can be achieved 12.16% when x^+=3833.3 and H^+=26.7 or can be achieved 11.1% when x^+=3666.6 and H^+=26.7, after that, the drag reduction climbs quickly for a further increase H^+, in particular at the smaller x^+.

The chord wise centerline pressure distributions can be observed in second effect, exemplified by Figure 9. The pressure distributions present that shock wave strength reduced and the shock wave position moved backwards when compared with the no control. Figure 9 also shows that the affected area is primarily close to the bump region. The bump shifts the normal wave to downstream approximately 4.7%*c*. Due to the upstream communication through the subsonic portion of the boundary layer, the upstream influence of bump is restricted to a small region upstream of starting position.

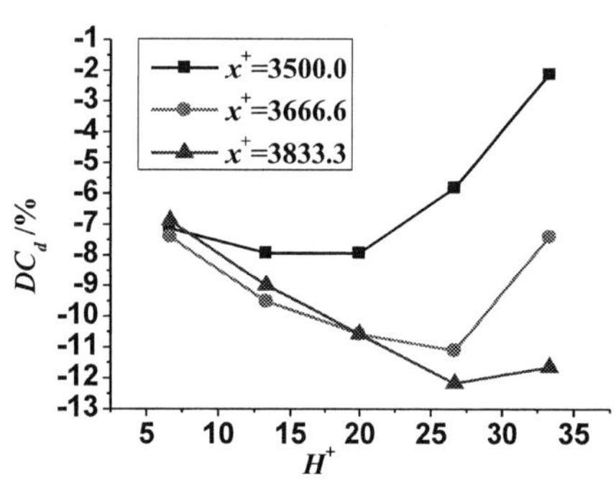

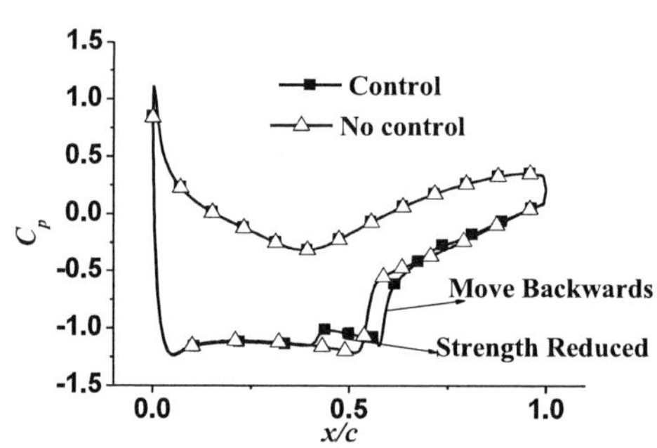

Figure 8 Effect of position and height. **Figure 9** Comparison of the pressure distribution.

The third bump effect is evident in the lift coefficients curves, shown in Figure 10. The bump control decreases the lift coefficients at low angle of attack. Lift coefficients are increased when the shock wave moves backwards along the airfoil, meanwhile, in the higher lift coefficient, the shock would interact with the bump. The bumps are not detrimental to lift, even enhance lift in the higher angle of attack, especially when the lift coefficient is exceeds 0.3403.

(2) Effect of height and length

According to the numerical results, the effect of the bump height has been research when the bump is fixed at x^+ = 3666.6 and the bump length L^+ = 666.7 or 1333.3 or 2000.0 or 2666.7. Figure 11 showed the effect of length and height of bump. At the same bump height, the mounts of drag reduction are increased when L^+<2000, and decreased when L^+>2000.0. If $L^+ \leq 2000.0$, the mounts of drag reduction increased first then decreased with the height, when the L^+=2000, and H^+=26.7, the maximum the mount of drag reduction is 12%. However, the mounts of drag reduction are decrease with the increasing of height when H^+>26.7. William E [16] examined the effect of discrete contour bumps on reducing the transonic drag at off-design condition on an airfoil. Their computational predictions indicated that when the bump was from 0.2*c* to 0.4*c*, the contour bumps generate significant drag reduction in the range of 12.0%-15.0%, at an off-design Mach number of 0.78. All these results indicate that shock weakening is closely related to the parameters of bump contour. As to 2D RAE2822, the optimum parameters L^+ is 2000.0 and H^+ is 26.7, the bump position is close to the shock. For other airfoils, the optimum parameters were L^+ at 1333.3 - 2000.0 and H^+ at 20.0-33.3. The results of research are consistent with reference 17.

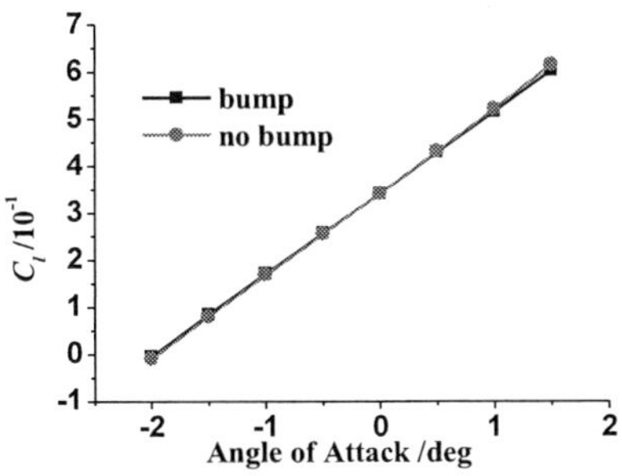

Figure 10 Comparison of the left coefficient.

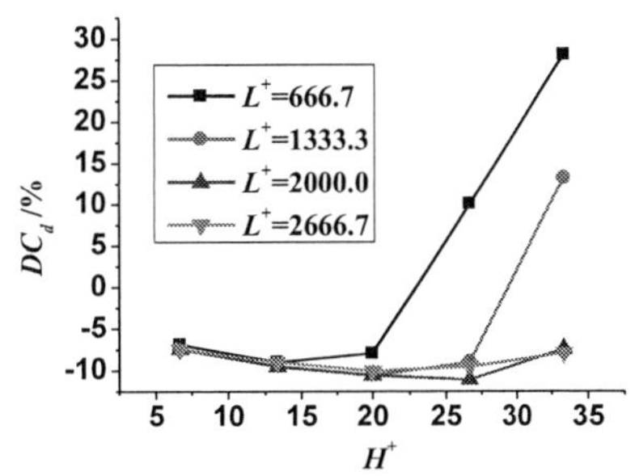

Figure 11 Effect of length and height of bump.

5. Three dimension model

Based on the results, a RAE2822 three-dimensional model has been tested and verified the control technology in the practical application. In this section, the L^+=2000.0, the H^+=26.7, crest of the bump x^+=3800.0 and the span of bump is one-tenth the span of limited 3D wing model. The span of the 3D wing model z^+ is 2×10^4. The Figure 12 and Figure 13 showed the change of the drag coefficients and lift coefficients of the three-dimensional wing model separately, when the span wise position (z^+) of bump is changed along the span of the wing. The maximum amount of drag reduction of three-dimensional model is dropped from 11.1% of the two-dimensional airfoil to 6.4% when the bump is located $z^+ = 1.5\times10^4$, at the same time, the lift coefficient is increased about 7%. Due to the wingtip effect, the result of drag reduction from three-dimensional model with bump is not better than 6.4%. Through appropriate adjustment and control, the shock strength can be reduced, and the position of shock can be move backward. The optimum parameters are L^+=2000.0, x^+=3800.0, H^+=26.7, and $z^+ = 1.5\times10^4$ for 3D models. These parameters are closed to the optimal values of 2D model. Based on the analysis, the control bump has many practical applications. The Figures demonstrate that the best control effect of limited wingspan model with local bump will be achieved when the bump control brought to the position where is near to the 80% of the wing tip.

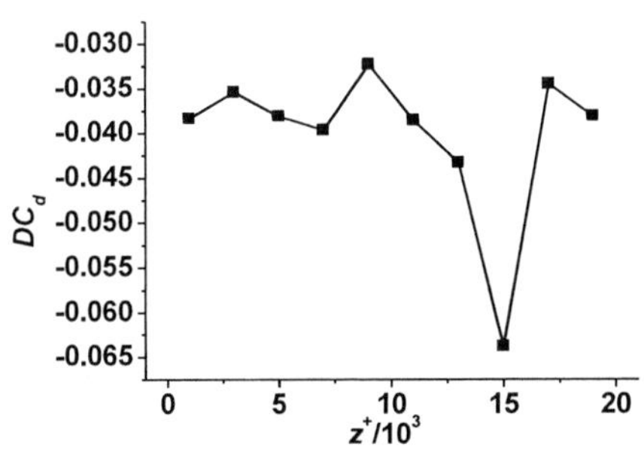

Figure 12 Drag coefficients.

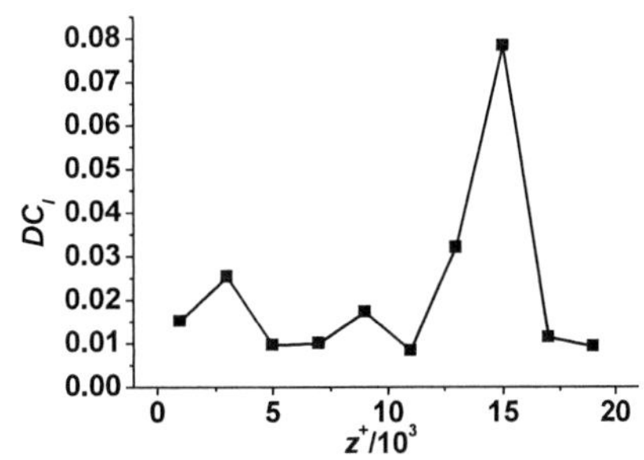

Figure 13 Lift coefficients.

The Figure 14(a-d) show the pressure coefficients on the chord wise section $z^+ = 1\times10^4$ when the bump is located on the different span position. The section is on the middle of the wing span. The span wise coordinates (z^+) of third bump is from 4×10^3 to 6×10^3 and the eighth bump is from 1.4×10^4 to 1.6×10^4. Both the bumps have same distance to the middle section. The Figure 14(a) shows the pressure coefficients when the wing is without the bump. The Figure 14(b) shows the pressure coefficient when the wing is with third bump. The Figure 14(c) shows the pressure coefficient when the wing is with eighth bump. The Figure 14(d) shows the pressure coefficient when the wing is with overall span bump. The Figure14 (a) and (b) indicate that the third bump has weakly impact on the pressure coefficients distribution. However, the effect of eighth bump is more obvious by comparing the Figure 14 (c) and (a). The Figure14 (c) indicate that area of negative pressure on the top surface moves backwards which will be beneficial to decrease drag. When the bump covers the whole wing, the zone of negative pressure on the upper left surface will be decreased, position of shock moves backwards, and the pressure coefficient on the bottom surface increase as show in Figure 14(d). These interactions are beneficial to increase lift and decrease the drag.

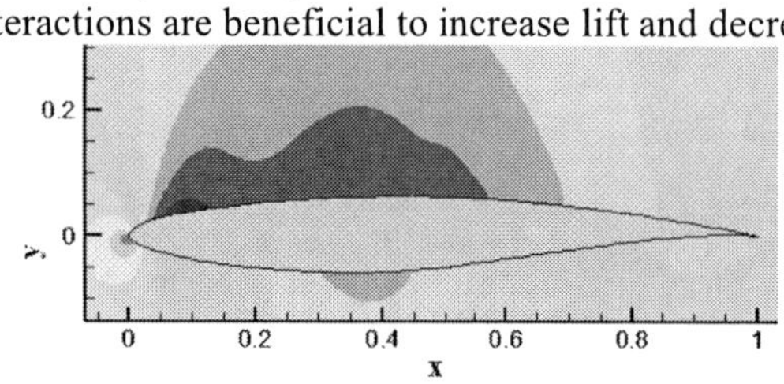

(a) No control with bump

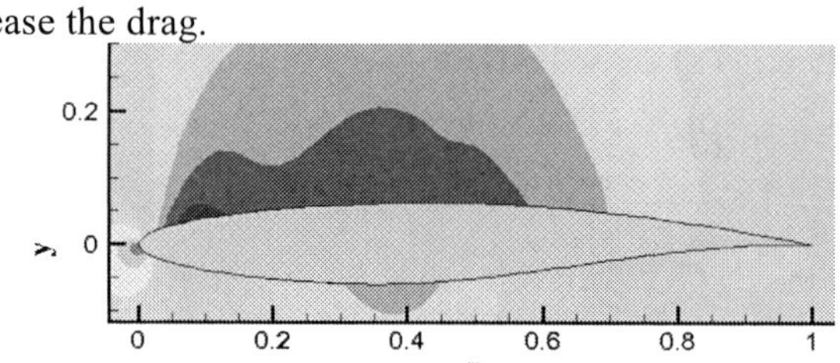

(b) Control with third bump

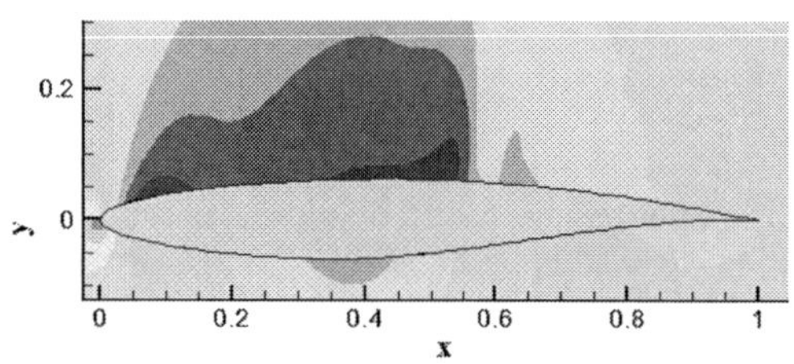

(c) Control with eighth bump

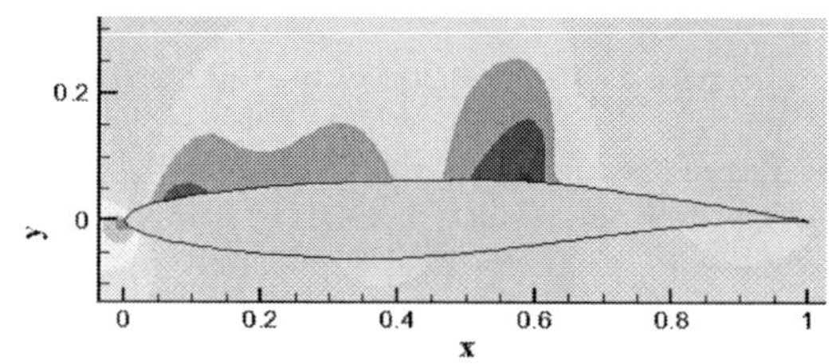

(d) Control with all bump

Figure14 The pressure coefficient

The simulational results of 3D wing model are more accurate and close to the actual condition than the 2D model. The Figure 14(a) indicates that the shock is not obvious on the top surface for the wing tip vortex. The Figure 14(b) has bump cotrol, but the position of bump locates inside the section, therefore, the third bump can not impact the tip vortex, therefore, the Figure 14(a) and (c) do not have big difference. The shock wave effect will turn up on the top of the surface when the bump position locates on the eighth part, as show in the Figure 14 (c). When the bump postion locates the outside of the section, the bump will resist the negtive effect because of tip vortex. The overall span bump will increase the relative thickness of airfoil which can increased the difficulty of flow over wing tip.

In spite of the overall span bump control can achieve better aerodynamic characteristic, but it has greater cost from structural damage and negative aerodynamic effect in off-design point. This paper has researched the partial bumps on the wing how to impact the aerodynamic characteristic of the wing. When the partial bump with one-tenth span of the wing is located 80% of the wing tip, the effect of control is better. The results should be able to guide engineering application.

6. Conclusion

The effects of bump control in reducing the transonic drag on two dimensional and three dimensional RAE 2822 airfoil have been examined. When the Reynolds number is 6.5million, the research focused on fully-turbulent flow conditions. The computational and experimental aerodynamic characteristics are made comparisons which prove the computing methods are accurate. The following conclusions can be drawn from the research:

1) The computational predictions indicated that the bump control can generate significant drag reduction, in the range of 2.0% - 12.0%. When the $Ma = 0.73$, the $H^+=26.7$, the $x^+= 3833.3$, and the normal shock wave located on $x^+=3666.7$, the bump control can achieve the maximum amount of drag reduction for RAE2822. The drag reduction occurred by modifying the strong normal shock wave on

the baseline airfoil, and creating a weaker shock wave pattern.

2) The computational results indicated that the bump has performed well in lift-drag ratio when the Ma = 0.73 for 2D RAE 2822. The results demonstrated that the lift-drag ratio can increase 13% when the L^+ = 2000.0, H^+ = 20.0 and x^+=3833.3. The bump would be likely required for an aircraft.

3) The CFD results predicted that the current bump contour will increase the drag coefficients if the normalized bump height is more than 30.0. The optimum parameter of drag reduction is L^+=2000, x^+ = 3800.0 and H^+ = 26.7 for 2D RAE2822.

4) Due to the wingtip effect, the maximum drag reduction of three-dimensional model with chord c_2 of 1m and span of $3c_2$ was 6.4%, which is not as good as two-dimensional airfoil. The best control effect of limited wingspan model with local bump will be achieved when the bump control brought to the position where is near to the 80% of the wing tip.

Acknowledgements

This research is supported by PAPD, the National Natural Science Foundation of China (11372133). Funding of Jiangsu Innovation Program for Graduate Education (CXLX12_0165), The Fundamental Research Funds for the Central Universities.

References

1. Schneider W. “The importance of aerodynamics in the development of commercially successful transport aircraft,” *Aerodynamic Drag Reduction Technologies*, edited by Peter Thiede, Vol. 76, Notes on Numerical Fluid Dynamics. 2000, Springer, New York, pp. 9–16.

2. Green JE. “Civil aviation and the environmental challenge,” *The Aeronautical Journal*, 2003, Vol.107, No.1, pp. 281-299.

3. Dargel G and Archambaud JP. “Assessment of the capability of drag reduction of the shock control device’ SC bump’ on airfoil flows and application aspects on wing,” *In IUTAM Symposium on Mechanics of Passive and Active Flow Control.* Editor: G. E. A. Meier and P. R. Viswanath, Kluwer Academic Publishers, 1998.

4. JH Qiu, DW Li. “Increased lift airfoil of vibration flow control technology,” *Journal of Nanjing University of Aeronautics & Astronautics*, 2012, Vol. 44, No.5, pp. 638-644.

5. Ashill PR, Fulker JL, Simmons MJ. “A review of research at DRA on active and passive control of shock wave,” *ICAS*, September, Sorrento, Italy, 1996.

6. Ashill PR, Fulker JL. “A review of flow control research at DERA. Mechanics of Passive and Active Flow Control,” *IUTM Symposium*, Gottingen, Germany, September 7-11, 1998: 43-56.

7. Zhu Y, N Qin D. “Numerical study of active shock control for transonic aerodynamics,” *International Journal of Numerical Methods for Heat and Fluid Flow,* 2004, Vol. 14, No. 4, pp. 444-466.

8. John V. Becker. “The high speed frontier,” *NASA*, SP-445, 1980: 36–59, 84–117.

9. Li PF, Zhang BQ. “Wave drag reduction of airfoil with shock control bump,” *Acta Aeronautica et Astronautica Sinica*, 2011, Vol. 32, No. 6, pp. 971-977.

10. Tao Y, Lin J. “Investigation on drag reduction with contour bump located on a supercritical airfoil,” *Acta Aerodynamica Sinica*. 2007, Vol. 25, No.1, pp. 116-119.

11. Cook, P.H., M.A. McDonald, M.C.P. Firmin. “Aerofoil RAE 2822- pressure distributions, and boundary layer and wake measurements,” *Experimental Data Base for Computer Program Assessment*, AGARD Report AR, 138, 1979.

12. Garnier Eric. “Large eddy simulation of shock/boundary-layer interaction,” *AIAA Journal*, 2002, Vol. 40, No. 10, pp. 1935-1944.

13. W. Kyle Anderson. “An implicit upwind algorithm for computing turbulent flows on unstructured grids,” *computers fluids,* 1994, Vol. 23, No. 1, pp. 1-21.

14. ANSYS 13.0 Help. // Theory Guide :: 0 // 4. Turbulence // 4.2. Spalart-Allmaras Model // 4.2.2. Transport Equation for the Spalart-Allmaras Model.

15. Zhang YF, Zhang XL. “Credibility analysis of RAE2822 airfoil transonic flow computation,” *aeronautical computing technique*, 2009, Vol. 39, No. 4, pp. 68-70.

16. William E, Milholen II*. “On the application of contourbumps for transonic drag reduction,” *43rd AIAA aerospace Sciences meeting and Exhibit*, 10-13 January 2005, Reno, Nevada, 2005.

17. Stanewsky E. “Drag reduction by shock and boundary layer control,” *New York: Springer-Verlag.* 2002:383-413.

Variable-stiffness skin concept for camber-morphing airfoils

Wolfram Raither[1*], Emian Furger[1], Manuel Zündel[1], Andrea Bergamini[2], Paolo Ermanni[1]

[1] Centre of Structure Technologies, ETH Zurich, 8092 Zurich, Switzerland
[2] Mechanics for Modelling and Simulation, Empa, 8600 Dübendorf, Switzerland

Abstract

Morphing airfoils promise advances in performance and efficiency when compared to conventional designs. However, the conflict of requirements between compliance and stiffness, that is characteristic of shape-adaptable structures, often leads to compromise-driven solutions. This article presents the concept of a variable-camber airfoil with adjustable stiffness. Smart elements in the airfoil skin permit to adapt the structural rigidity to the system's operational states, resulting in more efficient morphing and potentially lighter designs. It is shown by a numerical study that, with respect to a reference configuration, the actuation energy is reduced by up to 97% and the structural mass can be lowered by up to two thirds. Furthermore, experiments on a scaled airfoil structure with a variable-stiffness skin based on electro-bonded laminates demonstrate effectivity and integrability of the proposed structural concept which permits changes in cambering stiffness by a factor of at least seventy.

1. INTRODUCTION

Metamorphosis, "morphing" [16, 2], of aeronautical structures has been proposed for a long time as an alternative design philosophy to overcome the general limitations of conventional rigid high-lift and control surfaces in terms of aerodynamic and lightweight qualities. Like in every shape-adaptable lightweight structure, the conflicting requirements of shape adaptivity, load-carrying capability and lightweight efficiency, however, constrain the design space for morphing aerostructures [3, 4] and generally lead to compromise-driven solutions: The stiffness that is required to carry operational loads without excessive deformation is detrimental during shape adaptations, for which it causes the need for large activation energy as well as load introduction issues. This design conflict finally results in a high structural mass of many morphing concepts, which limits their applicability.

Time-variable structural stiffness [12] has been suggested for semiactive morphing applications, where the structural response of an airfoil can be varied by changing certain stiffness components in order to induce, for example, a twist that can be exploited for controlling the lift [11, 5, 1, 2, 14]. In the present contribution, in contrast, adaptive stiffness is intended to be used in the skin of an actively cambering airfoil. Here, a state of high rigidity can be set to carry operational loads when no force is exerted by the actuators, while well-directed reductions in the skin's stiffness are put into effect during cambering (in both directions) in order to lower the energy required to deform the airfoil. Concerning the realization of adjustable structural stiffness, several smart materials like electro-bonded laminates (EBLs) [7], temperature-variable polymers [8, 10], shape-memory alloys [15] or adaptively pressurized porous materials [17] can be considered, in principle. The following investigations are however limited to material concepts based on polymer glass transition and electro-bonding.

2. CONCEPT OF ACTIVE AIRFOIL WITH VARIABLE-STIFFNESS SKIN

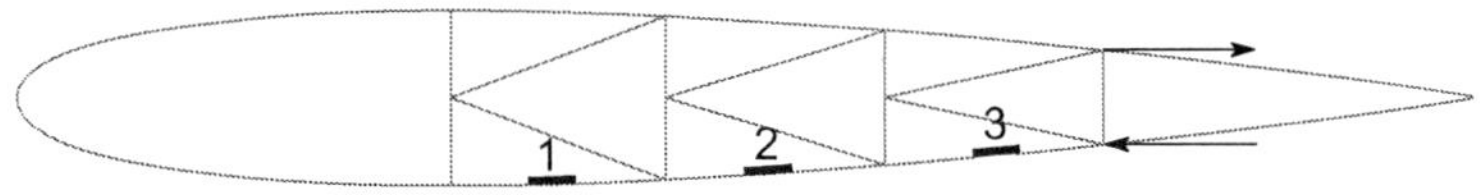

Figure 1. Concept of variable-camber morphing airfoil with adaptive-stiffness elements (1, 2, 3) in the skin (internal structure and actuation based on [9])

Figure 1 shows the concept of the variable-camber morphing airfoil with variable-stiffness skin constituting the subject of this work. It is based on the structural concept without variable stiffness reported in [9] which has been used as a reference configuration and which is shown in the undeformed and in the morphed configuration in figure 2. The rigid nose, the compliant internal structure and the two actuator forces, that can in principle be put into practice by any kind of linear actuator, have been adopted from this baseline configuration and extended by interfaces with adaptive shear stiffness in the skin. The design of the compliant structure itself is not part of the work on hand, so that, intentionally, a concept that is documented in detail in literature has been chosen as a baseline for the performance assessment of the presented idea and a detailed description of the compliant system is set aside here. However, its basic working principle is worthwhile mentioning, as it is relevant for the further discussion: In this concept, the pronounced cambering compliance is made possible by the nodes of the structure which are generated by the intersections of the internal ribs and the skin and which serve as compliant hinges with rotary kinematics for the cells of the airfoil.

In the example presented in figure 1 three variable-stiffness elements are placed in the lower part of the skin, but also different distributions are considered and will be discussed later, as well as an alternative internal structure.

While the adaptive elements are set to a state of high stiffness in the initial configuration in order to carry operational loads with sufficient rigidity and stability, they can be softened as soon as the actuators are activated for morphing. Remaining in the compliant state throughout the shape adaptation process, they allow for reduced actuator work in comparison with a structure with constant stiffness. Once a morphing cycle is completed, the adaptive interfaces can be reset to a state of high rigidity in order to restore the airfoil's original load-carrying capabilities for operational states without assistance of the actuator stiffness.

As long as the amount of energy that has to be spent for changing the stiffness by activating and resetting the adaptive elements is smaller than the one saved on the side of the actuators, the variable-stiffness concept provides an energetic advantage over the baseline configuration. Moreover, a design allowing for a certain change in camber can be realized with smaller actuators in this case by implementing a configuration with adaptive stiffness. This leads to weight savings on the actuators and, given the typical snowball effect of lightweight structures, on additional components, so that substantial savings in system mass and increases in efficiency can be expected.

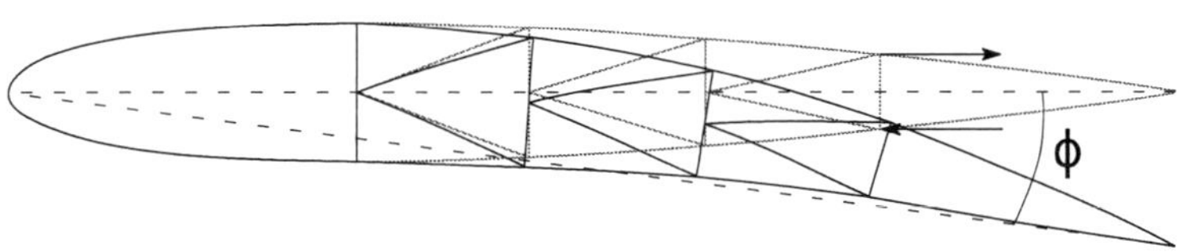

Figure 2. Reference configuration without adaptive stiffness based on [9] in undeformed and in cambered state

Figure 3 illustrates the working principles of the two kinds of implementations of variable-stiffness elements considered in the present work: On the one hand, the concept of a polymer layer with controllable temperature which allows for thermally-induced changes in stiffness across its glass transition is represented by a schematic, along with the temperature plot of its elastic modulus.* On the other hand, the schematic of an electro-bonded laminate (EBL) is shown, in which changes in voltage across the layers of an electric capacitor, and thus in attraction in through-the-thickness direction, are transformed into changes in shear stiffness by friction at the laminate's interface.

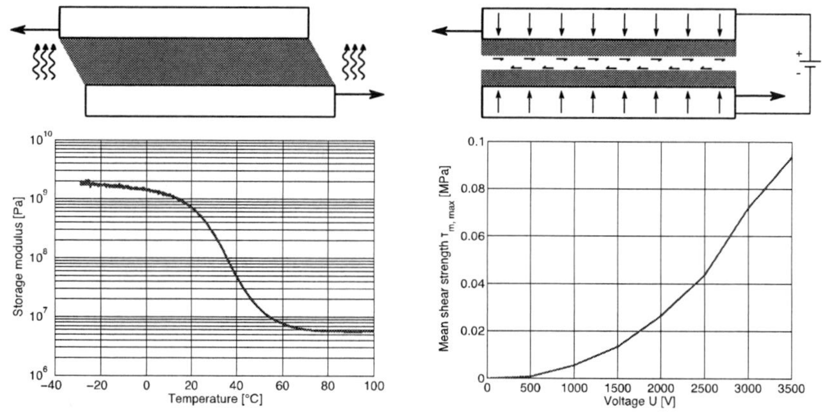

Figure 3. Working principles of two kinds of variable-stiffness elements: (a) Schematic of temperature-variable polymer layer and temperature plot of its elastic modulus over the glass transition and (b) schematic of EBL and plot of its shear strength in a double-lap shear test w.r.t the applied voltage

As illustration of the quantity representative of the EBL's adaptive capability, its strength (defined as the stress at the onset of sliding) in a double-lap shear experiment is plotted as a function of the applied voltage. Whereas the first material concept provides a continuous variation of the shear modulus by roughly two orders of magnitude, the second one is limited to discrete changes between the two states of static friction and sliding, at least in a first approximation and for its simplest implementation with one adaptive interface.

3. NUMERICAL SIMULATION

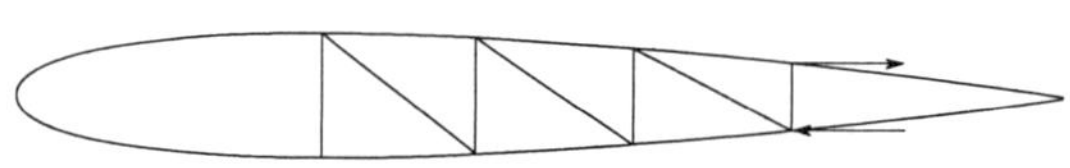

Figure 4. Second configuration of internal structure (shown in the baseline configuration without variable-stiffness elements)

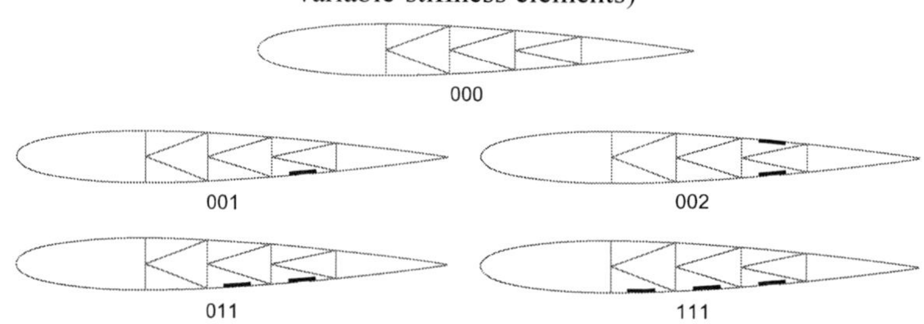

Figure 5. Arrangements of variable-stiffness elements (first internal structure as an example)

In order to assess the potential of the variable-stiffness concept, the influence of the adaptive skin on the mechanical behaviour of the active airfoil is investigated by means of finite element simulations.

---

* This temperature plot of the elastomer "DYAD 609" by "The Soundcoat Company" has been determined by means of dynamic mechanical analysis (DMA) using a frequency of 1 Hz and a heating rate of 2 K/min.

Whilst a NACA 0012 shape has been assumed for the skin of all models, the geometry depicted in figure 4 has been considered for the internal structure in addition to the one illustrated in figure 2. Concerning the arrangement of variable-stiffness elements, the configurations illustrated in figure 5 for the example of the first internal structure have been examined and the nomenclature system introduced for element arrangements is obvious from the figure. The applied geometrical parameters are defined in figure 6 for the overall airfoil and in figure 7 for the variable-stiffness elements and are quantified in table 1, while the wall thicknesses resulting from the sizing process described in section 3.1 are reported in section 5.1.

Modelling the case of a wing of infinite span, two-dimensional (plane strain) finite-element representations have been implemented using ANSYS. Whereas the airfoil's skin and substructure are discretized by means of two-node BEAM188 elements (three degrees of freedom at each node), membrane elements of the type SHELL41 (four nodes, each with three degrees of freedom) are applied for the polymer layers of the variable-stiffness elements simulating a thermal realization, which are subjected to shear deformations. Also the dielectric layers of electro-bonded laminates are represented using SHELL41 elements. The average length of the one-dimensional elements (and also the bigger dimension of the two-dimensional elements, on average) is approximately one percent of the chord. The unit depth relevant for some quantities of the model is 1 cm.

Table 2 summarizes the material properties used for the numerical model and the temperature plots of elastic modulus and specific heat capacity of the elastomer assumed for the thermal interfaces ("DYAD 609" by "The Soundcoat Company") are shown in figures 3 and 8, respectively. Concerning the thickness of the temperature-variable polymer layers, $t_{v,P}$ = 6 mm has been assumed. For the two operational states of the thermal concept, temperatures sufficiently below (T_1 = 0 °C) and above (T_2 = 60 °C) the polymer's glass transition have been chosen.

The geometric and material properties of the experimental airfoil structure (and the respective numerical models) are listed in tables 1 and 3, respectively.

As a detailed simulation of the electro-elastic coupling of the EBL material is not within the scope of this work, a simpler representation has been implemented for the numerical study: In order to include the limited shear strength of the EBL in its closed state, the EBL is modelled by a layer having the elastic properties of the dielectric which allows to determine the shear stress at the interface for comparison with the material's shear strength measured experimentally. The thickness of this layer was chosen to be $t_{v,EBL}$ = 0.4 mm, which allows to accommodate four EBL interfaces in parallel (for an exemplary arrangement see figure 9): If, like in this example, the two outermost electrodes are considered to be integrated in the adjacent overlapping parts of the airfoil skin, so that they do not have to be taken into account for the EBL thickness, and if, further, the four dielectric layers and the four inner electrodes are assumed to exhibit single thicknesses of 25 µm and 50 µm, respectively, their thickness sums up to 300 µm and the remaining 100 µm can be used for the insulation between the two innermost electrodes.[†] In order to simulate the compliant state of the EBL in a simplified way, the respective layer is removed from the model. This way, the small shear stress transfer due to sliding in this state is neglected, justified by the substantially lower normal stress when compared to the case when the EBL is closed. For the numerical study, the distance between the overlapping parts of the airfoil skin is held constant along the overlap in this configuration, corresponding to a guiding mechanism which might be required in a real design for ease of reclosing the interface.

The airfoil's rigid nose is implemented in the finite element models by a constraint of all the degrees of freedom of the respective nodes. The aerodynamic pressure distribution that is to be applied to the structure in the original as well as in the deformed configuration is determined by means of the airfoil theory based software "XFOIL", presuming an angle of attack of α = 5°, a flow speed of V = 100 m/s and an air density of ρ_A = 1.225 kg/m^3. Although neglecting viscous, compressible and three-dimensional effects, this way is considered appropriate at the present conceptual stage of this work. The performed analyses have also shown that the aeroelastic coupling of the considered airfoil is weak, as the

[†] Counting the electrodes to the geometrical thickness of the EBL model, but neglecting their elastic contribution is considered conservative with respect to the evaluation of the EBL in its stiff state, as the stiffness of the dielectric assumed for the EBL layer is considerably lower, since the electrodes usually consist of metal. In any case, the effect of this discrepancy on the results can be expected to be small.

aerodynamic loads have only a small influence on the deformation. This is a result of the specific sizing process applied which is described in the following subsection 3.1. Accordingly, a direct aeroelastic coupling in the numerical model has been dispensed with.

The simulation of the experimental airfoil structure, which differs in some points from the model of the purely numerical study presented in the previous paragraphs, is described in section 3.6.

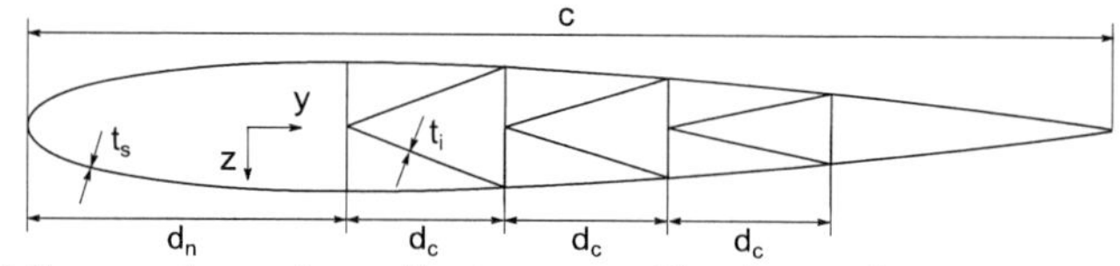

Figure 6. Airfoil geometry and coordinate system (first internal structure as an example)

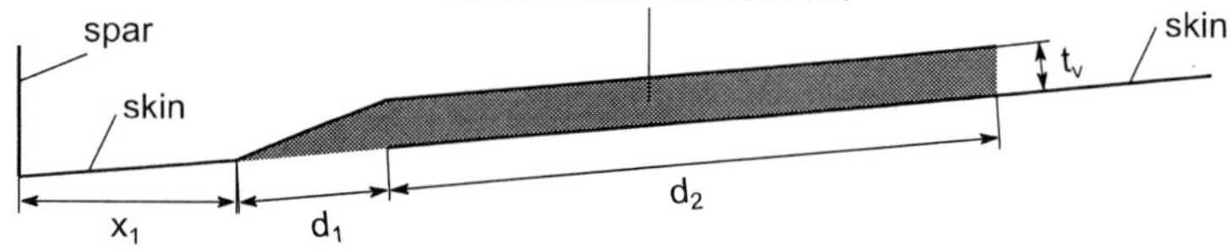

Figure 7. Geometry of adaptive interface

Table 1. Geometric parameters

Parameter	Symbol	Value, simulations	Value, experiment
Chord	c	1'000 mm	490 mm
Nose width	d_n	300 mm	150 mm
Cell width	d_c	150 mm	
Width of non-tapered part of variable-stiffness element	d_2	75 mm	
Width of tapered part of variable-stiffness element	d_1	$d_2/4$	
Thickness of variable-stiffness layer (temperature-variable	$t_{v,P}$	6 mm	
Thickness of variable-stiffness layer (EBL)	$t_{v,EBL}$	0.4 mm	
Thickness of dielectric layer	t_D	25 µm	
Offset of variable-stiffness element	x_1	$d_2/4$	20 mm
Skin thickness	t_s	(variable)	1 mm

Table 2. Material properties for simulations

Property	Symbol	Unit	Value
Young's modulus of aluminium	E_{Al}	GPa	70
Poisson ratio of aluminium	ν_{Al}	–	0.35
Density of aluminium	ρ_{Al}	kg/m^3	2'700
Young's modulus of polymer "DYAD 609" at 0°C	E_P	GPa	1.45
Young's modulus of polymer "DYAD 609" at 60°C	E_P	MPa	7.0
Poisson ratio of polymer "DYAD 609"	ν_P	–	0.499
Density of polymer "DYAD 609"	ρ_P	kg/m^3	1'250
Young's modulus of dielectric	E_D	GPa	3.2
Poisson ratio of dielectric	ν_D	–	0.499
Relative permittivity of dielectric	ε_r	–	3.59
Density of dielectric	ρ_D	kg/m^3	1'400

Table 3. Material properties for experiment

Property	Symbol	Unit	Value
Layup of GFRP laminate	–	°	[0/90; ±45]$_s$
Axial stiffness of GFRP laminate	a	N/mm	19'422
Flexural stiffness of GFRP laminate	d	Nmm	2'142

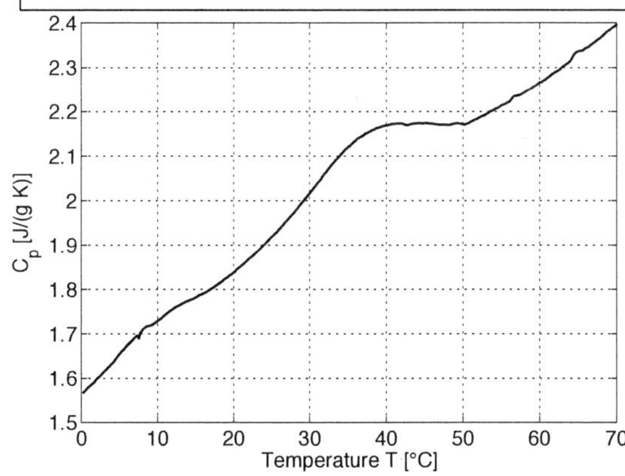

Figure 8. Specific heat capacity of polymer "Soundcoat DYAD 609" as a function of temperature, determined by differential scanning calorimetry (DSC)

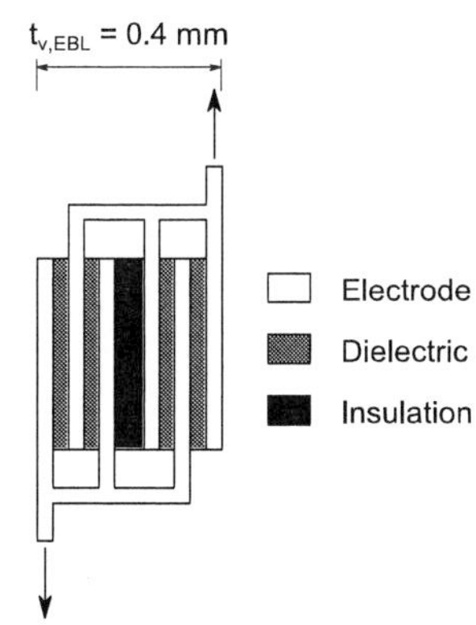

Figure 9. Schematic of an example of a parallel EBL arrangement

3.1 Sizing

If findings of general significance are of interest, the considered configurations of airfoil structures have to be comparable among each other and also with respect to designs evolving under real applicational conditions. This is ensured by applying the same representative sizing process to all configurations. In this procedure, which is visualized in figure 10, the actuator load is always equal to 80% of the critical load related to the first buckling mode, so that the structure's minimum buckling factor (ratio of buckling load to actuation load) is S = 1.25. Under this condition, wall thicknesses are then varied until the airfoil exhibits a cambering angle of $\varphi = 2°$ (defined, as evident from figure 2, as the angle between the airfoil's chord lines in the undeformed and in the cambered state). While the aerodynamic loading is held constant according to the aforementioned conditions, the actuator loads, on the other hand, vary in magnitude due to their dependence on the buckling stability.

In order to yield conservative results, the sizing procedure is performed in the more critical state of low stiffness.

With regard to a sufficiently wide design space, the ratio of wall thicknesses of skin and internal structure $\Theta = t_s/t_i$ is considered as additional degree of freedom.

It might appear counterintuitive that under these conditions an increase in wall thickness of the airfoil results in higher cambering deformation, which is caused by the coupling of the actuator load to the buckling load. Naturally, this practice does not produce especially compliant configurations, but for reasons that are further explicated in section 6 it can be considered to ensure the generation of comparable and realistic designs. In particular, it will turn out that the proposed variable-stiffness approach makes designs possible that would not be feasible by conventional means due to the stability requirements of the considered design problem.

Due to the completely opened skin in the configurations with EBLs in the soft state of the airfoil, their cambering compliance is substantially higher than the one of the airfoils based on the thermo-mechanical concept, for which the variable-stiffness interface is still closed by softened material. The discussed sizing process would therefore result in very thin skins and substructures for EBL-featured airfoils, leading to raised local deformation of the skin under aerodynamic loading. In order to counteract this effect, the smallest buckling factor has been required to be $S = 2$ for all configurations with EBL.

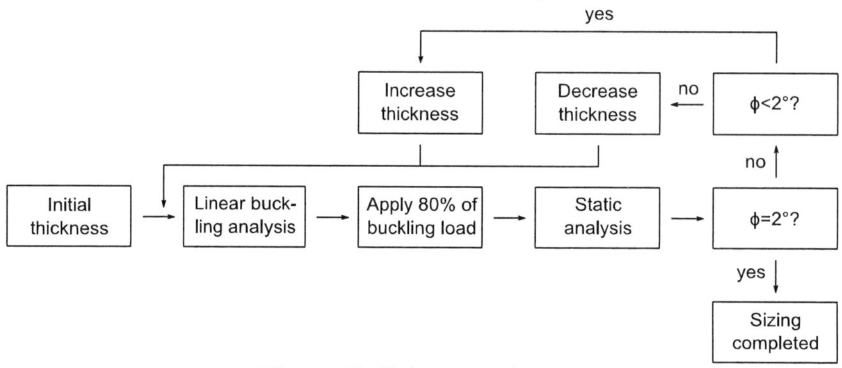

Figure 10. Sizing procedure

3.2 Mass Balance

The impact of the integration of variable-stiffness elements on the structural mass of the morphing airfoil can be decisive for the potential of future application of this design concept. Here, potential weight savings on the structure and on the actuators owing to the decrease in actuation energy are confronted with additional mass associated with the adaptive interfaces and with the infrastructure required for their energy supply and control.

Heating elements for the polymer layers can in principle be directly integrated as load-carrying structural members [14], so that no mass addition has been considered. Concerning the mass estimation of the EBL's electrodes, the same density as the one of the aluminium structure is assigned to these components.

3.3 Actuator Work

Following the main goal of the contribution on hand, the effectivity of the designs based on a variable-stiffness skin in changing their properties between a stiff and a compliant configuration has to be evaluated. As a representative performance measure in this sense, the work that has to be exerted by the actuators in order to achieve a certain cambering angle is extracted from the finite element models and compared for different designs. A linear force-displacement relation is used for this purpose, which has been confirmed to be sufficiently accurate by comparison with iterative analyses.

3.4 Energy Balance

In order to assess the potential of the variable-stiffness skin concept, an overall energy balance has to be performed, taking into account both the savings in actuator energy resulting from the increased compliance during morphing and the energy expenses related to the activation of the variable-stiffness interfaces.

Concerning the estimation of the latter, the following assumptions have been made: For the activation energy of the thermo-elastic implementation of variable-stiffness elements, on the one hand, only the heat related to the respective temperature changes of the polymer layers has been considered, corresponding to a system without thermal losses which would be approached by a design with very good thermal insulation of the adaptive layers. For this quantification, the specific heat (C_p) data of figure 8[‡] has been used and the density specified in table 2 has been assumed. The energy demand of the electro-elastic interfaces, on the other hand, is supposed to match its capacitive electrical energy, thus again assuming a

[‡] This temperature plot has been determined by means of differential scanning calorimetry (DSC) using a heating rate of 10 K/min.

system without losses, in this case of electric nature and mostly related to leakage current through the dielectric [6]. While these simplifications are not appropriate for a detailed energetic investigation of adaptive interfaces applied in a specific design, they permit an efficient general estimation of the order of the required energy, which is regarded as adequate for the conceptual conclusions aimed at by this work.

Based on this, the relevant expressions for thermal and electrical energy are given by $Q_{\text{th}} = m_{\text{P}} \int_{T_1}^{T_2} C_{\text{p}}\, dT$, m_P representing the mass of the polymer layers, and $Q_{\text{el}} = \frac{1}{2}\varepsilon_0 \varepsilon_{\text{r}} \left(\frac{U}{t_{\text{D}}}\right)^2 V_{\text{D}}$, respectively, ε_0 denoting vacuum permittivity, ε_r the dielectric's relative permittivity, U the voltage and V_D the volume of a single dielectric layer. In the latter formula, Q_{el} represents the electrical energy stored in one dielectric layer.

3.5 Aerodynamic Considerations

As the last of the important requirements which can decide about the concept's eligibility already at the present state, the aerodynamic properties of the airfoils with variable-stiffness skin under the conditions specified in the first part of section 3 are investigated qualitatively by means of the airfoil theory program "XFOIL", focusing on the aforementioned scope of this work. For this purpose, the geometries of the different configurations in their deformed state under combined aerodynamic and actuation loading as determined by structural finite element analyses have been handed over to "XFOIL" for aerodynamic characterization.

3.6 Simulation of Experimental Airfoil Structure

In addition to the purely numerical studies described in the previous subsections, the experimental electromechanical wing structure discussed in section 4 has been simulated with finite elements, using the properties given in tables 1 and 3[§]. Whereas the numerical study is based on linear simulations, the experimental structure has been simulated nonlinearly. As a fully coupled electromechanical modelling would imply a huge numerical effort, only the limiting cases of zero voltage and sufficiently high voltage have been considered and represented in the simulation by an ideally open and an ideally closed interface, respectively. In contrast to the model used for the purely numerical studies, the EBL interface of the experimental airfoil is composed of only two electrodes and one dielectric layer.

The purpose of the experimental airfoil structure is to demonstrate the effectiveness of the structural concept in static mechanical tests. No aerodynamic loads are hence applied in the respective simulation and therefore the sizing procedure used for the numerical studies was not followed for this simulation.

4. EXPERIMENT

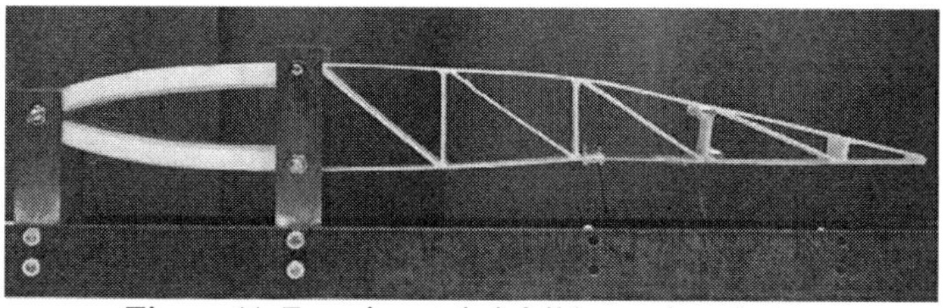

Figure 11. Experimental airfoil in cambered state

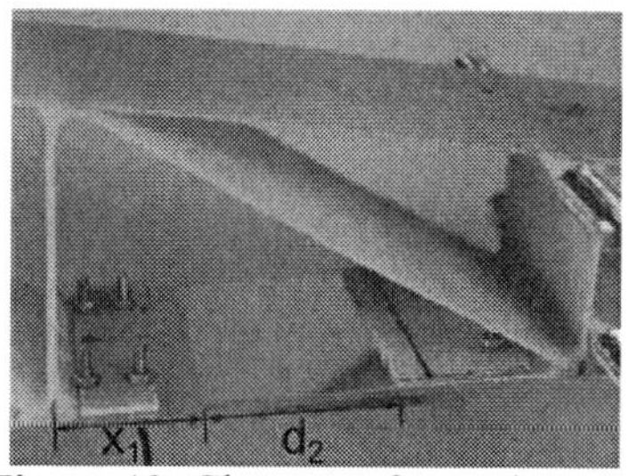

Figure 12. Close-up of experimental variable-stiffness interface

In order to validate the numerical results, an experimental airfoil structure with substructure 2 and electromechanical variable-stiffness element in the skin has been manufactured and characterized by static mechanical testing.

The geometric and material properties of the experimental structure made of GFRP have already been presented in section 3 and a photograph of this experimental structure attached to the test rig and set to a

[§] For a complete description of the properties of the experimental airfoil see also section 4.

morphed configuration is shown in figure 11. As it can be seen, the infinitely rigid leading edge considered in the simulation has been approached by a sandwich design using a foam core in this region, so that the experimental structure's leading edge is much stiffer than the rest of the airfoil. A detailed image of the EBL interface is given in figure 12, from which the EBL configuration with two steel electrodes and one dielectric layer made of polyimide is evident.

For a mechanical characterization, the leading edge of the experimental airfoil is bolted to a steel test rig placed in a Zwick/Roell Z005 tensile tester with 100 N load cell. Controlled forces in positive z-direction (see figure 6 for the definition of the coordinate axes) are then introduced in the skin at the location of the rearmost rib (distributed over the entire depth of the airfoil) and the z-wise deflection of the structure is recorded by a Micro-Epsilon ILD1700 laser triangulation sensor on the skin at the chordwise location of the second rearmost vertical rib.

5. RESULTS

5.1 Sizing

Figure 13 shows the influence of the thickness ratio Θ on the cambering angle for substructure 1 in the baseline configuration 000 without adaptive stiffness. Each curve in these plots is related to a certain structural mass. The respective curves representing the designs that evolve under the conditions of the specified sizing procedure are highlighted. For substructure 2, very similar curves evolve and the corresponding plots for the actuator force show the same qualitative trend as the one presented for the cambering angle.

The effect of an increase in flexural stiffness EI of skin and internal structure on the skin thickness resulting from the sizing process is presented for the example of configuration 000 in figure 14, where EI_0 represents the bending stiffness in the original situation corresponding to an isotropic and homogeneous beam with rectangular cross section. Thickness ratios of $\Theta = 1.00$ (substructure 1) and $\Theta = 0.85$ (substructure 2), which have been shown to lead to the highest respective cambering angles by the previous investigation, have been assumed for these and all the following analyses. Here, the actuator force shows the same qualitative behaviour as the presented skin thickness.

Figure 15 contains the values of skin thickness of the designs obtained by the discussed sizing procedure for a stiffness ratio of 20. The electro-elastic concept (EBL) for variable skin stiffness has only been investigated for configuration 001 and substructure 1.

Further aspects that have to be considered for the proof of concept of the variable-stiffness skin are constituted, first, by the maximum strain in the thermally-activated polymer layers in their soft state and, second, by the maximum stress in the closed EBL interfaces when the airfoil is subjected to elastic restoring loads and aerodynamic forces in the cambered state. Concerning the first point, maximum shear strain values ε_{xy} of the order of 100% are observed in the shell elements representing the polymer layer. Referring to the second issue, the average shear stress of all elements of the EBL interface in the critical state amounts to $\tau_{xy} = 0.62$ MPa. The maximum shear stress (averaged over the area) which could be transferred in recent lap shear experiments on EBLs with multilayer dielectrics at a voltage of 5 kV is around 0.2 MPa (cf. [7]), so that it is shown by the result above that an EBL with four interfaces in parallel, as assumed for the simulation, in principle provides the required load-carrying capability.

5.2 Mass Balance

Figure 16 presents the structural mass of the designs with and without variable stiffness that have been reported in the previous paragraph.

5.3 Actuator Work

Actuator force and work of all considered configurations are given in figure 17. The figure contains the results of both thermal and EBL variable-stiffness representations, where the latter has only been evaluated for substructure 1 and configuration 001, as mentioned before.

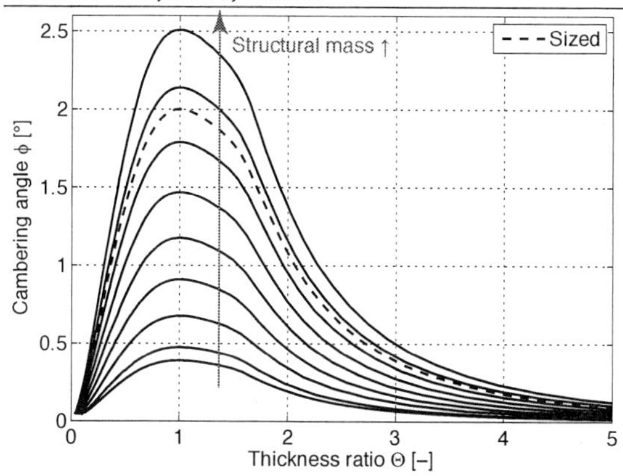

Figure 13. Cambering angle with respect to thickness ratio (array parameter: structural mass between 363 g and 1000 g, steps of 91 g between solid lines) in configuration 000 with substructure 1

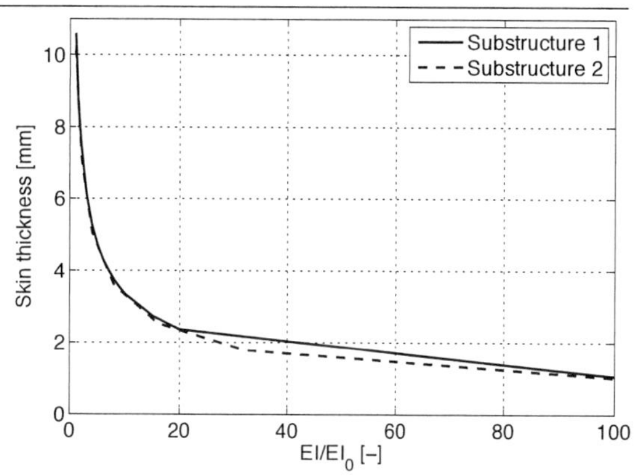

Figure 14. Skin thickness with respect to stiffness ratio for configuration 000

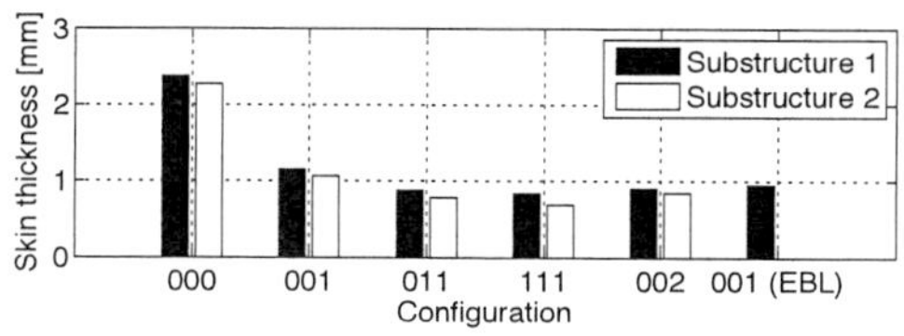

Figure 15. Skin thickness of the investigated airfoil configurations. Stiffness ratio $EI/EI_0 = 20$

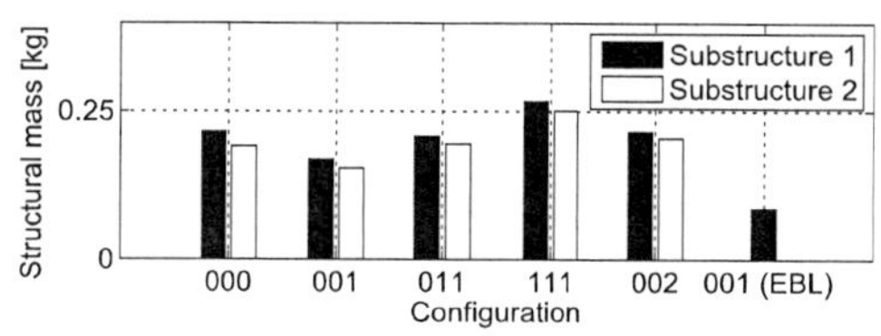

Figure 16. Structural mass for $EI/EI_0 = 20$

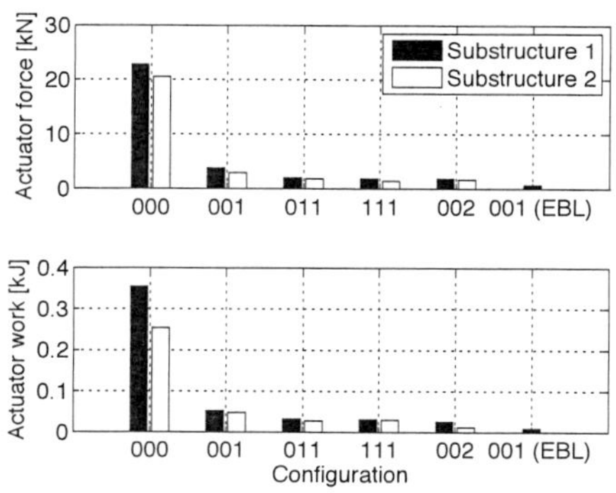

Figure 17. Actuator force and work for $EI/EI_0 = 20$

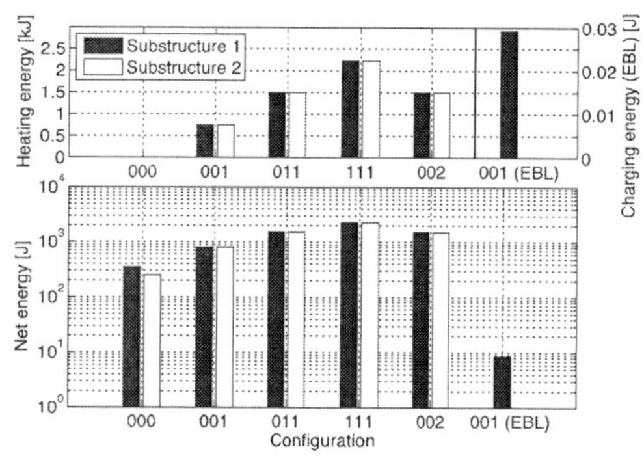

Figure 18. Energy balance accounting for actuator work (see figure 17) and heating/charging energy, $EI/EI_0 = 20$

5.4 Energy Balance

The results of the balance of the energies required for actuation and stiffness change (heating energy for the thermal and charging energy for the electrostatic concept) are shown in figure 18 for the different configurations, again considering a stiffness ratio of 20. Also here, the EBL results are limited to the conditions mentioned in the previous paragraph.

5.5 Aerodynamic Considerations

The pressure distributions of the airfoil with baseline configuration 000 and of the most compliant one with a variable stiffness skin featuring EBL are contrasted in figure 19, which also contains the cambered geometry of the EBL design. Table 2 summarizes the values of lift coefficient of all analyzed airfoils with substructure 1** in the cambered state. In the unmorphed situation, the lift coefficient amounts to c_l = 0.603. All aerodynamic results are based on inviscid calculations and an angle of attack of $\alpha = 5°$.

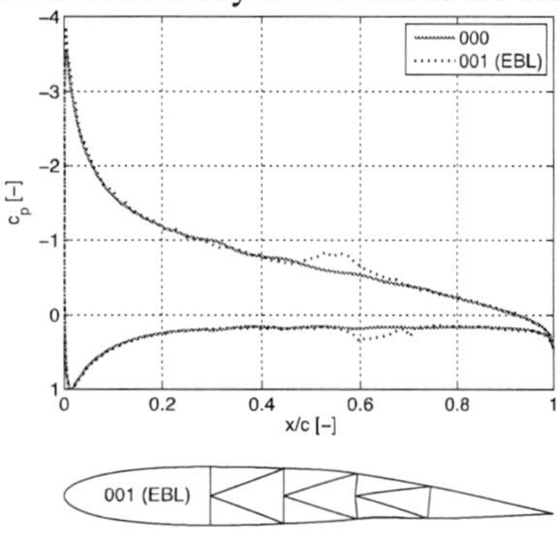

Figure 19. Pressure coefficients of baseline and variable-stiffness airfoil in the cambered state and deformed shape of the latter

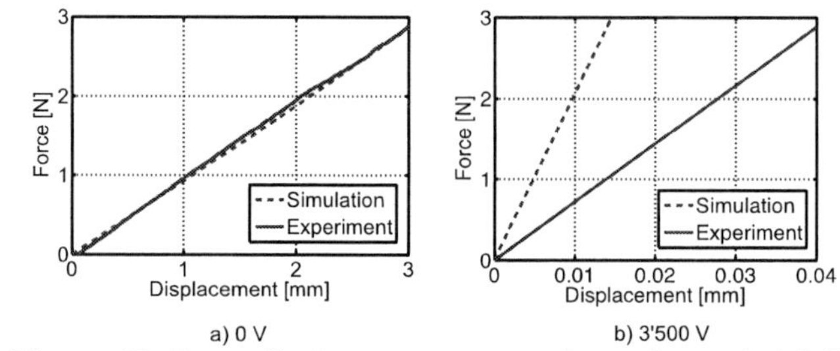

Figure 20. Force-displacement curves of experimental airfoil structure at voltages of 0 V and 3'500 V. Comparison between experiment and simulation

Table 2. Values of lift coefficient for different configurations with substructure 1 in the cambered state

Configuration	000	001	011	111	002	001 (EBL)
Lift coefficient c_l [–]	1.0175	1.0294	1.0172	1.0033	1.0405	1.0725

5.6 Experimental Results

Figure 20 shows the force-displacement curves of the experimental airfoil structure for the considered cambering loading starting from the uncambered configuration. The results are shown for interface voltages of zero and 3'500 V and the measurement data are confronted with the simulation results representing open and closed state, respectively.

6. DISCUSSION

As illustrated by the parametric study related to the structural sizing of the baseline configuration without adaptive stiffness, the thickness ratio Θ has a substantial influence on the compliance of the resulting designs: Θ values considerably smaller and bigger than one lead to low buckling loads due to small thickness of the skin and of the internal structure, respectively, and thus to lower actuation forces and cambering angles. For both types of substructure, the thickness ratios for which actuator force and cambering angle reach their maximum are relatively close to one.

It is obvious that in the applied stability-driven sizing process increased stiffness ratios EI/EI_0 permit lower wall thicknesses for both kinds of internal structure. On the other hand, the decoupling of flexural and axial stiffness is limited in reality by available geometries and materials. However, the elevated value of $EI/EI_0 = 20$ which most of the presented results are based upon can be put into practice by application of, for example, a sandwich structure. At this point, it should also be stated that the sizing procedure chosen for this work is regarded as realistic in terms of transferability of the considered two-dimensional airfoil sections to a wing with spanwise extension, as especially thin-walled designs that might be dimensioned sufficiently for the two-dimensional case, but tend to be undersized for requirements of the

** For substructure 2, similar values can be expected.

three-dimensional design problem like spanwise bending stiffness and buckling stability are penalized by the stability criterion applied.

The pronounced shear strain in the elastomer layers of the thermal solution might pose demands on the material selection and the detailed design of a variable-stiffness airfoil for a certain application, which are beyond the scope of this conceptual work. The EBL solution, on the other hand, has been shown to comply with the strength requirement in the critical state in section 5.1.

Under the considered conditions, all airfoils with variable-stiffness skin exhibit substantially lower wall thicknesses than the baseline design 000 and this effect is even more pronounced for a stiffness ratio of 1, for which the results have not been shown. Correspondingly, actuator force and work are largely reduced in the configurations with adaptive skin rigidity. Due to the high amounts of energy that have to be spent for heating the polymer, however, no positive energy balance of the thermo-mechanical variable-stiffness solution can be achieved for $EI/EI_0 = 20$. This is different for a stiffness ratio of 1, where for configuration 001 the net energy can be lowered by 48% (substructure 1) and 29% (substructure 2) with respect to the baseline configuration. The EBL, in contrast, provides a considerable energy saving of 97% for $EI/EI_0 = 20$ due to its low energy demand.

The mass balance reveals that for the higher stiffness ratio of 20 only the EBL configuration exhibits a considerable mass benefit, due to the lower thickness of the reference configuration in this case. At a stiffness ratio of 1, on the other hand, also the thermomechanical designs permit mass savings of the order of 60%, resulting from the reductions in wall thickness.

A comparison of the wall thicknesses shows that substructure 2 is characterized by a higher cambering compliance than substructure 1, especially for the variable-stiffness configurations. This is due to the fact that, in the state of low stiffness, in case of substructure 2 each of the airfoil's cells mainly rotates about one compliant hinge, while this degree of freedom is more constrained in case of substructure 1 due to the additional nodes inside the structure. For configuration 111 the difference between the two kinds of internal structures is thus particularly big because of the higher stiffness contribution of the rigid leading edge in case of substructure 1.

It should be added here that the energy balance can be improved by targeted selection or design of a polymer material with more favourable heat characteristics, but, on the other hand, thermal losses as neglected by the energy balance performed in this work cannot be avoided completely in an application. Moreover, it has to be noted that the sizing procedure followed in this investigation leads to relatively thick-walled designs of the baseline configuration and thus to a high actuation energy in this case, so that the benefit of the variable-stiffness skin might be bigger here than under different conditions. However, unrealistically compliant solutions can be excluded this way, as explained before.

While variations in wall thicknesses have been considered in the mass balance, downsizing of the actuators resulting from lower activation forces and savings in system weight due to amplification effects on various components have not been taken into account, but can be additionally expected to contribute substantially to the benefit of the variable-stiffness airfoils, if these are integrated in an aircraft system. On the other hand, certain weight additions come along with the power supply infrastructure required for the variable-stiffness interfaces. Based on previous experience, however, this influence is deemed to be comparably low, especially for the electrostatic concept.

The weight which has to be added for actuating the morphing airfoil largely depends on the type and integration of actuators, which both go beyond the scope of this work. If, though, a very general relation between actuator force and mass as proposed in [13] is assumed, the actuators of the configurations without variable stiffness turn out to be much too heavy for a feasible implementation. Under this condition, the novel variable-stiffness design thus permits to enter a design space that is inaccessible with conventional methods.

As indicated by the results, the higher compliance of the variable-stiffness skin in its soft state leads to more pronounced local bending of the structural elements and, compared to the baseline configuration, to a less smooth deformation. Accordingly, based on the elementary aerodynamic model applied, higher gradients of the pressure coefficient are observed at some locations. The resulting lift coefficients are still very similar to the ones of the baseline configuration without stiffness adaptivity, so that the morphing performance, the variation in c_l between about 0.6 and 1.0, is not affected considerably by the presented skin concept. In this context, it should be recalled that the wall thickness is constant for the whole substructure, as well as for the whole skin. It is considered as a task of structural optimization, which is not included in the present general investigation, to locally vary the single elements' thicknesses (and/or material properties) and possibly to adjust the EBL composition accordingly in favour of a more continuous deformation.

The mechanical testing shows that the experimental airfoil structure permits changes in cambering stiffness by a factor of more than seventy between the states of zero voltage and 3'500 V. Under the latter condition, the experimental structure can hold camber deformations of at least ca. 2°, showing for the design point of 2° camber angle the capability of the simulated parallel EBL arrangement of sustaining not only the elastic restoring load, but also additional loads like aerodynamic forces. It has to be mentioned, however, that the elastic properties of the experimental structure are not directly comparable to the ones of the structures investigated for the numerical study, since different sizing procedures have been applied.

At zero voltage, the measured stiffness of the experimental airfoil structure agrees very well with the simulated one, while at 3'500 V large deviations of the numerical results from the experimental ones are observed. This indicates that, first, the EBL is far from completely closing the interface and, second, the airfoil's cambering stiffness is very sensitive to this influence. The fact that no substantial nonlinearities in terms of force-displacement relation have been observed in the mechanical response of the experimental structure gives rise to the explanation that the parts of the interface which are being closed by voltage application stay in the regime of static friction in the range of forces considered, but a certain part of the interface, most likely in the region of the rear edge, remains always open and causes the discrepancy from the conditions assumed for the simulation.

7. CONCLUSION

The concept of a variable-stiffness skin in a camber-morphing airfoil has been shown to be effective by simulation and experiment. The performed numerical studies demonstrate that the additional design parameter of variable skin stiffness permits to access design spaces that remain closed to structures with time-constant stiffness. For the configurations considered in these simulations, adaptive skin rigidity leads to substantially increased efficiency, enabling reductions in actuation energy by up to 97% and considerable savings in structural weight of around 60%.

Whereas the benefit of a thermo-mechanical design of the variable-stiffness skin is basically limited to the mass balance for the lower one of the two considered stiffness ratios due to the considerable energy expenses related to the required temperature changes, the high energy efficiency of electro-bonded laminates (EBLs) permits large energy savings with respect to the reference configuration without variable stiffness. In addition, the general advantage of the electro-mechanical concept over the thermo-mechanical one in terms of activation speed, which can be decisive in many applications, has to be stressed.

Besides the numerical studies, an experimental airfoil structure with variable-stiffness skin based on an electro-bonded laminate has been manufactured and tested. The measurements reflect, at least qualitatively, the respective results of the simulation and demonstrate that the cambering stiffness of the experimental airfoil can be varied by a factor of about seventy.

The findings of this work lay the foundation for the application of the idea in morphing wings. A higher-fidelity, electro-mechanically coupled numerical modelling of the EBLs, a complete experimental validation of the presented energy balance and a more detailed aerodynamic model are considered to be the most important following steps that have to be approached in order to bring the concept forward to higher maturity.

References

1. Amprikidis, M. and Cooper, J. E., "Experimental validation of wing twist control using adaptive internal structures," *45th AIAA/ASME/ASCE/AHS/ASC Structures, Structural Dynamics & Materials Conference*, 2004.
2. Barbarino, S., Bilgen, O., Ajaj, R. M., Friswell, M. I., Inman, D. J., „A Review of Morphing Aircraft," *Journal of Intelligent Material Systems and Structures*, 2011, Vol. 22, No. 9, pp. 823–877.
3. Campanile, L. F., "Smart shape control: using compliant and active materials to adapt structural geometry. Challenges and good reasons," *14th International Conference on Adaptive Structures and Technologies (ICAST)*, 2003.
4. Campanile, L. F., "Lightweight Shape-Adaptable Airfoils: A New Challenge for an Old Dream," in Wagg, D., Bond, I., Weaver, P., Friswell, M. (ed.s), "Adaptive Structures: Engineering Applications," John Wiley & Sons, 2007, pp. 89–135.
5. Chen, P. C., Sarhaddi, D., Jha, R., Liu, D. D., Griffin, K., Yurkovich, R., "Variable Stiffness Spar Approach for Aircraft Maneuver Enhancement Using ASTROS," *Journal of Aircraft*, 2000, Vol. 37, No. 5, pp. 865–871.
6. Di Lillo, L., Carnelli, D. A., Bergamini, A., Busato, S., Ermanni, P., "Quasi-static electric properties of insulating polymers at a high voltage for electro-bonded laminates," *Smart Materials and Structures*, 2011, Vol. 20, No. 5, p. 057002.
7. Di Lillo, L., Raither, W., Bergamini, A., Zündel, M., Ermanni, P., "Tuning the mechanical behaviour of structural elements by electric fields," *Applied Physics Letters*, 2013, Vol. 102, No. 22, p. 224106.
8. Gandhi, F., Kang, S.-G., "Beams with controllable flexural stiffness," *Smart Materials and Structures*, 2007, Vol. 16, No. 4, pp. 1179–1184.
9. Gandhi, F., Anusonti-Inthra, P., "Skin design studies for variable camber morphing airfoils," *Smart Materials and Structures*, 2008, Vol. 17, No. 1, p. 015025.
10. Gordon, B. O., Clark, W. W., "Morphing Structures by way of Stiffness Variations," *48th AIAA/ASME/ASCE/AHS/ASC Structures, Structural Dynamics, and Materials Conference*, 2007.
11. Griffin, K. E., Hopkins, M. A., "Smart stiffness for improved roll control," *AIAA 36th Structures, Structural Dynamics and Materials Conference*, 1995.
12. Kuder, I. K., Arrieta, A. F., Raither, W. E., Ermanni, P., "Variable stiffness material and structural concepts for morphing applications," *Progress in Aerospace Sciences*, in press.
13. Marden, J. H., "Scaling of maximum net force output by motors used for locomotion," *Journal of Experimental Biology*, 2005, Vol. 208, No. 9, pp. 1653–1664.
14. Raither, W., Heymanns, M., Bergamini, A., Ermanni, P., "Morphing wing structure with controllable twist based on adaptive bending-twist coupling," *Smart Materials and Structures*, 2013, Vol. 22, No. 6, p. 065017.
15. Siler, D. J., Demoret, K. B. J., "Variable stiffness mechanisms with SMA actuators," *Proc. SPIE 2721, Smart Structures and Materials 1996: Industrial and Commercial Applications of Smart Structures Technologies*, 1996.
16. Sofla, A. Y. N., Meguid, S. A., Tan, K. T., Yeo, W. K., "Shape morphing of aircraft wing: Status and challenges," *Materials and Design*, 2010, Vol. 31, pp. 1284–1292.
17. Vos, R., "Mechanics and Applications of Pressure Adaptive Honeycomb.," PhD thesis, University of Kansas, 2009.

Embeddable Variable Stiffness Multi-stable Composites

A. F. Arrieta[1*], I. K. Kuder[1], T. Waeber[1] and P. Ermanni[1]
[1] **Centre of Structure Technologies, ETH Zurich, Leonhardstrasse 27, Zurich, CH-8092 Switzerland[1]**

Abstract

The possibility to achieve shape adaptation gives structural systems the potential to adjust for optimal operation in a wide range of conditions. Yet, several challenges remain to be addressed for the implementation of this promising technology. In particular, the conflicting requirements posed by shape adaptation for achieving compliance, while maintaining sufficient structural strength and light-weight continue to be unresolved. The formidable challenges posed by shape adaptation, and in particular in morphing applications, can be potentially addressed through the development of distributed compliance systems featuring highly directional stiffness properties. These characteristics can be further enhanced by embedding in such systems elements featuring variable stiffness. In this paper, a novel type of multi-stable composites capable of having two opposing edges restricted is presented. Multi-stability arising from residual thermally induced stresses due to unsymmetrical lamination is used in the present study. A tailored lay-up exhibiting regions of unsymmetric and symmetric lamination is implemented to achieve multi-stability even when restricting two opposing edges. The design space of the achievable statically stable states for the embeddable multi-stable composites is explored by varying dimensional parameters using the Finite Element Method (FEM). In addition, the difference between the structural responses of the obtained multiple stable shapes is numerically investigated. The restoring force for each stable state is examined yielding a significant variability in the stiffness. Experimental specimens are manufactured showing good agreement with the numerical results, hence validating the model.

1 INTRODUCTION

The ability to operate optimally in a diverse range of conditions has resulted in the idea of adapting the properties of systems, resulting in the field of adaptive structures. A particularly difficult type of property adaptation are those concerned with shape variation of structural elements. This problem is exemplified by the concept of conformally adapting aerospace structures which is widely known as morphing. Morphing requires structures exhibiting highly directional properties to address the conflicting requirements of high load-carrying capabilities along the direction of the loads, high compliance in the directions of shape adaptation, while maintaining light-weight characteristics. Distributed compliance structural systems are a promising solution to the formidable challenges presented by shape adaptation requirements, particularly those posed by morphing applications. A new trend to significantly augment the structural directionality of such systems is to embed elements capable of varying their stiffness [1]. The ability to change the structural response of certain components in distributed compliance systems allows for a higher degree of decoupling between the

[1*] Corresponding author, e-mail, andresar@ethz.ch

compliance and strength requirements. Composite laminates exhibiting multiple statically stable shapes are good candidates to serve as variable stiffness elements embedded within distributed compliance structures. Each statically stable shape exhibits significantly different directional structural response. Multi-stable structures have been up until now the focus of extensive research due to their capabilities to exhibit several statically stable states. Several types of systems including biological [2], and synthetic [3–6] exhibit multi-stability as a means to achieve unattainable performances with conventional structures. Amongst multi-stable structures, composite materials tailored to exhibit large differences between their statically stable shapes configurations have been considered for morphing applications [7–10]. Although such structures featured promising results, the discrete nature of the achievable deflections shapes associated to multi-stability restricts the range of applicability of multi-stable structures as a means to obtain large deflections in load carrying elements. The possibility offered by multi-stable structures to tailor different structural properties for each of the attainable stable configuration of a component allows for the realisation of load-carrying variable stiffness elements. Exploiting this characteristic multi-stable elements can be designed to exhibit variable stiffness properties arising from the different structural response associated to the stable configurations. Furthermore, as multi-stability can be achieved through passive means, the capability to show significant variation in the structural response can be realised in a robust compact manner.

In this paper a novel class of multi-stable composites featuring significantly different variable stiffness characteristics associated to each of the stable configurations is introduced. The multi-stable components herein presented are specifically designed to be embeddable into other structural systems, thus providing the complete system with passive variable stiffness capabilities. In particular, the designed multi-stable laminates allow for integration into a structure without fixation parts, which is particularly suitable to enhance the structural directional response of distributed compliance systems through the added capability of stiffness adaptation. A tailored lay-out featuring several regions with different stacking sequences is designed achieving both a significant difference in the structural response and the capability to embed the components into a wider system. Thermal stresses induced during the cool-down process arising from regions of unsymmetric lamination are exploited to realise the desired multi-stable behaviour in the composite structural elements. The FEM is employed to simulate and design the variability of the stiffness characteristics associated to each stable configuration. The room temperature shapes of the designed multi-stable components predicted with the FEM are in good agreement with manufactured specimens. The structural response of the stable configurations is studied both numerically and experimentally by axially loading tests multi-stable specimens. Numerical results show significant variability between the stiffness featured by each stable configuration of the designed laminates. Tests on experimental specimens showing a close match with the FEM results validate the used models. The realisation of purely elastic, passive, embeddable load carrying variable stiffness elements can be potentially used to significantly alter the structural response of a load-carrying system in a simple and robust fashion. The variable stiffness offered by the presented multi-stable components can be exploited to enhance the directionality of the structural characteristics of systems for which structural efficiency is paramount importance, as for instance in morphing structures. This can result in a reduction of the actuation requirements to achieve large deflections, or in altogether passive morphing exploit the external forces on the structure to induce desired shape adaptation. Furthermore, the stiffness variability can be implemented on applications where structural hysteric cycles are required as for instance in energy dissipation applications.

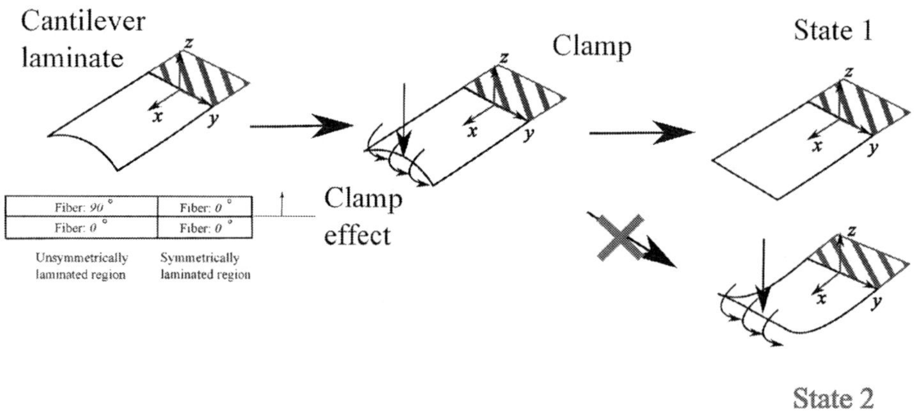

Figure 1: Effect of clamping the free end of a cantilevered bi-stable laminate configuration.

2 LAY-OUT DESIGN

The lay-outs for the multi-stable embeddable variable stiffness elements are designed mainly to fulfil two derised characteristics. First, the resulting structural behaviour of the obtained composite components should be so that significant stiffness variability is attained. Equally important, the resulting lay-out should allow for the designed components to be integrated into a wider structure without compromising the obtained variable stiffness characteristics. Furthermore, the integration should be realised such that a distributed compliance structure requiring no conventional fixation components, such as rivets or bolts, is achieved.

Previously, multi-stable laminates in a cantilevered configuration, i.e. with one edge completely clamped, have been obtained using a symmetric and unsymmetric stacking sequences [11, 12]. However, the possibility of embedding a laminate by connecting through clamping two ends into a larger structure was not achieved. Simply clamping a curved end of the laminate results in a moment being applied to the laminate restricting the deflection required to achieve the second configuration rendering it mono-stable, as schematically shown in Fig. 1. Therefore, a smooth transition reducing the curvature of the edge to be clamped is necessary on the resulting shape for all the attainable shapes of the multi-stable component. Starting from a rectangular bi-stable region, appropriate elastic boundaries allowing for a smooth reduction of the curvatures in both stable states are added to achieve the objective of embedding of the component. Adding symmetrically laminated regions to the short ends (parallel to the y-axis) of the rectangular unsymmetric region reduces smoothly the curvature along the short edges. However, these regions impose high bending moment opposing the straightening of the edges impeding the adoption of stable state 2 as shown in Fig. 1, reducing dramatically the parameter space for which such the complex lay-out would exhibit bi-stability. In fact, only a highly directional symmetric part being compliant in the y-direction of the short edge, while maintaining sufficient stiffness to smoothen the curvature in the x-direction could result in a multi-stable element. However, the mechanical properties of such a solution would not be satisfactory for load carrying purposes. An alternative for achieving the design objectives to simply adding symmetric regions on either side of the main rectangular unsymmetric region is achieved by inserting two additional small unsymmetric sections to the lay-out. These are designed to show curvatures in y-direction opposing the deformation in the (longitudinal) x-direction imposed by the main rectangular unsymmetric part smoothening adequately the behaviour of the edges at room

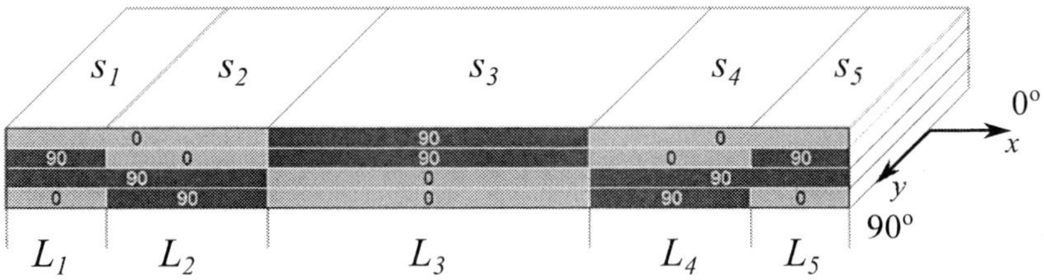

Figure 2: Tailored lay-out showing regions (s_i) with specific lay-ups of length L_i.

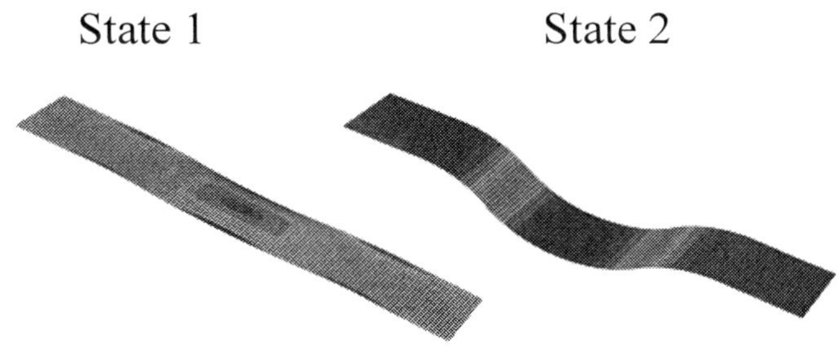

Figure 3: Example of stable shapes obtained with the introduced multi-region lay-out.

temperature. An example of the introduced five-section model is shown in Fig. 2. As described before, the unsymmetric sections s_2 and s_4 have the stacking sequence of the unsymmetric section s_3 mirrored about the mid-plane of the laminate.. The results from an FE model featuring the five-section lay-out show stable straight (state 1) and curved (state 2) shapes, as can be seen Fig. 3.

3 STATIC SHAPES

To study the capabilities of a multi-stable embeddable laminates to achieve purely elastic variable stiffness components, first the existence of multi-stability for a given lay-out design needs to be investigated. For the type of multi-stable laminates herein studied the multi-stability is mainly controlled by thermally induced stresses during the cool-down process from the curing temperature due unsymmetric lamination. Hence, the cool-down process for the designed lay-out featuring regions with different stacking sequences, i.e. the specific lay-up of the region, is simulated with FEM. Specifically, finite element analysis (FEA) performed in ABAQUS® including of nonlinear effects of large displacement is carried out for the cool-down process from the curing temperature in the autoclave. The S4R shell elements are used, the sides of which are chosen to be between 2.5 mm and 7.5 mm and the aspect ratio is kept between 0.5 and 2. The width and length are divided at least into 20 and 92 finite elements, respectively. To avoid rigid body displacement the edges of the laminate during cool-down are restricted from translation and rotation. The temperature field on the laminate during cool down is applied in to the FE model by setting ramp-like input from 140 to 0°C, simulating a curing temperature of 160°C and a room temperature of 20°C. The material properties used for the numerical and experimental specimens are given in Tab. 1. A slightly modified lay-out shown in Fig. 4 featuring 2 additional regions to ensure fibre continuity,

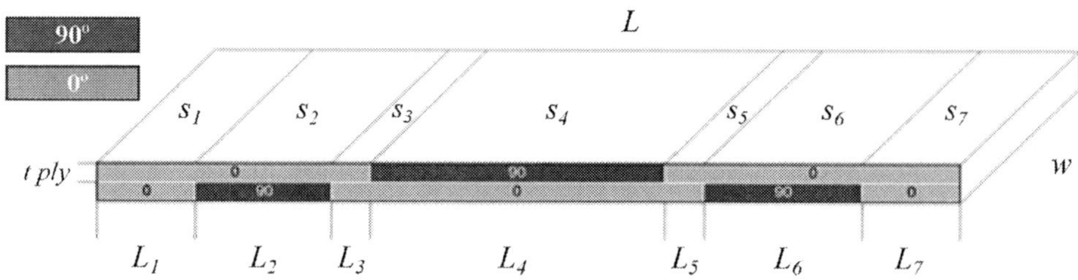

Figure 4: Tailored lay-out introducing fibre continuity regions for minimising failure at the interfaces of the specifically laminated regions.

thus reducing the possibility for failure at the interfaces is employed for all calculations hereafter. The geometrical properties of the numerically and experimentally specimens studied in this work are given in Tab. 2.

Fibre vol. [%]	E_{11} [GPa]	E_{11} [GPa]	G_{12} [GPa]	$\nu_{12} = \nu_{21}$ [-]	Density $[\frac{kg}{m^3}]$	α_{11} C^{-1}	α_{22} C^{-1}	α_{22}^{id} C^{-1}
60	161	10	4.4	0.3	1570	-1.8E-8	3E-5	2.5E-5

Table 1: Material properties for a typical ply of carbon fibre reinforced polymer prepreg used to manufacture the multi-stable experimental specimens.

The cool-down shapes of specimen 1 for the straight and curved states are shown in Figs. 5(a) and 5(b), respectively. It can be seen that a very good match between numerical and experimental results is obtained. Here an identification process of the effective thermal expansion coefficient given in Tab. 1, α_{22}^{id}, is conducted. Due to moisture pickup after the room temperature shapes are attained, the effect of the mismatch between the theoretical thermal expansion coefficients in the longitudinal and, α_{11} and transverse, α_{22}, directions, leading to the appearance of multi-stability in unsymmetrically laminated composites is diminished. This reduction is accounted for by lowering the used value of the theoretical thermal expansion coefficient of the matrix, α_{22}, in the FE calculation to fit the adopted shape. Such a procedure is more practical than including hygrothermal terms in the equations, due to the difficulty to obtain reliable data for the hygrothermal coefficient [13]. Furthermore, including hygrothermal terms in the FE calculations precisely results in a reduction of the effective thermal coefficient of the laminate. Notice that such a parameter identification method does not compromise the accuracy of the modelling as only one parameter is varied, while all other quantities involved in the calculations are taken from data sheets.

It is also interesting to notice from Fig. 5(a) that the straight configuration exhibits an angle between its edges and the central flat part of its geometry (x-axis). This angle is a critical parameter to be controlled to maintain the multi-stability when clamping the opposing short ends. In the curved configuration, a similar angle is described by the edges with the x-axis, however it is much less critical due to its small value as seen from Fig. 5(b). In the next section a specimen showing a reduced angle is studied in detail to investigate the stiffness variability offered by the introduced multi-stable composite components.

Specimen	L [mm]	w [mm]	L_1 [mm]	L_2 [mm]	L_3 [mm]	L_4 [mm]	L_5 [mm]	L_6 [mm]	L_7 [mm]
1	300	64	55	30	5.0	120	5.0	30	55
2	270	64	55	15	5.0	120	5.0	15	55

Table 2: Geometric properties of the regions composing the introduced lay-outs for the studied specimens. Refer to Fig. 4 for the schematic representation of the given parameters.

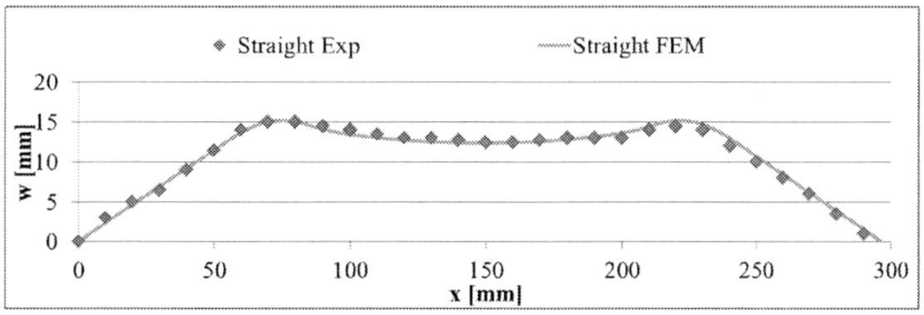

(a) State 1

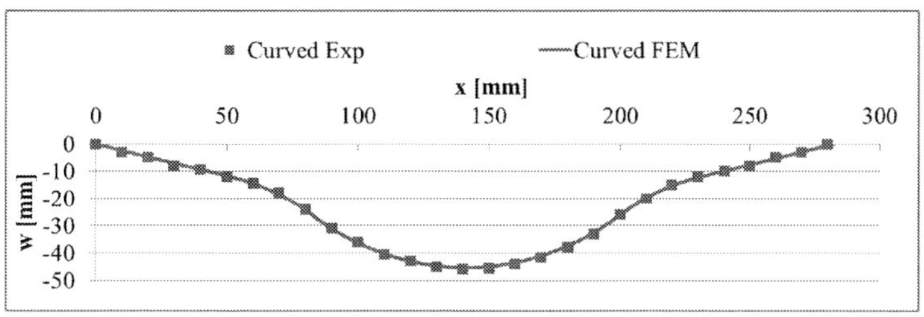

(b) State 2

Figure 5: Comparison between experimental and simulated shapes for embeddable multi-stable variable stiffness component, specimen 1. a) State 1 b) State 2.

4 STRUCTURAL VARIABLE STIFFNESS CHARACTERISTICS

To test the variability in the structural response offered by the designed embeddable multi-stable elements, numerical and experimental displacement controlled compression tests for both stable shapes are performed. In this study the focus is placed on the longitudinal stiffness of the components as this are envisioned to work mostly supporting in-plane forces when integrated into a larger structure. However, it should be noted that, although not shown here, considerable variability is also obtained for the bending stiffness of the designed laminates. The load case for testing the axial variability in the stiffness is schematically shown in Fig. 6 and the following boundary conditions are set. After the cool-down is finished, the x-coordinates of the nodes at the left edge of the laminates are fixed at their position simulating the effect of a clamp. The axial force is introduced to the laminate following a displacement controlled procedure. As the displacement is changed, the increments in the reaction forces, F_a, in x-direction of the nodes at left edge are stored and plotted in Fig. 7.

The test rig used for the experimental axial loading study showing the clamping jaws used to

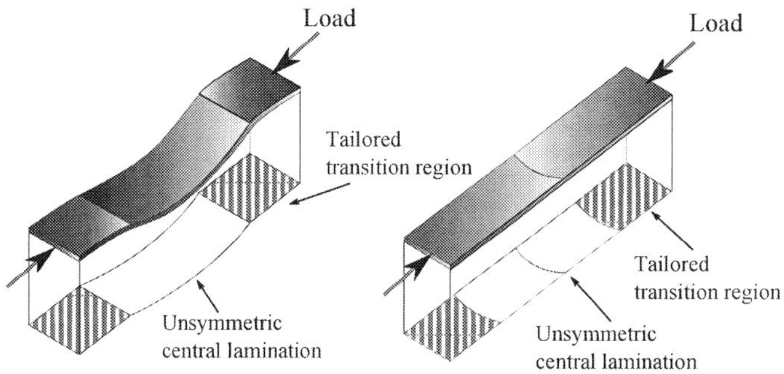

Figure 6: Schematical representation of the axial loading process applied to the multi-stable components on each stable state.

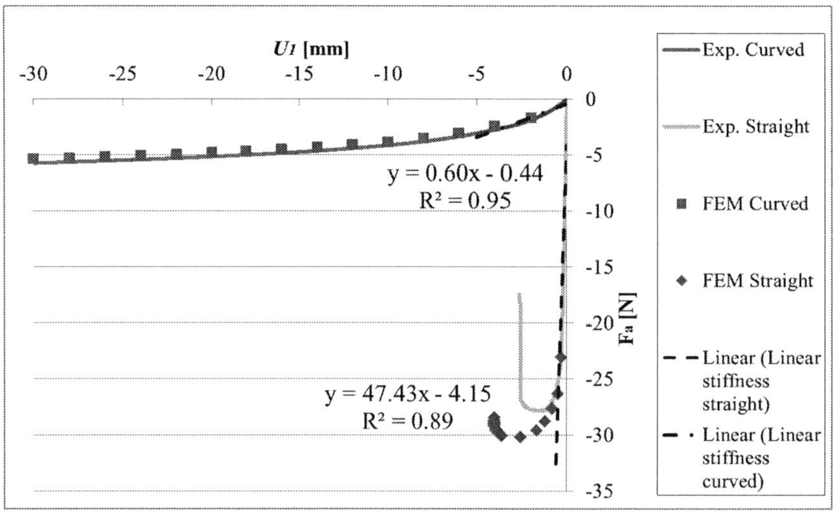

Figure 7: Numerical and experimental restoring force plot obtained from the axial loading test for both stable configurations. The linear regression lines obtained for a region of small displacements about the stable equilibrium shapes is added to highlight the obtained stiffness variability.

simulate the integration into a wider structure is shown in Fig. 8. A Zwick/Roell Z005 testing machine combined with a 100 N force transducer capable of maintaining high accuracy for loads of less than 1 N is used for the measurements. The test is conducted by controlling the displacement with a test speed of 20 mm/min. For the conducted axial tests, specimen 2 is used.

The static nonlinear FE results are in good agreement with the measured experimental data, as can be seen from in Fig. 7. The restoring force experimentally obtained from the axial test for both stable configurations is closely matched by the FE results. The conducted FE simulations also capture quite accurately the maximum force before cracks are heard for the axial loading of the component straight state. The deformed shapes of each stable states while being loaded exhibit very similar behaviour in the simulation and in the experiment as shown in Fig. 9(a). The stiffness variability of the designed component can be observed by conducting a linear regression about a small deflection range, 5 and 0.5 mm for the curved and straight states respectively. The slopes obtained from the linear regression lines associated to each stable configuration for the studied

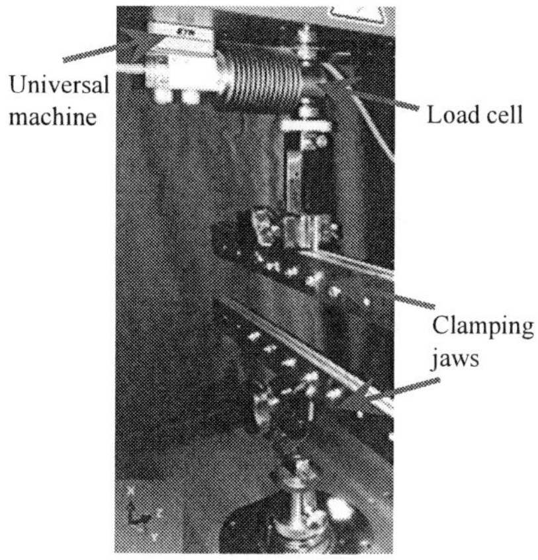

Figure 8: Test rig used for axial loading tests on multi-stable variable stiffness laminates.

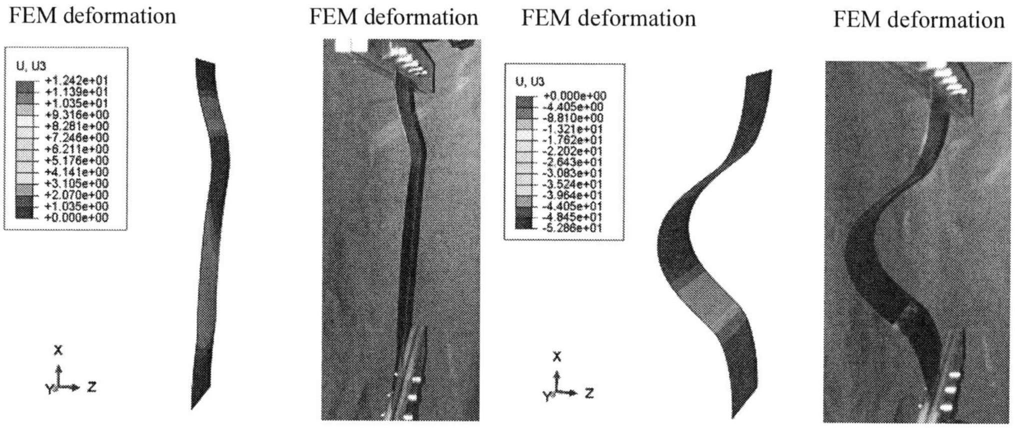

(a) State 1 (b) State 2

Figure 9: Comparison between experimental and simulated shapes at a deformed state during an axial loading test for a) State 1, and b) State 2.

component result in a ratio of $47.43/0.60 = 79.1$. Therefore, the linear stiffness variation exhibited by the component when changing from the straight to the curved state amounts to a nearly 80-fold variation. Studying the complete range for the restoring force shown in Fig. 7, it can be seen that the stiffness variation is amplified for larger deflections due to the softening of the axial stiffness for the curved state. It is worth noticing from the restoring force graph that the multi-stable embeddable laminate exhibits other desirable characteristics. Namely, the straight state is capable of carrying significant amounts of loads before failure, and the curved state provides the possibility of achieving large deflections of the component with a relatively low actuation requirement. These characteristics are interrelated to the already discussed stiffness variability, however, this can be further tailored for specific purposes via optimisation techniques.

5 CONCLUSIONS

Multi-stable composite laminates exhibiting variable stiffness characteristics obtained by introducing a novel lay-out are presented. The innovative design proposed combines different regions of symmetric and unsymmetric lamination specifically designed to allow for clamping two opposite edges. This boundary constraint capability permits embedding the devised variable stiffness elements into a wider structure. The purely elastic nature of the introduced components enables simple implementation as no external devices are required to obtain the stiffness variability of the laminates. Furthermore, the purposely designed lay-out facilitates integration without the need for conventional fixation devices such as rivets or bolts. This monolithic integration potential is particularly important when embedding the presented multi-stable components into distributed compliance structures. This allows for a monolithic integration, which is particularly important when integrating the presented multi-stable components into distributed compliance structures. In view of the composite materials used to obtain the passive stiffness variability, the elimination of the necessity of drilling holes in the components for fixation represents an additional advantage over conventional solutions which introduce discontinuities and thus increased failure risk of the laminates. Numerical results obtained with FEM showed the feasibility of the devised complex lay-out. Very good agreement between numerical and experimental results for the stable configuration shapes of the designed variable stiffness laminates is obtained. The available stiffness variation exhibited by the clamped configurations is studied both numerically and experimentally. The results show a very good match exhibiting an 80-fold stiffness variation between the structural response in axial compression associated to the straight and curved stable shapes. On going work focusing on the optimisation of the difference in stiffness between the states and the structural performance of the components when integrated into a wider structure show encouraging preliminary results. Finally, multi-stability can be achieved by other means to induce a residual stress field, such as with fibre pre-stressing procedures, which could potentailly further increase the design space for the achievable variability of the embeddable variable stiffness components.

Acknowledgements

The authors would like to thank the support of the ETH Research Commission and the Marie Curie Actions Cofund Program; Dr. A. F. Arrieta is funded through an ETH Postdoctoral Fellowship.

References

[1] I. K. Kuder, A. F. Arrieta, W. E. Raither, and P. Ermanni. Variable stiffness material and structural concepts for morphing applications. *Progress in Aerospace Sciences*, Available online 3 August 2013, 2013.

[2] Y. Forterre, J. M. Skotheim, J. Dumais, and L. Mahadevan. How the venus flytrap snaps. *Nature*, 433:421–425, 2005.

[3] E. Kebadze, S.D. Guest, and S. Pellegrino. Bistable prestressed shell structures. *International Journal of Solids and Structures*, 41(11-12):2801–2820, 2004.

[4] S. Daynes, K. D. Potter, and P. M. Weaver. Bistable prestressed buckled laminates. *Composites Science and Technology*, 68(9):3431–3437, 2008.

[5] X. Lachenal, P. M. Weaver, and S. Daynes. Multi-stable composite twisting structure for morphing applications. *Proceedings of the Royal Society A*, pages 1–22, 2012.

[6] A.F. Arrieta, T. Delpero, A. E. Bergamini, and P. Ermanni. Broadband energy harvesting with cantilevered bi-stable composites. *Applied Physics Letters*, 102(17):173904, 2013.

[7] M. W. Hyer. Some observations on the cured shapes of thin unsymmetric laminates. *Journal of Composite Materials*, 15(15):175–194, 1981.

[8] A. F. Arrieta, O. Bilgen, M. I. Friswell, and P. Hagedorn. Passive load alleviation bi-stable morphing concept. *AIP Advances*, 2(3):032118, 2012.

[9] C. G. Diaconu, P. M. Weaver, and F. Mattioni. Concepts for morphing airfoil sections using bi-stable laminated composite structures. *Thin-Walled Structures*, 46(6):689–701, 2008.

[10] S. Daynes, P.M. Weaver, and J.A. Trevarthen. A morphing composite air inlet with multiple stable shapes. *Journal of Intelligent Material Systems and Structures*, 22(9):961–973, 2011.

[11] F. Mattioni. *Thermally induced multi-stable composites for morphing aircraft applications.* PhD thesis, University of Bristol, 2009.

[12] A. F. Arrieta, O. Bilgen, M. I. Friswell, and P. Ermmani. Modelling and configuration control of wing-shaped bi-stable piezoelectric composites under aerodynamic loads. *Aerospace Science and Technology*, 29(1):453–461, 2013.

[13] S. W. Tsai and H. T Hahn. *Introduction to Composite Materials.* Technomic Publishing Company, Inc., 1980.

ICAST2013 #Put paper number here

Aeroelastic Control of Non-Rotating and Rotating Wings using the Dynamic Stiffness Modulation Principle via Piezoelectric Actuators

Fred Nitzsche[1*], Douglas D'Assunção[2], Carlos De Marqui Jr.[2]

[1] Department of Mechanical and Aerospace Engineering, Carleton University, Ottawa, Canada
[2] Department of Aeronautical Engineering, EESC / University of São Paulo, São Carlos, Brazil

Abstract

Carleton University's Rotorcraft Research Group is working since 2003 in the development of an active rotor control system that incorporates a mechanism for helicopter blade pitch dynamic stiffness modulation at the root, the Active Pitch Link (APL). We believe that this system will vastly expand the capability of smart material systems to overcome their intrinsic stroke limitations and to attain superior performance in performing Individual Blade Control for helicopter rotor-induced vibration reduction. The proposed system was recently tested at the whirl-tower facility built by Carleton University on the grounds of the National Research Council Canada (NRCC). This paper first reports the remarkable achievements obtained with a dynamically similar, 1-meter span hinged rotor blade model and further discusses its results on physical and mathematical grounds. It was verified that up to 100% reduction in the transmitted loads occurred at the target 2/rev frequency when the blade was excited by a transversal jet to mimic the asymmetric flow existent in the helicopter rotor forward flight condition. This remarkable result was obtained using a simple open-loop control algorithm that can be optimized to a target higher-harmonic frequency of the rotor, minimizing its impact on the rotor fundamental cyclic control frequency at 1/rev. In another experiment performed at University of São Paulo, semi-passive control techniques using shunted piezoelectric materials were investigated for the aeroelastic control of a non-rotating wing. Piezoelectric coupling was added to the plunge spring suspension of a typical section. In the experiments, the Synchronized Switch Damping on Inductor technique (SSDI) was employed. It was demonstrated that both flutter and limit-cycle oscillations of the typical section were controlled out over a range of airflow speeds. Finally, the similarity between the APL and the SSDI technique is discussed and recognized that they are both based on a dynamic stiffness modulation control principle.

1. INTRODUCTION

The use of smart materials in vibration control problems has increased in the last years. The main reasons are the sensing and actuation capabilities of piezoelectric material due to the direct and converse piezoelectric effects, respectively, as well as ease of application.

In general, the control techniques using piezoelectric materials can be divided as passive and active control. In active control applications, piezoelectric material is used as actuator (converse piezoelectric

* Corresponding author, fred_nitzsche@carleton.ca

effect). An input voltage is applied to the piezoelectric material and mechanical strain is produced in order to reduce unwanted vibrations (Fuller and Elliot, 1996). Sensors and external source of energy are required. In passive case, the piezoelectric material is shunted to a passive electrical circuit where the mechanical energy converted to electrical energy (direct piezoelectric effect) is dissipated (Uchino and Ishii, 1988; Forward 1979; Hagood and von Flotow, 1991; Wu, 1996).

The literature of aeroelastic control also presents piezoelectric-based vibration control investigations. Heeg (1993) reports the use of piezoelectric elements as actuators for flutter suppression of a rigid wing mounted on a flexible system. The flutter speed of the wind tunnel device was increased by 20% when the control loop was closed. The control of dynamic aeroelastic phenomena was demonstrated in the Piezoelectric Aeroelastic Response Tailoring Investigation (PARTI) conducted at NASA Langley Transonic Tunnel (McGowan *et al.*, 1998). A composite plate-like wing with 36 piezoceramic patches was used and an increase of 12% in flutter dynamic pressure as well as a 75% reduction of gust bending moment was achieved. Another widely known program is the Actively Controlled Response of Buffet Affected Tails (ACROBAT). Different authors investigate the use of piezoelectric actuators to damp tail buffeting of the F/A-18 aircraft (Moses, 1997; Moses, 1999; Hopkins *et al.*, 1998; Dürr *et al.*, 1999).

Although the literature demonstrates the good performance using piezoelectric actuators in *active* aeroelastic control some issues need to be addressed, namely: the large amount of power required, the added hardware as well as the control law design and implementation. A few papers report the use of purely *passive* control schemes to damp aeroelastic oscillations. Among them, McGowan examined the performance of shunted piezoelectrics to reduce the aeroelastic response of a two-degree-of-freedom typical airfoil section (McGowan, 1999). An alternative approach was presented in another publication to to damp one single mode or a number of modes (Fleming and Moheimani, 2005). In such a case, an external source of energy is not required and a simple electrical circuitry can be used. A modal-based modeling and analysis of the effectiveness of resistive-inductive shunted piezoelectric materials to damp aeroelastic systems has been reported (Agneni *et al.*, 2003). Although a weak performance of the passive controller for improving the stability margin (flutter envelope) of the composite wing of an unmanned glider was stated in this work, the authors showed the ability of the passive devices to reduce the gust response amplitude of the wing in the neighborhood of the flutter speed.

Since the frequencies of aeroelastic oscillations are very low (typically under 20Hz), the required inductances in strictly passive control techniques are extremely large and not practical. The use of synthetic inductance circuits (Riordan, 1967) would be a good alternative to address this issue; however the internal resistance reduces the performance of passive controls (Park and Inman, 2003). The SSD techniques are *semi-passive* methods that can address both the issues of passive controllers associated with the large inductance required in aeroelastic applications and the complexities of active controllers by introducing a nonlinear treatment of the voltage output of the piezoelectric elements and by so allowing an increase in the efficiency of the mechanical to electrical energy conversion. Exemples are the SSD (Synchronized Switch Damping) class of techniques (Richard *et al.*, 1999, 2000; Clark, 2000; Guyomar *et al.*, 2007; Corr and Clark, 2002). In these semi-passive methods the piezoelectric material is kept in open-circuit condition (maximum voltage output) except for a brief period of time when the voltage either 1) goes to zero due to a switch to a small resistance – as in the case of SSDS (Synchronized Switch Damping on Short-circuit) technique or 2) is inverted due to a switch to an inductor – as in the case of SSDI (Synchronized Switch Damping on Inductor). In both cases the switching is performed synchronously with the mechanical displacement. In the SSDS case electrical energy is dissipated during the voltage cancelation in the short-circuit condition, resulting in increased damping effect. In the SSDI case, the required inductance is small when compared to the passive techniques because it is associated to the electrical and not to mechanical system control frequency.

Another example where aeroelastic control can be achieved with a *semi-passive* technique is shown in the conceptual drawing in Fig 1, optimized to exploit the large force delivered and the frequency bandwidth of piezoelectric materials to perform an indirect-active vibration control of the axial loads generated at the "structure" that are transmitted to the "base." The primary advantage of this patented system (nicknamed "Smart Spring", Nitzsche *et al*, 1999) when compared to other piezoelectric actuator based approaches is that the device does not rely on the piezoelectric actuators to achieve high stroke and force simultaneously (meaning high power and high required voltages). Rather, the device only requires the actuators to produce micro displacements to generate relatively high actuation forces.

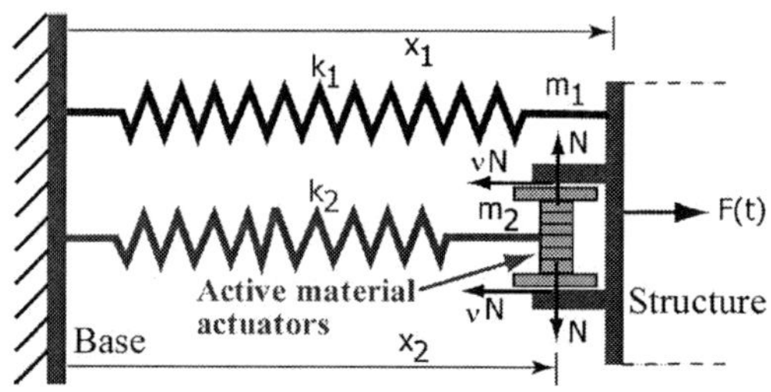

Figure 1. "Smart Spring" conception.

2. A SEMI-PASSIVE TECHNIQUE TO CONTROL THE TRANSMISSIBILITY OF THE AEROELASTIC LOADS PRODUCED BY ROTATING WINGS

The "Smart Spring" device was already extensively reported in other publications as a mean of reducing rotating wing vibrations. In the SHARCS (Smart Hybrid Active Rotor Control System) project, the "smart spring" is inserted between the blade horn and the rotor swashplate (Fig. 2), hence replacing the conventional pitch link to reduce the transmitted vibration between the helicopters rotating to the non-rotating frames (Nitzsche *et al*, 2013). It is important to understand that this Active Pitch Link (APL) based on the "Smart Spring" concept relies on delivering power not to actuate against the aerodynamic loads but to change the *effective spring constant* of the device (or, more precisely, its *impedance*). It is the *change of the flexural torsional characteristics of the blade* (throughout is boundary conditions) that allows the control of the aeroelastic response of the system. Energy is redistributed in the spectrum of vibration, removed from the frequencies were the transmissibility is desired to be low (i.e. at the frequencies integer multiples of the number of blades, *n*/rev and they immediate neighbors) and displaced it at the other harmonics known to have lower transmissibility in the helicopter rotor. Therefore, such a device is able to control the blade aeroelastic response in an *indirect* way. The primary advantage in the APL system, compared to other systems designed with the same purpose, is that the APL does not rely on the actuators to achieve high stroke and force simultaneously. Rather, the device only requires the piezoelectric actuators shown in Fig. 1 to produce micro displacements to close the initial gap between the stacked actuators and the sleeve connected to the "structure." The gap is kept as low as possible to maximize the actuation force, N that produces the friction force νN between the two surfaces. This friction force ideally should be enough to guarantee that there is no sliding between the actuators and the sleeve and, therefore, $x_1 = x_2$ at the times when the springs are perfectly engaged, $k = k_1 + k_2$. However, even when some sliding exists the device is able to perform stiffness modulation because a dynamic (complex-valued) equivalent spring constant is created due to the coulomb friction between the two surfaces, $k = \mathrm{Re}(k) + i\,\mathrm{Im}(k)$. This additional "damping effect" is beneficial to the system stability because part of the mechanical energy is converted into heat. In the experiments, instability of the system was never observed. Nevertheless, it must be noted that for security reasons (fail safeness) in the practical implementation of the device the situation $k = k_1 + k_2$ occurs when the applied voltage is zero. Hence, the actuators are used to *disengage* the two springs *rather than engaging* them (assume in Fig. 1 that the actuators contract rather than expand for non-zero voltages, opening the gap and leaving only the primary spring, k_1 in the load path). Furthermore, in order to modulate the dynamic response of the blade, the device is evidently more effective if placed near the root to actively alter its boundary conditions, which may instantaneously change from a perfect inelastic connection to the swashplate (if $k_2 \rightarrow \infty$) to some finite elastic value, $k = k_1$.

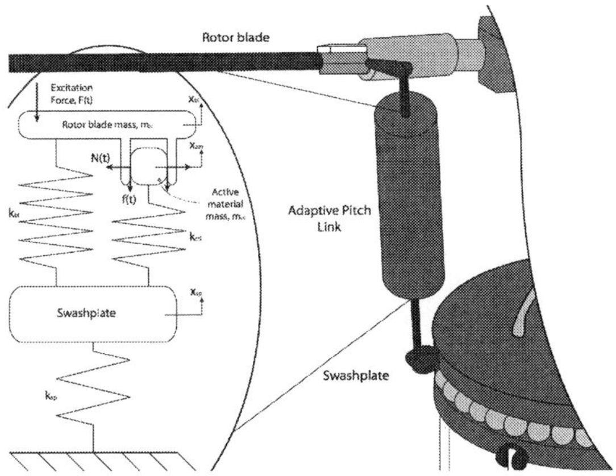

Figure 2. Active Pitch Link (APL) control device.

2.1 The Active Pitch Link hardware development

Over the past ten years, three prototypes Active Pitch Links (APL) were developed. The first prototype, developed in 2003, was a large-scale system to prove the concept, Fig. 3(a). Hence, no effort was made to meet any design requirement such as size and fail safeness. The second generation of APL was manufactured in 2006, Fig. 3(b). This design was already very compact (0.108m long) with adjustable preload settings and a total mass of 0.196kg to fit the 1-m SHARCS blade. After testing this unit in both non-rotating and whirl tower tests the present third generation of the APL prototype was designed and built in 2010, Fig. 3(c). It incorporates a clamp mechanism whose operation is largely independent from the centrifugal loads for improved performance in the rotating tests and has the necessary fail-safeness characteristics. There are three modes of operation for the APL and the input voltage sent to the piezoelectric actuator controls them. The first is the "solid" link mode, which is theoretically achieved with an input voltage in the range of 0-60V. In this mode, the stiffness of the APL is virtually infinity and it acts as the traditional solid pitch link. The second mode of operation is in "transition" range, where the APL stiffness varies from its maximum to its minimum value. An input voltage in the range of 60-120V controls this mode. The last mode of operation is the "soft" link mode, controlled by an input voltage of 120-150V. In this mode, the load path is carried though the APL softer spring resulting in the design stiffness of 160kN/m.

2.2 Active Pitch Link characterization

The APL stiffness curve is a plot of the calculated resultant stiffness as a function of the piezoelectric input voltage. To create this plot, an external dynamic load was applied axially to the APL by an electromagnetic shaker. The force magnitude F = 37N was kept constant in the tests with a frequency of ω = 15.7Hz. This procedure was repeated for several different piezoelectric input voltages and data was collected for ten input voltages, ranging between 15 and 150V. For each data point the displacement time-history (measured by a Hall sensor) was recorded. This data was used to determine the magnitude of the "Smart Spring" displacement for each corresponding input voltage. The equivalent stiffness was then calculated using the ratio of force amplitude to the spring displacement amplitude for each input voltage. As the APL was designed to use a friction clamp to vary the stiffness of the system, the force required to overcome a frictional force is much less for already moving parts than for initially stationary parts. For this reason, the initial conditions of the friction clamp were verified to be very important when determining the experimental stiffness curve of the APL. In order to show the effect of the initial conditions of the friction clamp, the experimental stiffness curve was developed twice: for 1) the initially stationary – "1st initial condition" and 2) the initially moving – "2nd initial condition" shown in Fig. 4, where the real parts of the effective stiffness k versus the actuator voltage curves for the two stated initial conditions are reported. The curves approach asymptotically the static values of k_1 and k_2 at their lower and higher voltage branches, respectively. However, no spring displacement data was collected when the piezoelectric actuator voltage applied was zero or near zero as in this condition the APL approaches the "solid" pitch link with very high stiffness k_2 and, therefore, the data points became scattered and not repeatable. Nevertheless, the curves capture relatively well the transition zone of the APL, where only partial clamping occurs and there is sliding between the actuators and the sleeve seen in Fig. 1.

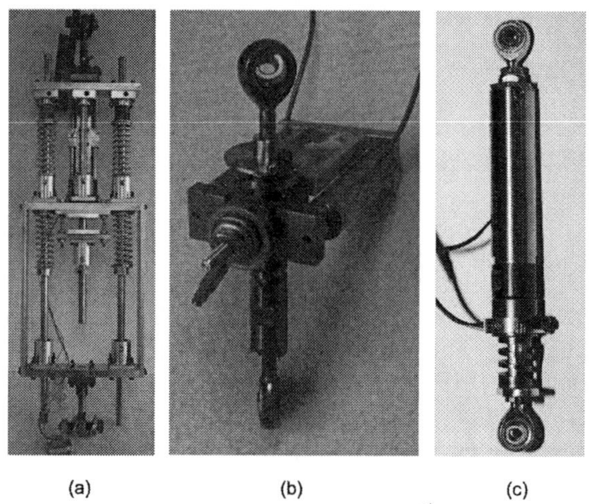

Figure 3. APL prototypes: (a) – 1st generation, 2003, (b) – 2nd generation, 2006 and (c) – 3rd generation, 2010.

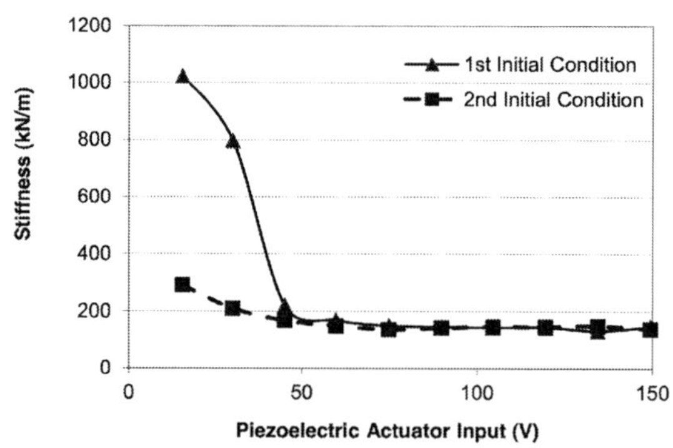

Figure 4. Real part of the APL dynamic stiffness variation with the input voltage for $\omega = 15.7$Hz.

2.3 Active Pitch Link rotating tests (in whirl tower)

The APL was subsequently tested at the Carleton University whirl tower. In the whirl tower, a single blade hinged rotor hub was installed. In the tests, the APL was positioned at its normal location, replacing the conventional pitch link, Fig. 5. The APL control voltage was fed into the rotating frame by slipping rings and the data was acquired using the wireless telemetry located inside the hub. A calibrated strain-gauge sensor at the base of the pitch link was used to collect the vibrational data that would be transmitted into the non-rotating frame (the swashplate in the helicopter case) from the blade. The tests were carried out at three spinning ratios, $\Omega = 600$, 750 and 1,000RPM.

Figure 5. APL installation in the rotor hub.

For these tests, a fan was positioned at the bottom of the whirl tower to provide a transverse flow with the intention to excite the blade with a periodic aerodynamic load, as it happens in forward flight. The excitation provided by the fan was considered sufficient for at least the first 3 harmonics of the rotor fundamental frequency for all spinning ratios, as seen in Fig. 7. The centerline position of the fan, whose relative dimension is also shown in the same figure, was taken as the reference azimuth for the control signal input (0 degree control phase). Since the purpose of the tests was to demonstrate the effectiveness of the APL to redistribute the vibration spectrum in the rotating condition, no optimization of the control law was pursued in this work. The APL design objective of not interfering with the fundamental frequency of the rotor, $p\Omega$ = 1/rev was, therefore, completely relaxed (where p is the harmonic index of the rotor frequency). Hence, the tests were conducted with two single harmonic (pure sinusoidal) control signals, namely at the $p\Omega$ =1/rev and 2/rev frequencies. For these two control frequencies, the control phase angle with respect to the fan centerline was then varied between ±180 degrees.

The best result obtained in the whirl tower tests show a perfect attenuation (100%) of the frequency 2/rev with the APL control performed at 2/rev and a phase angle of approximately –45 degrees (Fig. 8). In this condition, the frequency response at 1/rev was almost unchanged as it should be the case for the helicopter applications because the vibrational control system provided by the APL must not interfere with the rotor cyclic control done at 1/rev. In all tested cases, although the control performance varied considerably, the results confirmed the vibration spectrum reshaping generated by the APL, extracting energy from one harmonic and shifting it at other harmonics. It is worthwhile to mention that the control of vibration was done without introducing significant changes in the blade angle of attack (less than 0.1 degree), indicating that the control was performed by strictly structural means, without generating additional aerodynamic loads as in the traditional methods used in rotating wings (i.e. individual blade control or high-harmonic control). This feature, together with the known helicopter rotor characteristics of low transmissibility at harmonics not multiple of the number of the blades can be explored to develop effective closed-loop control algorithms to decrease the vibration transmissibility from the rotating to the non-rotating frames.

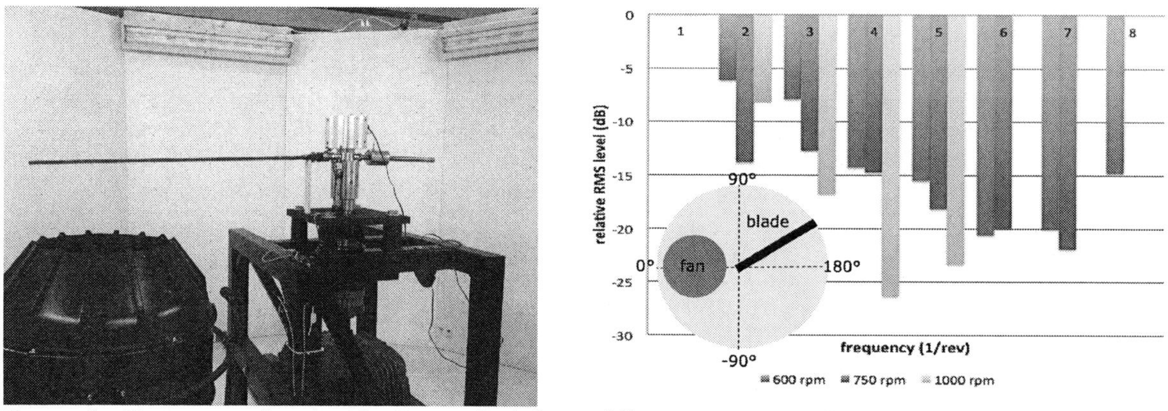

Figure 6. Test set-up showing the fan on the left providing a transversal flow to excite the blade (on the left) and the blade first 8 harmonics attenuations with respect to the rotor fundamental frequency for different rotor spinning ratios (on the right).

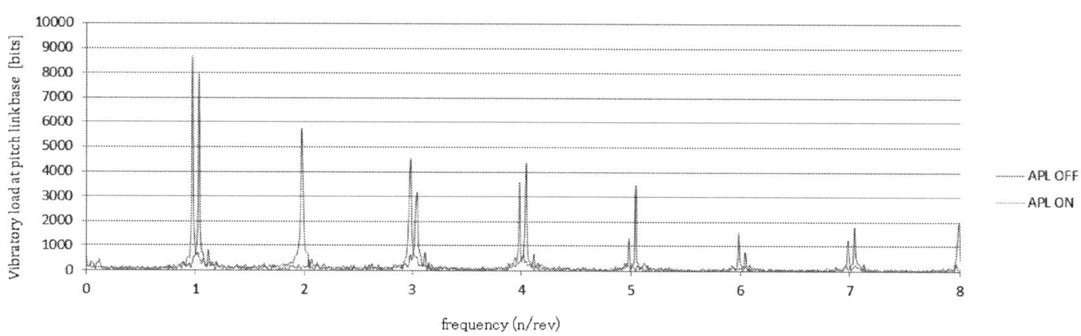

Figure 7. Transmissibility from the blade to the pitch link base at the first $p = 8$ harmonics of the rotor spinning frequency ($\Omega = 600$ rpm).

3. OTHER SEMI-PASSIVE METHODS TO CONTROL AEROELASTIC PHENOMENA

Semi-passive methods have been introduced in attempt to overcome the previously mentioned drawbacks of purely passive and active controllers (Richard *et al.*, 1999, 2000; Clark, 2000; Guyomar *et al.*, 2007; Corr and Clark, 2002). In this section, the SSDS and SSDI methods are presented and discussed. Different semi-passive methods also reported in the literature (Clark, 2000; Corr and Clark, 2002) are not included in this section.

The *SSDS method*, originally presented by Richard *et al.* (1999), consists in leaving the piezoelectric element in open-circuit condition (the electrical boundary condition that gives the maximum voltage output) except when a local maximum voltage is detected. At this point, the system is switched to short-circuit condition for a small period of time. In the *SSDI method*, also presented by Richard *et al.* (2000), when a local maximum is detected the system is switched to an inductor, L and an inversion on the voltage of the piezoelectric element occurs. This inversion is due to a change in the electrical frequency that takes place when the inductor is briefly connected to the capacitance of the piezoelectric element, C_p. During the voltage inversion a resonant circuit LC_p with resonance frequency ω_n is generated. The inversion time is half the pseudo-period, Δt_i of the electrical network (Lallart *et al.*, 2008):

$$\Delta t_i = \pi/\omega_n = \pi\sqrt{LC_p} \tag{1}$$

Therefore, the inversion time is directly proportional to the inductance L, which in practice must be chosen in such a way that the resonance frequency of the electrical circuit obtained during the switch is 10 to 50 times the target control frequency of the structure (Mohammadi, 2008). In the SSDI case, the voltage after the inversion has an absolute value slightly lower than the initial voltage due to losses. The *voltage inversion efficiency* is measured by defining the inversion coefficient, γ, defined as follows:

$$V_{after} = -\gamma V_{before} \tag{2}$$

The negative sign in (2) indicates that the voltage is inverted. The inversion coefficient is related to the electric quality factor, Q of the switching network, which behaves as a transient R_iC_pL circuit in series:

$$Q = \frac{L\omega_n}{R_i} = \frac{1}{R_i}\sqrt{\frac{L}{C_p}} \tag{3}$$

where R_i is the resistance of the switching network, including the inductor. Therefore, the amplitude envelope of the R_iC_pL circuit transient response gives the voltage amplitude after the inversion time:

$$V_{after} = -V_{before}\exp(-\zeta\omega_n\Delta t_i) \tag{4}$$

where ζ is the damping coefficient of the circuit, $\zeta = R_i/2\sqrt{C_p/L}$. The minus sign is introduced again to indicate the voltage inversion. Using this expression into the previous equations:

$$V_{after} = -V_{before}\exp(-\pi/2Q) \tag{5}$$

and the relationship between the inversion coefficient and the quality factor is obtained:

$$\gamma = \exp(-\pi/2Q) \tag{6}$$

From (6) it is clear that the SSDS may be seen as the particular case of the SSDI when the quality factor is zero and, consequently, the inversion coefficient is also zero (Mohammadi, 2008).

3.1 The experimental set-up

In this section, the semi-passive SSDS and SSDI techniques defined in the previous section are experimentally investigated for aeroelastic control cases of a fixed wing. Figure 8 shows the schematics of a linear 2-DOF typical section. The plunge and pitch Degrees of Freedom (DoF) are denoted by h and α, respectively. The plunge displacement is measured at the elastic axis, i.e., at point P (positive downwards) and the pitch angle is measured about the elastic axis (positive clockwise). In addition, b is the semi-chord of the airfoil section, x_α is the dimensionless chord-wise offset of the elastic axis from the centroid,

C, k_h is the stiffness per unit length in the plunge DoF, k_α is the stiffness per unit length in the pitch DoF, d_h is the damping coefficient per unit of length in the plunge DoF, d_α is the damping coefficient per unit of length in the pitch DoF and U is the airflow speed. In this work, the electromechanical coupling is added to the plunge DoF. Hence, the schematics of the electro-aeroelastic section are also shown in Fig. 8 (right side). Therefore, the strain of the piezoelectric patches due to the plunge motion produces the electrical output.

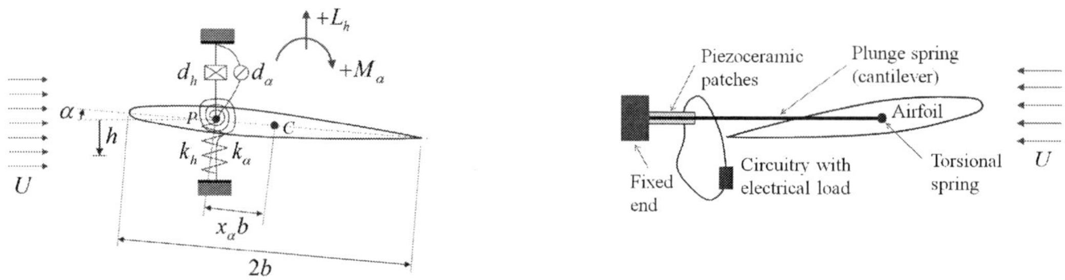

Figure 8. Test set-up showing the aeroelastic typical section under airflow excitation (on the left) and the tested electro-mechanical system with the piezoelectric patches on the plunge DoF (on the right).

Figure 9 shows the experimental setup used for investigating the aeroelastic behavior of the typical section instrumented with piezoelectric patches. This setup is a realization of the typical section presenting electro-mechanical coupling as shown in Fig. 8 (on the right). The plunge stiffness is due the four elastic beams with clamped-clamped end conditions. The free ends of the elastic beams are connected to metal plates at the top and the bottom. A shaft (pitch axis) is mounted to the upper and the lower plates through a pair of bearings. The pitch stiffness is given by a spring wire clamped onto the shaft (at the elastic axis), shown in the enlarged view of Fig. 9 (on the right). The free end of the wire is simply supported on the top plate. Two piezoceramic patches (type QP10N from Mide Technology Corporation) are attached symmetrically to the root of the two bending stiffness members and their electrodes are connected in parallel to an external resistive load.

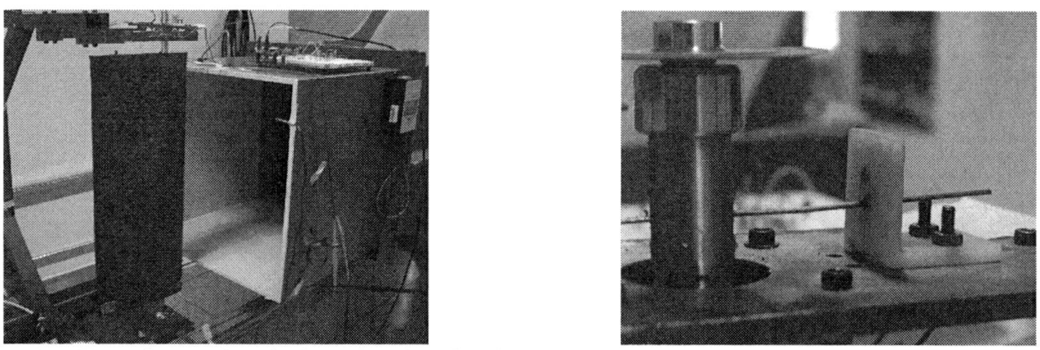

Figure 9. Experimental typical section (on the right) and detailed view of the pitch spring (on the left).

The plunge displacement, h was measured by using a strain gauge bonded on the beam, on a ¼-bridge configuration. The pitch displacement, α was measured with a digital encoder (US model HEDS-9000-T00) attached to the aluminum axis of the wing. The acquisition of all data was performed with a dSPACE®DS1104 system. Table 1 lists the experimental parameters of the typical section used in this paper. Note that in the Table 1 m_e is the added mass to the airfoil of mass m due to bearings, flanges, etc. and θ is the electro-mechanical coupling coefficient.

Table 1. Experimental parameters.

Parameter	Value	Units
b	1.25 x 10^{-1}	m
$x_a b$	3.2 x 10^{-1}	kg/m
m	1.542	kg/m
m_e	2.548	kg/m
I_a	7.2 x 10^{-3}	kg.m
k_a	5.08	N/rad
k_h	4.2 x 10^{3}	N/m^2
d_a	6.35 x 10^{-2}	N.s/rad
d_h	1.8146	N.s/rad^2
θ	1.55 x 10^{-3}	N/V
C_p	85 x 10^{-9}	F

3.2 The electronic switching circuit

The electronic circuit used in this paper is based on the circuit breaker proposed by Richard *et al.* (2007) and Zhu *et al.*, (2012). Passive electronic components were used and the three basic functions of the *maximum circuit breaker* were addressed by the envelope detector (R1, C1, D1), the comparator circuit (R2, D2, Q1) and the switch circuit (D3, R3 and Q2), where R is a resistor, C a capacitor, D a diode and Q a transistor, as shown in Fig. 10. It is important to stress that this is an *autonomous* switching circuit that results in a *self-powered flutter controller*.

The envelope detector is a low-pass filter whose output amplitude should be very similar to the voltage piezoelectric signal with a small phase delay. The comparator circuit (R2, D2, Q1) compares the filter output signal on C1 with the voltage of the piezoelectric material. Thus, while the voltage amplitude of the envelope is smaller than the voltage of the piezoelectric the transistor Q1 is blocked. When the piezoelectric voltage is smaller than the voltage of the envelope and the difference between the envelope voltage and piezoelectric voltage is larger than the threshold of the transistor Q1, the latter conducts. Consequently, the transistor Q2 is triggered and the switching process starts. Accordingly, the *minima circuit breaker* is performed in a similar way with the corresponding function components (R6, C2, D6), (R5, D5, Q4) and (D4, R4, Q3) but having different polarities of diodes and transistors, as shown in Fig. 10.

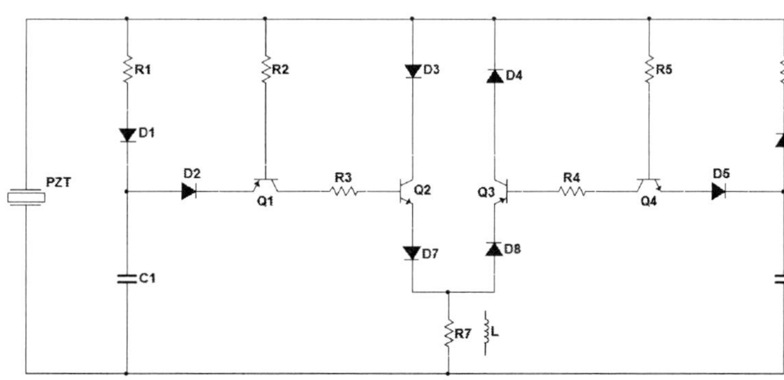

Figure 10. Circuit diagram of the self-powered electronic circuit breakers.

In this paper, the same electronic circuit breaker is used for both the SSDS and SSDI control cases. The differences are also shown in Figure 10, where the resistor R7 of 100Ω was used for the SSDS case and the inductor L of 8H (with internal resistance of 505Ω) replaced it for the SSDI case. Diodes D7 and D8 were used to guarantee the right switching for the maximum and minima detectors. Table 2 shows the values of all electrical components used in the self-powered electronic breaker circuit.

Table 2. Electrical components values of the self-powered electronic breakers.

D1 to D8	Diodes fast soft recovery	BYW95C
Q1 and Q3	Transistor – PNP high voltage amplifier	MPSA92
Q2 and Q4	Transistor – NPN high voltage amplifier	MPSA42
R1 and R6	Resistor	1 kΩ
R1 to R6	Resistor	10 kΩ

3.3 Aeroelastic control results for the fixed-wing tests

In this section, the SSDS and SSDI methods are employed to control the typical section at the flutter and post-flutter conditions.

The flutter speed of the electromechanically coupled typical section was first experimentally determined considering the piezoelectric in open-circuit condition. The flutter speed was estimated by testing the system at various airflow speeds. An initial condition was applied to the linear displacement DoF (plunge) and the free response of the system was measured. Figure 11 presents on the left the time histories of voltage, pitch and plunge displacements for an airflow speed of 11.7m/s. Note that the time histories exhibit a decaying behavior. The same figure on the right presents the corresponding time histories for the airflow speed of 11.9m/s. At this airflow speed the typical flutter behavior with persistent oscillations and increasing amplitudes is observed. Therefore, the airflow speed of 11.9m/s was taken as the flutter speed in the open-circuit condition.

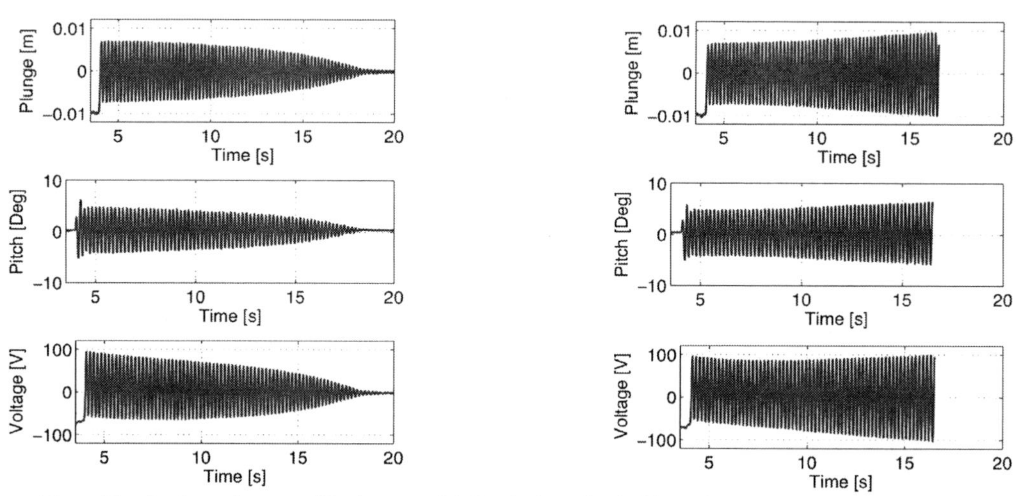

Figure 11. Typical section oscillations at 11.7m/s (on the left) and at 11.9m/s (on the right) at the open-circuit condition.

The experimental investigation of flutter suppression of the typical section by using the SSDS and SSDI schemes is presented next. In the wind tunnel experiments, an initial displacement was applied to the plunge DoF for the flutter speed. In the two cases (SSDS and SSDI), the piezoelectric patches were in open-circuit condition for the 10 initial seconds, when they were connected to the self-powered electronic breaker. Figure 12 shows the time histories of pitch, plunge and voltage output for at the open-circuit flutter speed using SSDS control scheme on the left. As can be observed that the mechanical and electrical oscillations decay and are suppressed within 8 seconds. The same figure on the right shows the corresponding time histories for the SSDI control scheme. The mechanical and electrical oscillations decay and are suppressed within 6 seconds, indicating that it is a more efficient control technique.

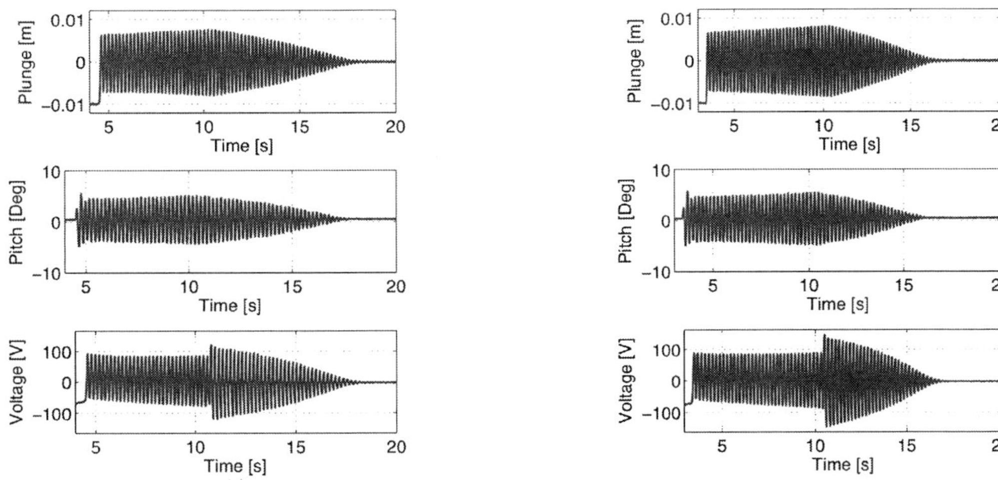

Figure 12. Typical section oscillations at 11.9m/s using the SSDS scheme (on the left) and using the SSDI scheme (on the right).

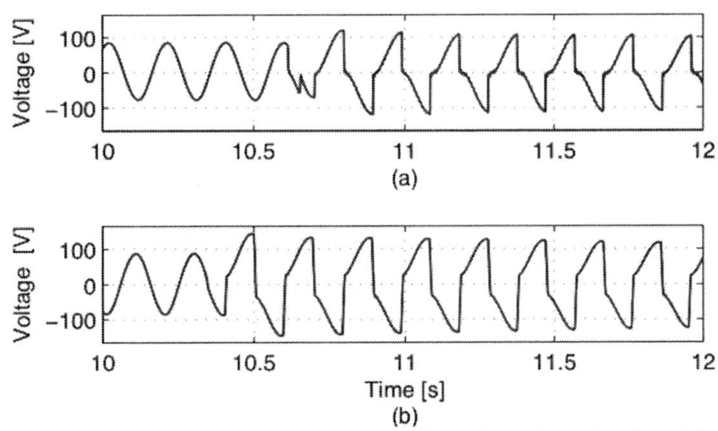

Figure 13. Detailed view of the electrical response after triggering the circuit breaker: (a) using SSDS; (b) using SSDI.

Figure 13 shows the detailed view of the voltage output for the two switching processes reported in Fig. 12. The piezoelectric patches are kept in open-circuit condition except for a brief period of time when the switch is closed. Figure 13(a) for the SSDS case shows that the switch is closed for one-quarter of the oscillating period, when the voltage reaches an extreme. While the switch is closed, the voltage drops to zero. The time history of voltage output for the SSDI case is shown in Fig. 10(b). One should note that the voltage inversion is not complete in this case. This is due to the internal resistances of the switching circuit and inductor. In fact, using (5), the electric quality factor Q = 1.13 was determined for the inversion coefficient measured in the experiments, γ= 0.25.

In another experiment the performance of the two semi-passive systems at the post-flutter condition was analyzed. In both cases (SSDS and SSDI techniques), the piezoelectric remained connected all the time to the circuit breaker and the systems were tested for airflow speeds larger than the open-circuit flutter speed of 11.9m/s.

When the SSDS technique was used, the flutter oscillations were suppressed until the airflow speed of 12.7m/s (Fig. 14, on the left). For airflow speeds larger than 12.7 m/s the typical flutter behavior was observed, as shown in the same figure (Fig. 14, on the right). Then, by using this technique, the closed-loop flutter speed was 8.5% larger then that of the open-circuit condition.

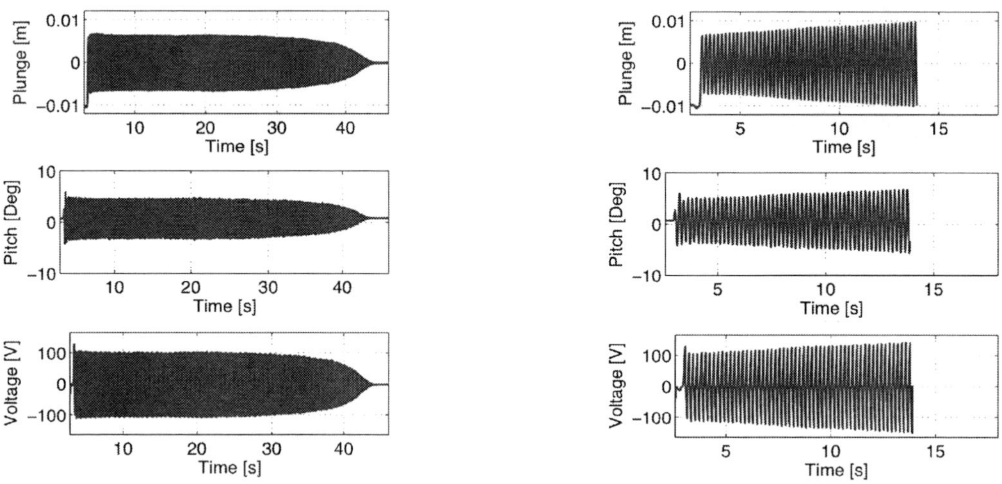

Figure 14. Typical section oscillations using the SSDS technique: at 12.7m/s (on the left) and at 12.8m/s (on the right).

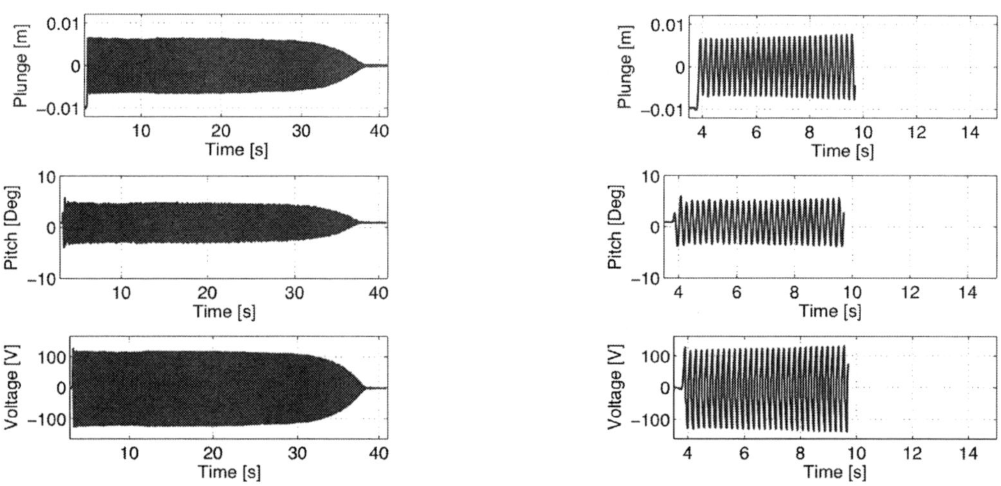

Figure 15. Typical section oscillations using the SSDI technique: at 13.1m/s (on the left) and at 13.2m/s (on the right).

When the SSDI technique was used, the flutter oscillations were suppressed until the airflow speed of 13.1m/s (Fig, 15, on the left). For airflow speeds larger than 13.1m/s the typical flutter behavior was observed (Fig. 15, on the right). It is important to note that when the SSDI control was used, the flutter speed was 11.4% larger then that the open-circuit flutter speed.

4. DISCUSSION ON THE RELATIONSHIP BETWEEN THE SMART SPRING AND THE SSDS AND SSDI SEMI-PASSIVE METHODS

In this paper one mechanical and two electro-mechanical semi-passive techniques to dampen structural vibrations are presented. It is, however, interesting to discuss the relationship between them. For this, it is necessary first to recall the analogy between mechanical and electrical circuits, where the mass, m spring constant, k and damping coefficient, c are commonly associated to the inductor, L inductance (inverse of the capacitance, C_p associated with the piezoelectric element), and resistor, R_i, respectively. For the elecric circuit:

$$\ddot{I}+\frac{R_i}{L}\dot{I}+\frac{1}{LC_p}I=0 \tag{7}$$

The mechanical counterpart is written as:

$$\ddot{x}+\frac{c}{m}\dot{x}+\frac{k}{m}x=0 \tag{8}$$

where the displacement of the mass, $x(t)$ in (8) replaces the current, $I(t)$ in (7). Hence, the resonance frequencies are $\omega_n=\sqrt{k/m}$ and $\omega_n=1/\sqrt{LC_p}$ for the mechanical and electrical systems, respectively. The damping ratio of the mechanical system is $\zeta=\frac{1}{2}c/\sqrt{mk}$, whereas in the case of the series R_iC_pL circuit the damping ratio is defined as $\zeta=\frac{1}{2}R_i\sqrt{C_p/L}$.

Examining the "Smart Spring" schematics in Fig. 1, it consists of two *load paths*: the *primary*, comprised by k_1 and m_1 and the *secondary*, defined by k_2 and m_2. The secondary load path is only added to the primary load path when the stack of piezoelectric actuators expand under an applied voltage, close the gap and apply the normal force N on the sleeves, creating the friction force νN that engages the secondary path and, thus, increases the resultant spring constant of the device from k_1 to $k = k_1 + k_2$. In the dynamic operation there is always some sliding in this engaging process and the resultant spring constant is not a pure real quantity, as discussed in Section 2.2. It possesses real and imaginary parts due to the Coulomb dry friction: $k=\mathrm{Re}\{k\}+i\,\mathrm{Im}\{k(N,\omega,\nu)\}$ for forced simple-harmonic motion at frequency ω. The real part can be identified to the spring constants, $\mathrm{Re}\{k\}=k_1+k_2$ when the spring is completely engaged. The imaginary part produces the losses due to the switching process, always observed in the operation of the "Smart Spring" that are a function of the contact normal force, N (itself a function of the applied control voltage to the piezoelectric elements) and the kinematic friction coefficient, ν between the contact surfaces. One can relate the "Smart Spring" Coulomb dry friction to the equivalent viscous damping using the corresponding complex frequency response functions at the operation frequency. Using (8) to describe the dynamic motion associated with the *secondary path when completely disengaged*: $x = x_2$, $\mathrm{Re}\{k\} = k_2$ and $m = m_2$. Therefore, the complex frequency response functions, $g(i\omega)$ of the "Smart Spring" secondary path for the actual and equivalent viscous damping cases are compared:

$$g(i\omega)=\frac{1}{-\omega^2 m_2+\mathrm{Re}(k)+i\,\mathrm{Im}(k)}=\frac{1}{-\omega^2 m_2+i\omega c_2+k_2} \tag{9}$$

yielding $k_2 = \mathrm{Re}(k)$ and $c_2 = \mathrm{Im}(k)/\omega$. Hence, for this secondary system the equivalent viscous damping and resonance frequency would be $\zeta = \frac{1}{2}\mathrm{Im}(k)/\omega\sqrt{m_2 k_2}$ and $\omega_n = \sqrt{k_2/m_2}$, respectively. Therefore, during its operation, the "Smart Spring" engages and disengages to a resonating system with frequency ω_n. This is also the case in the SSDI control system operation. The "Smart Spring" would then be a formal *mechanical realization of the SSDI*, where the equivalent quality factor is a function of the exciting harmonic frequency:

$$Q = \frac{1}{2\zeta} = \frac{\omega}{\mathrm{Im}(k)}\sqrt{m_2 k_2} \tag{10}$$

It is worthwhile to note that a "Smart Spring" with zero damping ratio ($Q \to \infty$) would be one where the friction is not an important parameter or, in other words, the engaging and disengage processes would occur perfectly, with no sliding (secondary path composed only by the mass m_2 and the spring k_2 as shown in Fig. 1), and the stiffness modulation would be perfect. Such a "Smart Spring" would be the *mechanical realization of the SSDS* following the discussion in Section 3.1. However, one must be cautious about this comparison because the SSDI an inversion in the displacement x_1 when the "Smart Spring" engages remains to be verified if practically feasible due to mechanical limitations that are irrelevant in electrical circuits. In this case, the SSDS case of the "Smart Spring", where no losses are observed, would be apparently its most efficient operation condition.

Another interesting observation is that the practical realization of the SSDI calls for a resonance electrical system that is 10 to 50 times the control target frequency of the associated mechanical system. Likewise, in a good "Smart Spring" design, the resonance frequency associated with the secondary path (internal resonance frequency) must be very high, well above the target control frequency, ω. For this reason, the "Smart Spring" is called a *stiffness* modulation system where, by definition, the resonance frequency is much higher than the exciting frequency ($\omega_n >> \omega$) and, thus, the system is defined to be *stiffness dominated* (notice that as $k_2 >> k_1$ during the brief switching process both (10) and $\omega_n = \sqrt{k_2/m_2}$ are still good approximations). Following this reasoning, both the SSDS and SSDI techniques could be seen as a class of *stiffness* modulation (or control) methods as well. The objective of the "Smart Spring", SSDS and the SSDI methods are to artificially increase the operative damping of an electro-mechanical system at a range of relatively low frequencies where the structure is primarily stiffness dominated. All these three techniques are able perform aeroelastic control in an indirect way, effectively using the stiffness modulation principle. In the case of the "Smart Spring a real variation in the structural stiffness is introduced by the piezoelectric-driven switching mechanism. On the other hand, in the case of the SSDS and SSDI techniques, it is the *apparent* stiffness variation of the coupled electro-mechanical system, introduced by the piezoelectric elements between their closed- and open-circuit conditions that are efficiently explored to control the vibrations.

5. CONCLUSIONS

In this paper, the SSDS (Synchronized Switch Damping on Short-circuit) and SSDI (Synchronized Switch Damping on Inductor) semi-passive techniques using self-powered switching circuits were studied to control aeroelastic phenomena. The flutter of a fixed wing was successfully suppressed during the experiments and no external energy was required by the control system. The uncontrolled flutter speed of an electro-mechanically coupled typical section was first measured with the piezoelectric elements in the their open-circuit condition. After, the flutter oscillations were successfully suppressed using both the SSDS and SSDI control methods. The flutter speeds of the typical section were increased by 8.5% and 11.4% for the SSDS and SSDI control cases, respectively. The semi-passive techniques presented in this paper for aeroelastic control are interesting alternatives to strictly passive piezoelectric controllers. This is because the requirement of very high inductance in the passive controllers electrical circuits can be relaxed by the use of semi-passive methods. Excellent aeroelastic control performance in wind tunnel tests was demonstrated in this paper with the semi-passive methods using practical electrical circuits.

Another semi-passive technique to control aeroelastic response, using a strictly mechanical system (the "Smart Spring" device), also demonstrated superior performance to control the vibratory loads from helicopter blades in rotating tests. The technique was able to alleviate 100% of the vibrations in the harmonic frequency twice the spinning frequency of the blade.

Finally, it was discussed the similarity between the three investigated control methods (SSDS, SSDI and the "Smart Spring") and it was recognized that the SSDI and SSDI can be seen as the electro-mechanical counterparts of the mechanical "Smart Spring" and, as such, can also be regarded as stiffness modulation techniques as well.

References

Agneni, A.; Mastroddi, F.; Polli, G.M. (2003). Shunted piezoelectric patches in elastic and aeroelastic vibrations. Computers & Structures, v. 81, n. 2, p. 91–105.

Clark, W.W. (2000). Vibration Control with State-Switched Piezoelectric Materials. Journal of Intelligent Material Systems and Structures, v. 11, n. 4, p. 263–271.

Corr, L.R.; Clark, W.W. (2002). Comparison of low-frequency piezoelectric switching shunt techniques for structural damping. Smart Materials and Structures, v. 11, n. 3, p. 370–376.

Dürr, J. (1998). Integration of piezoceramic actuators in fiber-reinforced structures for aerospace applications. Journal of Intelligent Material Systems and Structures, v. 3326, p. 81–92.

Fleming, A.J.; Moheimani, S.O.R. (2005). Control orientated synthesis of high-performance piezoelectric shunt impedances for structural vibration control. IEEE Transactions on Control Systems Technology, v. 13, n. 1, p. 98–112.

Forward, R L. (1979). Electronic damping of vibrations in optical structures. Applied Optics, v. 18, n. 5, p. 690.

Fuller, C.C.; Elliott, S.; Nelson, P.A. (1996). Active control of vibration. 1st Ed.: Academic Press, San Diego, CA.

Guyomar, D.; Richard, C.; Mohammadi, S. (2007). Semi-passive random vibration control based on statistics. Journal of Sound and Vibration, v. 307, n. 3-5, p. 818–833.

Hagood, N.W.; Von Flotow, A. (1991). Damping of structural vibrations with piezoelectric materials and passive electrical networks. Journal of Sound and Vibration, v. 146, n. 2, p. 243–268.

Heeg, J. (1993). Analytical and Experimental Investigation of Flutter Suppression by Piezoelectric Actuation. NASA Technical Paper 3241.

Hopkins, M.A *et al.* (1998). Active vibration-suppression systems applied to twin-tail buffeting. In: J. M. Sater (Ed.); SPIE. v. 3326, p.27–33. San Diego, CA.

Lallart, M.; Lefeuvre, É.; Richard, C.; Guyomar, D. (2008). Self-powered circuit for broadband, multimodal piezoelectric vibration control. Sensors and Actuators A: Physical, v. 143, n. 2, p. 377–382.

McGowan, A.M.R. (1999). An examination of applying shunted piezoelectrics to reduce aeroelastic response. CEAS/AIAA/ICASE/NASA Langley International Forum on Aeroelasticity and Structural Dynamics 1999. p.553–572.

McGowan *et al.* (1998). Aeroservoelastic and structural dynamics research on smart structures conducted at NASA Langley Research Center. Proceedings of SPIE - The International Society for Optical Engineering. v. 3326, p.188–201.

Mohammadi, S., (2008). Semi-passive vibration control using shunted piezoelectric materials. Thesis: Department of Electrical Engineering, Institut National des Sciences Appliquées de Lyon.

Moses, R.W. (1997). Vertical-tail-buffeting alleviation using piezoelectric actuators: some results of the actively controlled response of buffet-affected tails (ACROBAT) program. Proceedings of SPIE: The International Society for Optical Engineering. v. 3044, p.87–98.

Moses, R.W. (1999). Contributions to active buffeting alleviation programs by the NASA Langley Research Center. Proceedings: 39th AIAA/ASME/ASCE/AHS/ASC Structures, Structural Dynamics and Materials Conference, v. 2, p.1034–1042.

Nitzsche, F.; Grewal, A.; Zimcik, D. (1999). Structural component having means for cyclically varying its stiffness to control vibrations – Smart Spring. Patents: USA (5,973,440), European Community (EP-996570-B1) and Canada (2,242,214).

Nitzsche, F.; Feszty, D.; Grappasonni, C.; Coppotelli, G. (2013). Whirl-tower Open Loop Experiments and Simulations with an Adaptive Pitch Link Device for Helicopter Rotor Vibration. Proceedings: 54th AIAA/ASME/ASCE/AHS/ASC Structures, Structural Dynamics and Materials Conference. Paper No. AIAA-2013-1711.

Park, C.H.; Inman, D.J. (2003). Enhanced piezoelectric shunt design. Shock and Vibration, v. 10, n. 2, p. 127–133.

Richard, C.; Guyomar, D.; Lefeuvre, E. (2007). Self-powered electronic breaker with automatic switching by detecting maxima or minima of potential difference between its power electrodes. WO Patent, Patent PCT/FR2005/003000, Publication number: WO/2007/063194.

Richard, C.; Guyomar, D.; Audigier, D.; Bassaler, H. (2000). Enhanced semi passive damping using continuous switching of a piezoelectric device on an inductor. In: V. V. Varadan; N. M. Wereley; R. O. Claus; et al. (Eds.); Proceedings of SPIE: The International Society for Optical Engineering. v. 3989, p. 288–299.

Richard, C.; Guyomar, D.; Audigier, D.; Ching, G. (1999). Semi-passive damping using continuous switching of a piezoelectric device. (V.V. Varadan, N.M. Wereley, Y. Bar-Cohen., Eds.) Proceedings of SPIE: The International Society for Optical Engineering, v. 3672, March, p. 104–111.

Riordan, R.H.S. (1967). Simulated inductors using differential amplifiers. Electronics Letters, v. 3, n. 2, p. 50.

Uchino, K.; Ishii, T. (1988). Mechanical damper using piezoelectric ceramics. Nippon Seramikkusu Kyokai Gakujutsu Ronbunshi/Journal of the Ceramic Society of Japan, v. 96, n. 8, p. 863–867.

Wu, S. (1996). Piezoelectric shunts with a parallel R-L circuit for structural damping and vibration control. Smart Structures and Materials 1996: Passive Damping and Isolation. v. 27201259, p. 259–269.

ICAST2013: 24th International Conference on Adaptive Structures and Technologies October 7-9, 2013, Aruba

Modeling Impact-Based Vibration Energy Harvesting with Application to Frequency Up-Conversion

Adam M. Wickenheiser[1*]

[1] Department of Mechanical & Aerospace Engineering, The George Washington University, Washington, U.S.A.

Abstract

Many vibration energy sources available in the environment are better approximated by a sequence of impulses, periodic or random, rather than by harmonic motion. Combustion engines, footfalls, and passing vehicles are a few examples of vibration sources that fall under this category. This paper presents a distributed-parameter model of a piezoelectric energy harvester subjected to impacts. A single mode equivalent spring-mass-damper model is developed and shown to be an accurate representation of the impact dynamics. This simplified model is then used to find an estimate of the total energy harvested per impact. To demonstrate the utility of this model, it is applied to the frequency up-conversion technique for low-frequency energy harvesting. This technique uses a magnetic tip mass inside a ferrous enclosure to create a sequence of potential wells. The transition of the tip mass between the wells induces a pluck followed by a free response, similar to an impact. The dynamics of the base motion and the free response are then decoupled by (1) predicting how often the plucks occur and what their initial deflections are and (2) calculating the subsequent free response to each pluck. The frequency response of this method and a purely linear system are compared to the full nonlinear response. Simulations show that the nonlinear response converges to the impact-based approach at low frequencies (compared to the fundamental natural frequency), whereas it converges to the purely linear system in the vicinity of the fundamental natural frequency. Hence, the impact-based approach shows utility for modeling low-frequency energy harvesting using frequency up-conversion.

1. INTRODUCTION

Resonant piezoelectric energy harvesters convert mechanical energy into electrical through the strain induced in the material by inertial loads. Typically, piezoelectric material is mounted on a structure that oscillates due to excitation of the host structure to which it is affixed. If a natural frequency of the structure is matched to the predominant excitation frequency, resonance occurs, where large strains in the piezoelectric material are induced by relatively small excitations. In order to take advantage of resonance, the natural frequency of the device must be matched to the predominant frequency component of the base excitation [1]. For many potential applications, ambient vibrations are low frequency, requiring longer length scales or a larger mass to match the resonance frequency to the excitation frequency [2–4]. This, in

* amwick@gwu.edu

turn, reduces the specific power or power density of the device and may make it impractically large or heavy for the application.

In this study, a technique known as frequency up-conversion is employed to generate strong off-resonant responses. This technique is similar to the repetition of a bi-stable system, first studied by Moon [5], to create a sequence of potential wells; the transition between them induces a "pluck" followed by a free response at the fundamental frequency. A similar concept has been pursued by Tieck et al. (2006), consisting of a rack placed transversely near the tip of the beam that would periodically pluck the beam as it vibrated. Other concepts utilizing mechanical rectification have been proposed for harvesting energy from buoy motion [6] and low-frequency, rotating machinery [7]. All of these techniques effectively decouple the input frequency from the natural frequency of the device, enabling miniaturization while maintaining good performance at low frequencies.

In the following sections, the equations of motion (EOMs) are stated for a uniform beam with a magnetic tip mass under periodic base excitation. The modal expansion of the solution is reduced to a single mode (the fundamental) in order to derive an approximate model for low frequencies well below the fundamental frequency. A simplification is derived based on neglecting the base excitation; this simplification leads to a model of the beam's excitation in terms of a sequence of plucks followed by free vibrations. It is demonstrated that modeling the magnets as an equivalent linear spring and the electrical dynamics as a linear spring and damper is an effective way of predicting the free response. A few case studies are presented to highlight the accuracy of this approximate model.

2. EQUATIONS OF MOTION

The layout of the piezoelectric, vibration-based energy harvester and the nearby magnetized structures used for mechanical rectification is presented in Fig. 1. For this study, a bimorph configuration is considered, in which piezoelectric layers are bonded to both sides of an inactive substructure. Electrodes are assumed to cover the upper and lower surfaces of each layer, and they are wired together in the "parallel" configuration, as depicted. It is assumed that the electrodes and connecting wires have negligible resistance and that the resistivity of the piezoelectric material is significantly higher than that of the external circuitry; thus, the transducer impedance is assumed to be purely reactive. A tip mass is connected to the free end of the beam, and its center of mass is displaced axially from the connection point by a distance d_t. Tip masses are traditionally added to decrease the natural frequency of the beam and to increase the strain due to base excitation. In this situation, the tip mass is considered to be a permanent magnet and is attracted to ferromagnetic structures placed in a line parallel to the y-axis with spacing d_m between them. These structures are not magnets themselves; rather, they become magnetized due to the proximity of the magnetic tip mass.

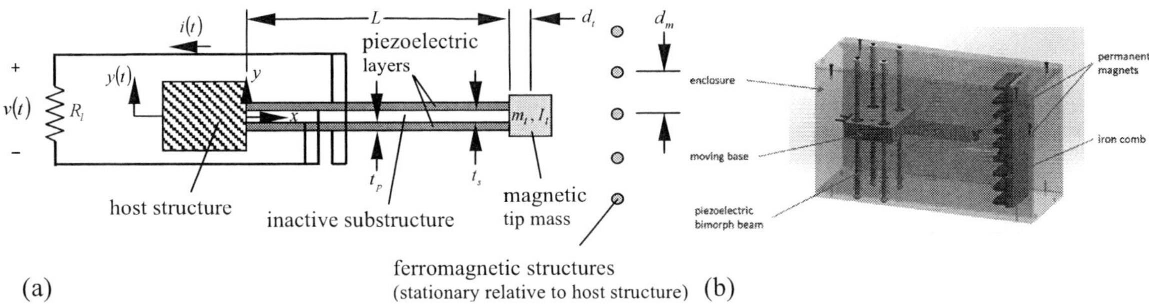

Figure 1. (a) Layout and geometric parameters of cantilevered vibration energy harvester in parallel bimorph configuration with magnetic tip mass. (b) CAD model of a prototype device for demonstrating magnetically induced frequency up-conversion energy harvesting.

In this model, the states of the electromechanical system are the following: $w(x,t)$ is the relative transverse deflection of the beam with respect to its base, $v(t)$ is the voltage across the energy harvester as seen by the external circuit, and $i(t)$ is the net current flowing into the external circuit. The input to the system is $y(t)$, the absolute transverse displacement of the base; therefore, $w(x,t)+y(t)$ is the absolute transverse deflection of the beam. The assumptions of Euler-Bernoulli beam theory [8] are adopted, and the following linearized constitutive equations for the piezoelectric layers are used:

$$\begin{aligned} T_1 &= c_{11}^E S_1 - e_{31} E_3 \\ D_3 &= e_{31} S_1 + \varepsilon_{33}^S E_3 \end{aligned} \tag{1}$$

where T is stress, S is strain, E is electric field, D is electric displacement, c is Young's Modulus, e is piezoelectric constant, and ε is dielectric constant. The subscripts indicate the direction of perturbation; in the cantilever configuration shown in Fig. 1, 1 corresponds to axial and 3 corresponds to transverse. The superscript $(\cdot)^E$ indicates a linearization at constant electric field, and the superscript $(\cdot)^S$ indicates a linearization at constant strain [9]. The use of Eq. (1) assumes the hypotheses of small deflections and plane stress, which is reasonable since the beams are not directly loaded in the other directions and the width-to-thickness ratio is large. The stress within the substrate layer is given simply by the linear stress-strain relationship $T_1 = c_{11,s} S_1$, where $c_{11,s}$ is Young's Modulus of the substrate material in the axial direction.

Under these assumptions, the transverse mechanical equation of motion is

$$\begin{aligned} &(\rho A)_{eff} \frac{\partial^2 w(x,t)}{\partial t^2} + (EI)_{eff} \frac{\partial^4 w(x,t)}{\partial x^4} + \vartheta \left[\frac{d\delta(x)}{dx} - \frac{d\delta(x-L)}{dx} \right] v(t) = \\ &-(\rho A)_{eff} \frac{d^2 y(t)}{dt} - \left[m_t \frac{d^2 y(t)}{dt} + f_{mag}(t) \right] \left[\delta(x-L) + d_t \frac{d\delta(x-L)}{dx} \right] \end{aligned} \tag{2}$$

where $(\rho A)_{eff}$ is the mass per unit length of the beam, and $(EI)_{eff}$ is the effective bending stiffness [10]. (Note that if the beam segment is monolithic, these constants are simply the product of the density and cross-sectional area or the Young's Modulus and the moment of inertia, respectively.) The constant multiplying the $v(t)$ term is defined as ϑ, the electromechanical coupling coefficient. Factoring into the load applied at the tip of the beam, m_t is the mass of the tip mass, $f_{mag}(t)$ is the magnetic force, and, as in Fig. 1, d_t is the displacement between the end of the beam and the tip mass centroid. Finally, $\delta(\cdot)$ is the

Dirac delta function. In this study, the magnetic force is assumed to be purely in the y-direction. Although there is a stiffening effect due to the axial attractive force, it is considered negligible. The negative sign on the magnetic force indicates that it is an attractive force.

The electrical EOM can be found by integrating the electric displacement over the surface of the electrodes, yielding the net charge $q(t)$ [9]:

$$q(t) = \vartheta \frac{\partial w(x,t)}{\partial x}\bigg|_{x=L} - C_0 v(t) \tag{3}$$

where the constant multiplying the $v(t)$ term is defined as C_0, the net clamped capacitance of the segment [10]. Eqs. (2–3) provide a coupled system of equations; these can be solved by relating the voltage $v(t)$ to the charge $q(t)$ through the external electronic interface.

One of the results of this study is to show that although the magnetic coupling is effective at low frequencies (compared to the beam's fundamental frequency), the effects of this coupling decrease to 0 in the vicinity of the fundamental frequency. In order to calculate the frequency transfer function, it is assumed that the electrical load can be represented by a resistor with value R_l, shown in Fig. 1. Eq. (3) can then be rewritten as

$$\frac{v(t)}{R_l} = i(t) = \frac{dq(t)}{dt} = \vartheta \frac{\partial^2 w(x,t)}{\partial t \partial x}\bigg|_{x=L} - C_0 \frac{dv(t)}{dt} \tag{4}$$

The system of coupled equations (2–3) can be solved by assuming that the transverse deflection of the beam can be written as a convergent series expansion of eigenfunctions, i.e.

$$w(x,t) = \sum_{i=1}^{\infty} \varphi_i(x)\eta_i(t) \tag{5}$$

where $\varphi_i(x)$ is the ith transverse mode shape function, and $\eta_i(t)$ is the ith modal displacement. The eigenvalues and mode shapes of the cantilevered beam with offset tip mass can be found in [11].

Substituting (5) into (2) and applying the orthogonality condition results in

$$\frac{d^2\eta_k(t)}{dt^2} + 2\zeta_k\omega_k \frac{d\eta_k(t)}{dt} + \omega_k^2\eta_k(t) + \Theta_k v(t) = -(\rho A)_{eff}\gamma_k \frac{d^2 y(t)}{dt^2} - \beta_k\left[m_t \frac{d^2 y(t)}{dt} + f_{mag}(t)\right] \tag{6}$$

at which point a modal damping term has been inserted. The kth modal short-circuit (i.e. $v(t) = 0$) natural frequency ω_k is given by

$$\omega_k = \sqrt{\frac{\lambda_k^4 (EI)_{eff}}{(\rho A)_{eff} L^4}} \tag{7}$$

for Euler-Bernoulli beams, where L is the length of the beam, and λ_k is the kth eigenvalue. Eq. (6) constitutes the EOM for the kth transverse vibrational mode. The modal influence coefficients appearing in Eq. (6) are given by

$$\Theta_k = \vartheta \frac{d\varphi_k(x)}{dx}\bigg|_{x=L}, \quad \gamma_k = \int_0^L \varphi_k(x)dx, \quad \beta_k = \varphi_k(L) + d_t \frac{d\varphi_k(x)}{dx}\bigg|_{x=L} \tag{8}$$

where Θ_k is the modal electromechanical coupling coefficient, γ_k is the modal influence coefficient of the distributed inertial force along the beam, and β_k is the modal influence coefficient of the concentrated force at the tip. A similar decoupling of the electrical EOM, Eq. (3), yields

$$\frac{v(t)}{R_l} = \sum_{i=1}^{\infty} \Theta_i \frac{d\eta_i(t)}{dt} - C_0 \frac{dv(t)}{dt} \tag{9}$$

The applied magnetic force is given by

$$f_{mag}(t) = F_{mag} \sin\left[\frac{2\pi}{d_m}\left(w(x_m,t) + y(t)\right)\right] \tag{10}$$

where x_m is the x-coordinate of the tip mass centroid, and d_m is the transverse spacing of the ferrous structures, as depicted in Fig. 1. The magnitude of this force F_{mag} is a complicated function of the material properties and geometry of the tip mass and the ferromagnetic structures [5,12]. The position of the tip mass centroid can be written in terms of the modal coordinates:

$$w(x_m,t) \approx w(L,t) + d_t \left.\frac{\partial w(x,t)}{\partial x}\right|_{x=L} = \sum_{i=1}^{\infty} \beta_i \eta_i(t) \tag{11}$$

Substituting Eq. (11) into Eq. (10) and inserting the result into Eq. (6) yields a system of coupled ordinary differential equations for the modal displacements $\eta_k(t)$.

3. APPROXIMATE MODELS

3.1 Linearization about fundamental frequency

Setting the magnetic force $F_{mag} = 0$ results in a linear system of equations (6,9). Consequently, if the base excitation is sinusoidal, the modal responses $\eta_k(t)$ and output voltage $v(t)$ are also sinusoidal at the driving frequency of the base excitation. If the excitation frequency is near the fundamental frequency, the dynamics of the fundamental mode dominate the response. In this case, the voltage frequency transfer function near the fundamental frequency can be accurately approximated by

$$\frac{V(\omega)}{Y(\omega)} \approx \frac{R_l \Theta_1^2 \dfrac{i\left[(\rho A)_{eff}\gamma_1 + m_t\beta_1\right]\omega^3}{\omega_1^2 - \omega^2 + i2\zeta_1\omega_1\omega}}{iR_l C_0 \omega + 1 + R_l \Theta_1^2 \dfrac{i\omega}{\omega_1^2 - \omega^2 + i2\zeta_1\omega_1\omega}} \tag{12}$$

where $i = \sqrt{-1}$, and ω is the base excitation frequency [13]. In the sequel, the 1 subscript is dropped and the fundamental natural frequency is written ω_n. Finally, the power frequency transfer function is given by $P(\omega) = V^2(\omega)/R_l$.

3.2 Free response approximation at low frequencies

In the presence of the magnetic field, the EOMs become nonlinear, and the analysis based off of the frequency transfer function detailed in the previous subsection is no longer valid. Instead, the vibrations induced by the spatially periodic magnetic field are interpreted as a series of plucks that occur each time the tip mass crosses an unstable equilibrium point between the ferrous structures. Each pluck is followed by a free response – underdamped in this case – superposed on the relatively slow base motion. An example response showing these two superposed motions is depicted in Fig. 2. The free response at the fundamental frequency of the beam, as opposed to the frequency of the base motion, drives the majority of the energy harvested.

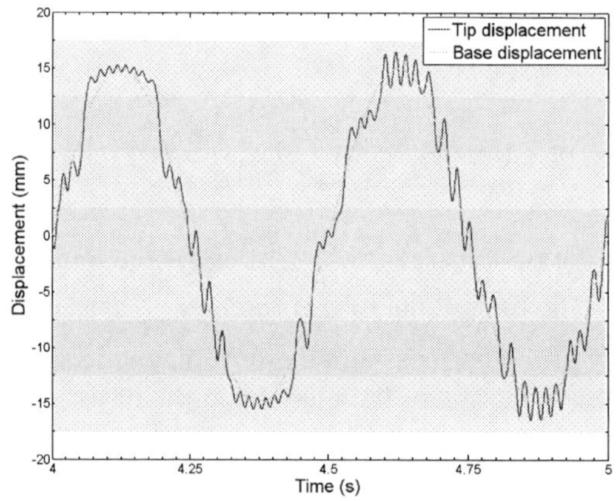

Figure 2. Absolute base and tip displacements: $d_m = 5\,\text{mm}$, $y(t) = Y\sin(\omega t)$, $Y = 15\,\text{mm}$, $\omega = 2\,\text{Hz}$. The shaded areas are the basins of attraction of the stable equilibria [13].

To analyze the energy harvested from a pluck, first the effect of the magnetic field strength on the free response is considered. To simplify the analysis, the inertial force due to base excitation is assumed to be negligible (it is assumed to happen on a much slower time scale than the beam's free response), and the effect of the energy dissipated by the resistor is approximated by an additional damping term. To estimate this additional damping, first the mechanical natural frequency, including the effects of the magnets, must be found. The modal EOM, Eq. (6), can be written as

$$\frac{d^2\eta(t)}{dt^2} + 2\zeta\omega_n\frac{d\eta(t)}{dt} + \omega_n^2\eta(t) = -F_{mag}\beta\sin\left[\frac{2\pi}{d_m}(\beta\eta(t) + y(t))\right] \quad (13)$$

where the electrical and base excitation terms have been dropped.

After each pluck, the beam oscillates freely about the center of the magnetic potential well. Assuming this well has a nearly parabolic potential energy profile means that the magnet acts like an additional spring-like force. To see this mathematically, Eq. (13) is linearized about the point $y(t) = kd_m$, $\eta(t) = 0$, where k is an integer, giving

$$\frac{d^2\eta(t)}{dt^2} + 2\zeta\omega_n\frac{d\eta(t)}{dt} + \left(\omega_n^2 + \frac{2\pi\beta^2}{d_m}F_{mag}\right)\eta(t) = 0 \quad (14)$$

Thus, the term in parentheses is the square of the natural frequency of the response including the linearized magnetic dynamics, to be labeled ω_{n+mag}. Since k is arbitrary, these dynamics govern the motion within any of the potential wells.

The amplitude of each pluck, and hence, the initial condition of the free response, is determined by the location of the unstable equilibrium between each pair of ferrous structures. In equilibrium,

$$\omega_n^2 \eta(t) = -F_{mag}\beta \sin\left[\frac{2\pi}{d_m}(\beta\eta(t) + y(t))\right] \tag{15}$$

which is derived from Eq. (14). Considering the case when the base is moving upward, i.e. $\dot{y}(t) > 0$, the pluck occurs when

$$\sin\left[\frac{2\pi}{d_m}(\beta\eta(t) + y(t))\right] = 1 \tag{16}$$

When this condition occurs, further vertical motion of the tip results in a decreased downward magnetic force. At this point, the beam has passed over a local maximum in the magnetic potential, and it begins accelerating towards the next stable equilibrium. By plugging Eq. (16) into Eq. (15), the amplitude of the pluck can be found:

$$\eta_0 = \frac{-F_{mag}\beta}{\omega_n^2} \tag{17}$$

If the times of the plucks are denoted t_k, then substituting Eq. (17) into Eq. (16) and solving for t_k, and using the fact that $y(t_k) = Y\sin(\omega t_k)$, yields

$$t_k = \frac{1}{\omega}\sin^{-1}\left(\frac{4k-3}{4}\frac{d_m}{Y} + \frac{F_{mag}\beta^2}{\omega_n^2 Y}\right),\ k = 1,\ldots,N,\ \text{ where } (N-1)d_m < Y \le Nd_m \tag{18}$$

A similar formula can be derived for the pluck times when $\dot{y}(t) < 0$:

$$t_{k+N} = \frac{\pi}{\omega} - \frac{1}{\omega}\sin^{-1}\left(\frac{4(N-k+1)-1}{4}\frac{d_m}{Y} - \frac{F_{mag}\beta^2}{\omega_n^2 Y}\right),\ k = 1,\ldots,N \tag{19}$$

It is assumed that the effect of the energy dissipated by the resistor is approximated by an additional damping term and an increase in the natural frequency. According to [14], these effects are functions of the electromechanical coupling and the impedance of the device. Respectively, these parameters are given by $k_e^2 = \Theta^2/(C_0\omega_{n+mag}^2)$ and $\alpha = R_l C_0 \omega_{n+mag}$. In the case of steady-state oscillations, the effective natural frequency can be approximated by [14]

$$\omega_{n,eff}^2 = \omega_{n+mag}^2\left(1 + k_e^2\frac{\alpha^2}{1+\alpha^2}\right) \tag{20}$$

This formula is validated for free oscillations in the sequel. Under the same conditions, the total effective modal damping ratio is written as

$$\zeta_{eff} = \zeta\frac{\omega_n}{\omega_{n,eff}} + \zeta_e \tag{21}$$

the sum of the mechanical and electrical damping, adjusted for the fact that these damping ratios are defined for different natural frequencies. The electrical damping term can be accurately approximated by

$$\zeta_e = \frac{k_e^2}{2\sqrt{1-k_e^2}}\frac{\alpha}{1+\alpha^2} \tag{22}$$

To calculate the voltage response, and hence the power, Eq. (9) is employed, which requires the derivative of the modal response. By substituting Eqs. (20–21) into Eq. (14), the derivative of the free response, assumed to be underdamped, can be found to be

$$\frac{d\eta(t)}{dt} == \eta_0 \frac{\omega_{n,eff}^2}{\omega_{d,eff}} e^{-\zeta_{eff}\omega_{n,eff}t} \sin(\omega_{d,eff}t) \tag{24}$$

where $\omega_{d,eff} = \omega_{n,eff}\sqrt{1-\zeta_{eff}^2}$ is the effective damped natural frequency, and the initial amplitude η_0 is given by Eq. (17). This solution can now be plugged into Eq. (9) to find the voltage response.

The energy harvested during the free vibrations can then be adequately approximated by the following formula:

$$E(t) = \frac{1}{R_l}\int_0^t v^2(\tau)d\tau \approx \frac{1}{2R_l}\int_0^t A^2 e^{-2\zeta_{eff}\omega_{n,eff}\tau} d\tau = \frac{A^2}{4\zeta_{eff}\omega_{n,eff}R_l}\left(1-e^{-2\zeta_{eff}\omega_{n,eff}\tau}\right) \tag{25}$$

where

$$A^2 = \frac{\Theta^2 R_l^2 \frac{\omega_{n,eff}^4}{\omega_{d,eff}^2}}{1-2\zeta_{eff}\alpha_{eff}+\alpha_{eff}^2}\eta_0^2, \text{ where } \alpha_{eff} = R_l C_0 \omega_{n,eff} \tag{26}$$

The accuracy of this approximate formula can be seen in Fig. 3. The results of the simulation of the original EOMs, Eqs. (6,9), are plotted using a solid line, whereas Eq. (25) is plotted using a dashed line. To arrive at Eq. (25), it is assumed that the initial transients and the oscillating terms in $v^2(t)$ integrate out to 0; hence, the result is a smooth exponential curve. Although instantaneously the approximate curve may not be accurate, it matches the overall growth of the exact solution. Hence, Eq. (25) is an accurate representation of the energy harvested from a free vibration with a non-zero initial deflection and a zero initial velocity.

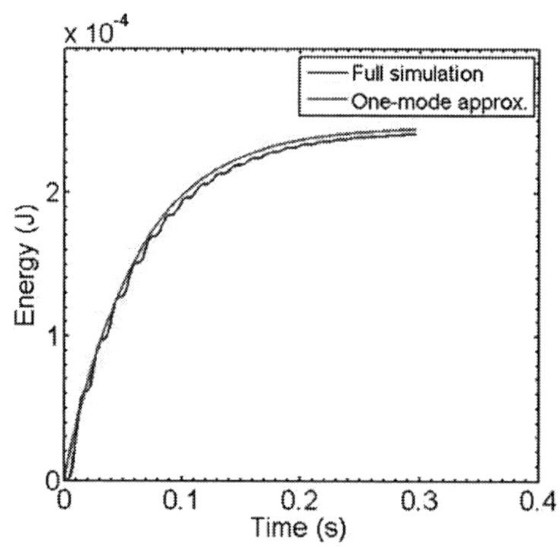

Figure 3. Energy harvested during free vibration after an initial pluck, comparison between simulation using exact EOMs and approximation using Eq. (28). Parameters are listed in Table 1.

For a sequence of plucks, which is what occurs with the frequency up-conversion technique, it is assumed that the plucks are instantaneous and that the deflection is "reset" to the value given by Eq. (17) after each pluck. Hence, the total energy harvested during a half cycle of the base excitation is

$$E_{total} = \sum_{k=1}^{N} E(t_{k+1} - t_k) \tag{27}$$

4. Simulated response to sinusoidal base excitation

In this section, the response of the system to sinusoidal base excitation, $y(t) = Y\sin(\omega t)$, is presented in both the time and frequency domains. The geometry and material properties used in the following simulations are listed in Table 1. The tip mass m_t is approximately one-third of the overall beam mass, and its moment of inertia I_t is calculated assuming the mass is roughly cube-shaped. The resistor value chosen for this study is the optimal value for energy harvesting at the fundamental frequency in the limit of small electromechanical coupling, i.e. $R = 1/(C_0\omega_1)$ [15].

The first three natural frequencies of the beam are $\omega_1 = 34.1\,\text{Hz}$, $\omega_2 = 271.6\,\text{Hz}$, and $\omega_3 = 806.8\,\text{Hz}$. Since frequencies around and below the fundamental frequency are of interest in this study, a three-mode expansion of the beam displacement is deemed sufficient. Furthermore, since the magnetic force is applied at the tip of the beam, only the fundamental mode is significantly excited by the plucking.

Table 1. Geometry and material properties

	Beam properties:	
L	length	100 mm
b	width	20 mm
t_s	thickness of substructure	0.5 mm
t_p	thickness of PZT layer	0.4 mm
ρ_s	density of substructure	8740 kg/m³
ρ_p	density of PZT	7800 kg/m³
$c_{11,s}$	Young's modulus of substructure	102 GPa
c_{11}^E	Young's modulus of PZT	66 GPa
e_{31}	piezoelectric constant	-12.54 C/m²
ε_{33}^S	permittivity	15.93 nF/m
ζ	modal damping ratio	6.4%
	Tip mass properties:	
m_t	mass	10 g
d_t	centroid displacement	5 mm
I_t	moment of inertia	1.7x10⁻⁷ kg-m²
	Derived properties:	
$(\rho A)_{eff}$	mass per length	0.20 kg/m
$(EI)_{eff}$	bending stiffness	0.25 N-m²
C_0	net clamped capacitance	160 nF
k_e^2	electromechanical coupling coefficient	0.049
R	resistance	29.3 kΩ

Wickenheiser and Garcia [13] have shown through simulations that the magnet has an increasing effect as the base excitation frequency decreases to 0. As is discussed in the sequel, at low frequencies relative to the fundamental resonance, the response converges to a sequence of free responses. In this regime, the inertial forces are negligible, and so the disturbances due to the magnetic force are relatively large. As the driving frequency increases, the beam tip has less time to oscillate in each potential well, and, thus, the power harvested is reduced. As $\omega \to \omega_1$, the frequency of the base motion approaches the frequency of the impulse response of the beam from the magnetic force. Hence, the time in which the beam is in free response decays to 0, and so linear, resonant behavior dominates. A discussion of the variation in frequency response with respect to the magnet parameters F_{mag} and d_m can be found in [13].

Fig. 4(a) compares two methods of simulating the tip deflection response: using the original EOMs and assuming a series of undamped free responses give by Eq. (27). The most notable difference that can be seen from the figure is that the free response assumption does not take into account the base motion, which causes the solution to the EOMs to drift upward or downward depending on the sign of the base velocity. Another difference between the two models is that the assumed time of the plucks, indicated by the vertical lines and given by Eqs. (18–19), generally occur before the plucks in the actual solution. This happens because the beam has enough inertia to resist the pull of the next magnet, i.e. to overcome the potential well barrier between magnets. Only after the beam's velocity decays sufficiently does it become trapped in the next potential well in the sequence.

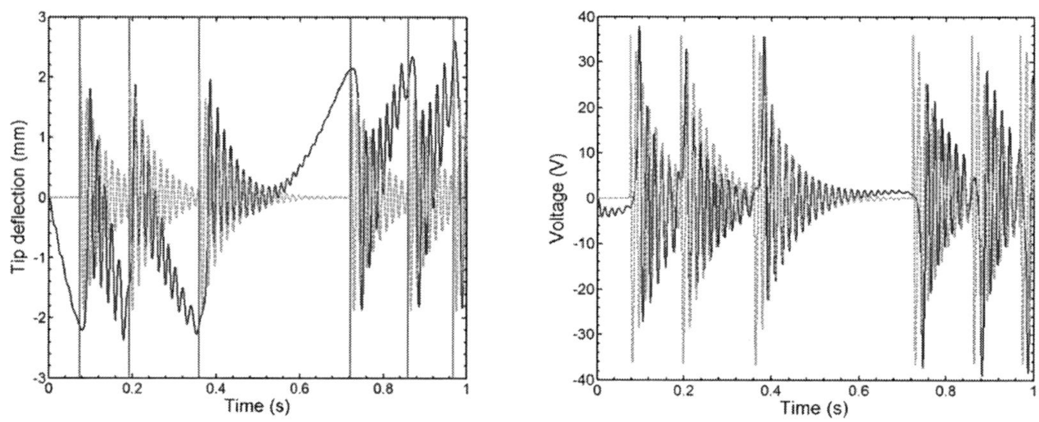

(a) (b)

Figure 4. Comparison of simulated tip deflection from EOMs (solid) and a series of plucks, Eq. (27) (dashed). Vertical lines indicate the estimated times of plucking according to Eqs. (18–19). (a) Time deflection vs. time. (b) Voltage vs. time.

Fig. 4(b) depicts the voltage response during the same simulation that has been plotted in Fig. 4(a). A comparison of the two curves plotted shows an excellent agreement between the simulation results and the predicted voltage from the sequence of plucks. There is a slight asymmetry in the curve representing the simulation results due to the base excitation. This discrepancy is much less pronounced than in Fig. 4(a) since the voltage is dominated by the velocity of the tip deflection, which is not affected directly by the base motion. The primary difference between the two curves in Fig. 4(b) is the error in predicting when the plucks occur, as discussed in the previous paragraph. This error causes an over-prediction in the number of cycles of free oscillation, but this error is small compared to the duration of the free response between plucks. The initial magnitude of the voltage free response is well predicted, however.

The comparison between the power frequency transfer functions of the simulation, the approximation by a series of plucks, and the linear system (without magnets) is shown in Fig. 5. This plot is generated using the same procedure as the one used to produce Fig. 4. The most striking feature of this plot is that the simulation results are seen to converge to the approximation by a series of plucks for low frequencies and converge to the linear system approximation at frequencies approaching the fundamental resonance. As previously mentioned, at frequencies around the fundamental resonance, the beam is not allowed to vibrate freely because the pluck frequency exceeds its natural frequency. Hence, there is no exponential decay in the amplitude of the tip deflection between plucks. In this case, the forced response (i.e. particular solution) dominates the motion, and so the frequency transfer function of the nonlinear system approaches that of the linear system. At low frequencies, the base excitation term becomes negligible, and so the mechanical EOM reduces to Eq. (13), the basis for the approximation. In this scenario, the magnetic force drives the excitation of the beam, whereas the inertial force due to the base excitation is negligible. This is manifested in the decrease in the linear response at low frequencies.

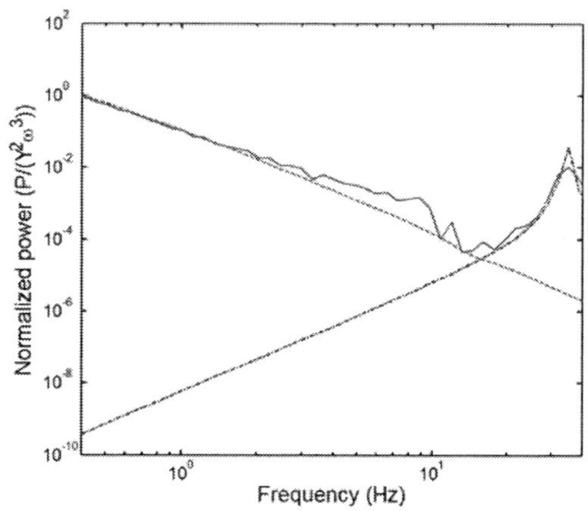

Figure 5. Comparisons of normalized power frequency transfer functions between EOMs (solid), a series of plucks (dashed), and the linear system (dash-dot).

6. Conclusions

This paper presents an accurate means of approximating the non-linear response of the frequency up-conversion technique as a series of free responses. This simplification is based on the assumption that the base excitation is negligible, and so it only holds at low frequencies compared to the fundamental resonance of the beam. This approximation, however, is useful in the design of energy harvesters utilizing this technique as it enables power to be generated at very low frequencies. This means that the device can be designed for a fundamental frequency much higher than the nominal base excitation frequency, which tends to result in smaller and lighter transducers. At low frequencies, the approximation derived herein is shown to agree well with the simulation results of the full non-linear equations of motion in terms of displacement, voltage, and power harvested. It is confirmed through analysis of the frequency transfer function that the non-linear system converges to the approximation by a series of free responses at low frequencies and to the linear system response at frequencies around the fundamental. Hence, a combination of analytical solutions can be used to predict the energy harvesting performance of this non-linear device in lieu of simulation of the full dynamics equations.

Acknowledgements

This work was supported by the Central Intelligence Agency [grant number CCNS20965F].

References

1. Anderson, B. and Wickenheiser, A., "Performance analysis of frequency up-converting energy harvesters for human locomotion", *SPIE Smart Structures and Materials & Nondestructive Evaluation and Health Monitoring*, San Diego, California, 2012, Vol. 8341, 834102.
2. Roundy, S., Wright, P. K., and Rabaey, J., "A study of low level vibrations as a power source for wireless sensor nodes", *Computer Communications*, 2003 Vol. 26, pp. 1131–1144.
3. Wickenheiser, A. and Garcia, E., "Design of energy harvesting systems for harnessing vibrational motion from human and vehicular motion", *SPIE Smart Structures and Materials & Nondestructive Evaluation and Health Monitoring*, San Diego, California, 2010, Vol. 7643, 76431B.
4. Wickenheiser, A. M., "Design Optimization of Linear and Nonlinear Cantilevered Energy Harvesters for Broadband Vibrations", *Journal of Intelligent Material Systems and Structures*, 2011, Vol. 22, pp. 1213–1225.
5. Moon, F. C., "Problems in magneto-solid mechanics ", *Mechanics Today*, 1978, Vol. 4, pp. 307–390.
6. Murray, R. and Rastegar, J., "Novel Two-Stage Piezoelectric-Based Ocean Wave Energy Harvesters for Moored or Unmoored Buoys", *SPIE Smart Structures and Materials & Nondestructive Evaluation and Health Monitoring*, San Diego, California, 2009, Vol. 7288, 72880E.
7. Rastegar, J. and Murray, R., "Novel Two-Stage Electrical Energy Generators for Highly Variable and Low-Speed Linear or Rotary Input Motion", *SPIE Smart Structures and Materials & Nondestructive Evaluation and Health Monitoring*, San Diego, California, 2010, Vol. 7643, 76430C.
8. Inman, D. J., *Engineering Vibration*, Pearson, Upper Saddle River: *Pearson*, 2007.
9. IEEE Standard on Piezoelectricity, New York: *IEEE*, 1987
10. Wickenheiser, A. M., "Analysis of Energy Harvesting Using Frequency Up-Conversion by Analytic Approximations", *Small-Scale Energy Harvesting*, Croatia: *InTech*, 2012, pp. 211–234.
11. Oguamanam, D. C. D., "Free vibration of beams with finite mass rigid tip load and flexural-torsional coupling", *International Journal of Mechanical Sciences,* 2003, Vol. 45, pp. 963–979.
12. Stanton, S. C., McGehee, C. C., and Mann, B. P., "Nonlinear Dynamics for Broadband Energy Harvesting: Investigation of A Bistable Piezoelectric Inertial Generator", *Physica D*, 2010, Vol. 239, pp. 640–653.
13. Wickenheiser, A. M. and Garcia, E., "Broadband vibration-based energy harvesting improvement through frequency up-conversion by magnetic excitation", *Smart Materials and Structures*, 2010, Vol. 19, No. 6, 065020.
14. Davis, C. L. and Lesieutre, G. A., "A modal strain energy approach to the prediction of resistively shunted piezoceramic damping", *Journal of Sound and Vibration*, 1995, Vol. 184, No. 1, pp. 129–139.
15. Wickenheiser, A. M. and Garcia, E., "Power Optimization of Vibration Energy Harvesters Utilizing Passive and Active Circuits", *Journal of Intelligent Materials Systems and Structures*, 2010, Vol. 21, No. 13, pp. 1343–1361.

ICAST2013: 24th International Conference on Adaptive Structures and Technologies October 7-9, 2013, Aruba

Comparison study of standard, conventional SSHI and self-powered SSHI interface circuits in piezoelectric energy conversion

Bin Li and Jeong Ho You*

Department of Mechanical Engineering, Southern Methodist University, Dallas, TX, USA

Abstract

Energy conversion efficiency of multiple identical piezoelectric oscillators has been studied in comparison between standard interface circuit, conventional parallel and series synchronized switch harvesting on inductor (P-SSHI and S-SSHI) interface circuits, and self-powered S-SSHI interface circuit. The simulation results indicate that the total output power increases linearly with the number of identical piezoelectric oscillators. For both conventional P-SSHI and conventional S-SSHI circuits, the requirements of displacement sensors and power supply significantly limit their applications. The self-powered S-SSHI interface circuit has been introduced to remove these requirements and it exhibits a higher harvesting efficiency than the standard interface circuit. The self-powered S-SSHI interface circuit is therefore considered as a good candidate for AC/DC conversion in energy harvesting using multiple piezoelectric oscillators.

1. INTRODUCTION

In our previous studies [1-3], a piezoelectric energy harvester using multiple piezoelectric cantilever plates are developed to harvest low frequency acoustic energy. In order to convert the AC output from the vibrating piezoelectric plates to DC to charge terminal electric loads such as an electrochemical battery, an AC/DC conversion external circuit should be added. Since the efficiency of piezoelectric energy harvesting system strongly depends on the design of external circuit, we investigate the effects of external circuits on the conversion efficiency of multiple piezoelectric oscillators.

Most previous studies for AC/DC conversion in piezoelectric energy harvesting system focused on a single piezoelectric oscillator [4-7]. The most basic AC/DC converter for piezoelectric oscillator is the standard interface circuit [8]. The standard interface circuit simply includes a full-bridge rectifier and a filter capacitor connected to the piezoelectric element. Later, nonlinear electronic interface circuits such as a synchronized switch harvesting on inductor (SSHI) interface circuit have been introduced to increase the energy conversion efficiency [9-13]. The principle of SSHI is inverting the piezoelectric element's voltage at the zero velocity or extremum displacement of the piezoelectric element to reduce the phase shift between the piezoelectric element's voltage and velocity which results in the increase of energy harvesting efficiency. Therefore, the conventional SSHI requires displacement sensors to measure the zero velocity moment and an external power supply to control the inversion switch. Implementation of displacement

* Corresponding author: jyou@lyle.smu.edu

sensors may be very difficult or even impossible due to the space limit of system. Also, the power consumption by the switches may not be negligible compared with the harvested power. In order to overcome these difficulties, a self-powered S-SSHI interface circuit has been developed to make the AC/DC conversion for a single piezoelectric oscillator without the uses of displacement sensors and power supply [14-16].

To the best of our knowledge, a self-powered S-SSHI circuit for multiple piezoelectric oscillators is not available. In this study, a self-powered S-SSHI circuit for multiple piezoelectric oscillators used in piezoelectric energy harvesting is introduced and studied theoretically and numerically using the circuit simulation software *Multisim*. The conversion efficiency of the self-powered S-SSHI circuit is compared with standard and conventional SSHI circuits. In Section 2, the mechanism of piezoelectric energy harvester is discussed. The equivalent circuit representation for a piezoelectric oscillator and the configurations of the interface circuits are introduced in Section 3. In Section 4, the circuit design for the multiple piezoelectric oscillators is presented. The simulation and theoretical results for the electric responses of interface circuits for multiple identical piezoelectric oscillators are discussed in Section 5. Finally, the conclusions are drawn in the last section.

2. IN-PHASE MULTIPLE PIEZOELECTRIC OSCILLATORS

In our previous studies [1-3], multiple parallel bimorph PZT piezoelectric cantilever plates (Stripe Actuator 40-2010 by *APC International, Ltd.*) were placed inside a 42 cm quarter-wavelength straight-tube resonator to harvest acoustic energy at a low frequency (~200 Hz), as shown in figure 1 (a) and (b). When the tube is excited by an external sound wave at its eigenfrequency, a resonant standing wave is developed inside the tube. The resonant standing wave excites the vibration motion of the cantilever plates, resulting in generating electricity. The piezoelectric plates are designed to have the same structural resonance frequency with the acoustic resonance frequency of the tube, so as to maximize the harvested energy. All piezoelectric plates are vibrated in phase inside the tube resonator.

The output voltage in an open-circuit condition (V_{open}) of a single piezoelectric plate placed at 5 cm from the tube inlet was measured as 1.51 V with the incident SPL of 100 dB at 199 Hz. Since the output voltage of piezoelectric plate is linearly proportional to the incident sound pressure, the output voltage will be 6.04 V with 112 dB SPL. Likewise, when the incident SPL is 112 dB, the output voltages of the piezoelectric plates *A*, *B* and *C* located near the tube inlet in the open-circuit condition are 5.44, 5.16 and 4.84 V, as shown in figure 1(c).

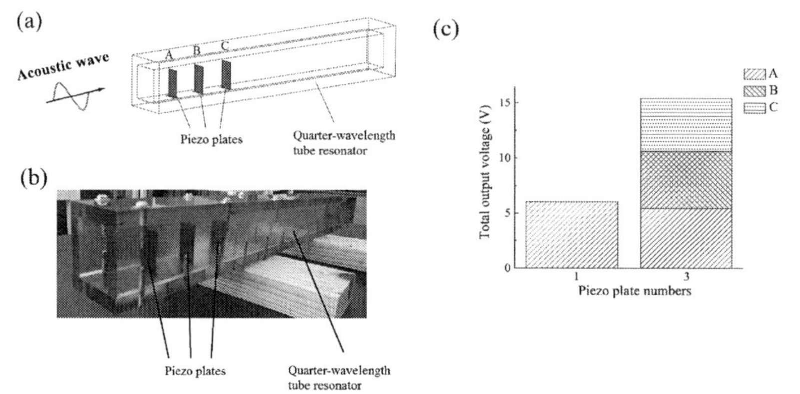

Figure 1. Sketch of (a) multiple piezoelectric oscillators used in acoustic energy harvesting. (b) Experimental setup. (c) Output voltages in open-circuit condition when placing one piezoelectric plate

and three piezoelectric plates.

3. INTERFACE CIRCUITS FOR MULTIPLE PIEZOELECTRIC OSCILLATORS

In order to design an interface circuit to convert AC to DC in multiple piezoelectric oscillators, the mechanisms for interface circuits for multiple identical piezoelectric oscillators need to be investigated. The parallel bimorph piezoelectric plate consists of a carbon fiber sandwiched by PZT-5X shwon in figure 2(a). Figure 2(b) shows the equivalent circuit for a single piezoelectric oscillator which is modeled as an equivalent current source I_{eq} in parallel connection with an internal capacitor C_p (C_p=75 nF for the piezoelectric plate used) [17]. I_{eq} can be calculated as $I_{eq} = j\omega C_p V_{open}$. With V_{open} of 6.04 V and the frequency of 199 Hz, the magnitude of I_{eq} is 566 μA. For n multiple piezoelectric oscillators, the equivalent circuit contains the current sources (I_{eq1}, I_{eq2}, ⋯, I_{eqn}) and the internal capacitances (C_{p1}, C_{p2},⋯,C_{pn}) connected in parallel.

A simple standard circuit typically composes of a full-bridge diode rectifier connected to the piezoelectric element, and a filter capacitor C_r which is used to maintain the rectified voltage stably, as shown in figure 2(c). The resistor R is used to match the impedance with the piezoelectric element to ensure a maximization of the harvesting efficiency. The most representative nonlinear electronic interface circuits are P-SSHI and S-SSHI, as shown in figure 2(d) and (e) respectively. In both circuits, an electronic switch S is added to the piezoelectric element before connecting to the rectified bridge. The only difference between P-SSHI and S-SSHI is the inductor L is connected to the piezoelectric element in parallel and in series. The switch is turned on when the displacement of piezoelectric oscillator reaches its extrema. The instant conduction of the switch at the extremum displacement of the piezoelectric element excites the LC oscillation and leads to the inversion of the piezoelectric voltage. In conventional SSHI circuits, displacement sensors and a power supply are required to detect the piezoelectric element's displacement and turn on the digital switch. In both conventional P-SSHI and S-SSHI, a quick inversion to the output voltage of single piezoelectric oscillator is generated due to LC oscillation with a period of $2\pi\sqrt{LC_p}$. Since the period of LC oscillation is set to be significantly smaller than the vibration period of the piezoelectric oscillator, the voltage inversion occurs instantly. The closing switch time is usually set as a half of the LC oscillation period $\pi\sqrt{LC_p}$. Then, the switch remains off until the next extremum of the piezoelectric oscillator's displacement. Compared with the standard circuit, the inversion of the piezoelectric voltages would usually significantly reduce the time shift between the mechanical velocity and the output voltage of the piezoelectric element, and also increase the piezoelectric voltage [11]. The reduction of the time shift allows more mechanical energy flowing into the external loading. The increase of the piezoelectric voltage confirms more energy extracted from the piezoelectric element, therefore improve the harvesting efficiency.

Originated from the conventional S-SSHI, self-powered S-SSHI interface circuit is able to automatically take the switching action without displacement sensors and power supply [14-16], as shown in figure 2(f). Self-powered S-SSHI circuit utilizes the voltage from the piezoelectric elements to drive the transistors acting as two symmetric switches $S1$ and $S2$. In the self-power circuit, the full-bridge rectifier in the standard and conventional SSHI circuits is also simplified to a pair of diodes to reduce the energy consumption in the circuit.

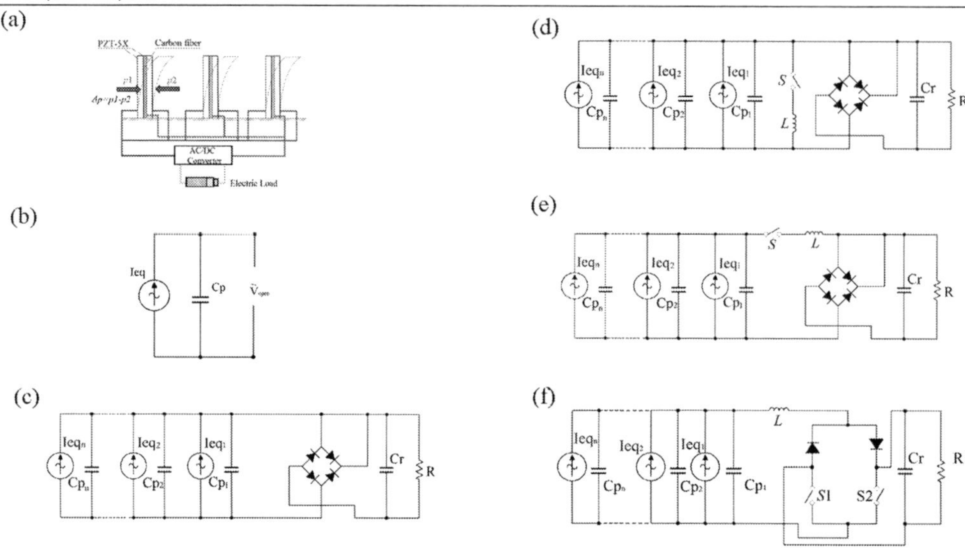

Figure 2. Sketch of (a) multiple piezoelectric cantilever plates energy harvesting system. (b) Open-circuit model of a piezoelectric oscillator. (c) Standard interface circuit, (d) conventional P-SSHI interface circuit, (e) conventional S-SSHI interface circuit, and (f) self-powered S-SSHI interface circuit.

4. CIRCUITS DESIGNS

The interface circuits for multiple piezoelectric oscillators are designed and simulated in the circuit simulation software *Multisim* 12.0. The circuit diagrams for three identical piezoelectric oscillators of conventional P-SSHI, S-SSHI and self-powered S-SSHI interface circuits are shown in figure 3. The mechanisms of self-powered S-SSHI for a single piezoelectric oscillator have been studied in previous studies [14-16]. All piezoelectric oscillators are assumed to vibrate in phase with I_{eq} of 566 μA. The inductor L in the conventional P-SSHI and S-SSHI, and the self-powered S-SSHI circuits is 100 μH. V_p is the voltage of the piezoelectric oscillators. V_{output} is the rectified output voltage to the external loading. i_s denotes the current flowing through the switch S in P-SSHI and S-SSHI. i_2 and i_3 denote the currents flowing through the transistors Q2 and Q3 in the self-powered S-SSHI circuit.

For both conventional P-SSHI and conventional S-SSHI, as shown in figure 3(a) and (b), the switch S is controlled by a pulse maker. The switch remains open for the majority of time. When I_{eq} equals to zero, the pulse maker would send a pulse signal to turn on the switch and excite the LC oscillation for a very short period $\pi\sqrt{nLC_p}$. Then the piezoelectric voltage V_p would suddenly drop down or boost up due to the conduction of the switch. Since the rectified bridge itself would consume energy when conducting, a diode 1N4004G with low forward voltage (~0.5V [15]) is selected in the circuit to minimize the energy loss in the operation.

The circuit design for self-powered S-SSHI interface circuit is shown in figure 3(c). Transistors Q1 to Q4 (TIP 31G and TIP 32G) are used to act as the switches in figure 2(f). Q1 and Q2 are initially off until V_p increases to the maximum. Then, Q1 and Q2 conduct quickly and excite the LC oscillation with the period about $\pi\sqrt{nLC_p}$ to make the inverse of V_p. Q1 and Q2 then are cut off after the half period of LC oscillation. Likewise, when V_p reaches the minimum, the transistors Q3 and Q4 will conduct and cut off due to the symmetry of the circuit.

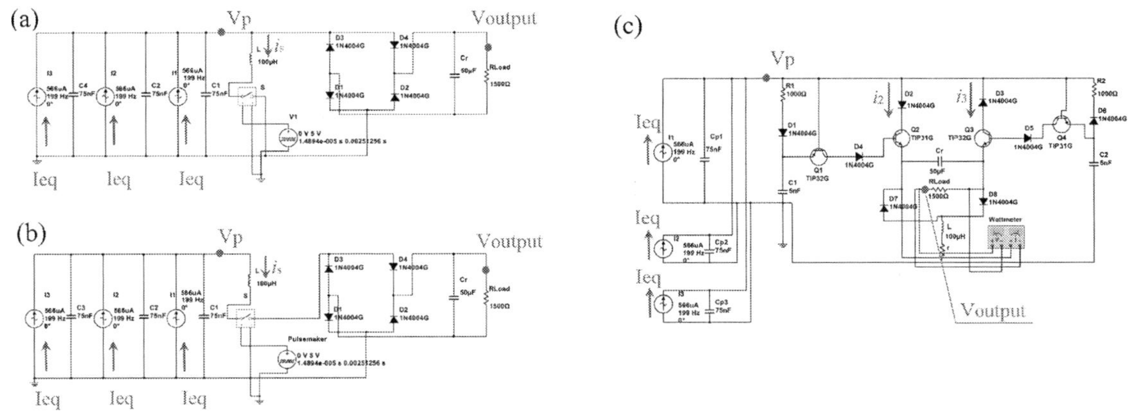

Figure 3. Circuit designs of (a) conventional P-SSHI, (b) conventional S-SSHI and (c) self-powered interface circuits for three identical piezoelectric oscillators in *Multisim*.

5. RESULTS AND DISCUSSIONS

The waveforms of i_s, i_2, i_3, V_p, I_{eq} and the rectified output voltages V_{output} for three identical piezoelectric oscillators in standard, P-SSHI, S-SSHI and self-powered S-SSHI interface circuits are shown in figure 4. figure 4(a) shows a typical waveform of standard interface circuit [8]. In figure 4(b) and (c), the wave shapes of V_p show the voltage of piezoelectric oscillators gets inverse for every half period of I_{eq}. The inversion amplitude of V_p is defined as factor q_I [9-10]

$$q_I = -\frac{V_{off} - V_{ref}}{V_{on} - V_{ref}}, \tag{1}$$

where V_{on} and V_{off} is the initial and terminal voltages of V_p in the voltage inversion. V_{ref} is the reference voltage. For P-SSHI, V_{ref} is zero; for S-SSHI V_{ref} is the rectified voltage. In this study, q_I is measured from the waveforms of the circuits. In Figure 4(b) and (c), the switch S controlled by the pulse maker would conduct every half period of I_{eq} when I_{eq} equals zero. Therefore i_s appears sharp positive and negative peaks sequently. During the conduction of the switch, LC oscillation in the circuit would suddenly inverse V_p. The wave shapes of V_p for both P-SSHI and S-SSHI show the voltages of piezoelectric oscillators getting inverse for every half period of I_{eq}. This phenomenon coincides with the previous studies in conventional SSHI [9-13].

Unlike the conventional SSHI, self-powered S-SSHI interface circuit takes use of the transitions and diodes driven by V_p to act as the switches [14-16]. As shown in figure 4(d) i_2 and i_3 which flow through Q2 and Q3 appear respective positive and negative sharp peaks for every half period of I_{eq}. This indicates Q2 and Q3 are conducting as closed switches. There is small amount of reverse currents in i_3 generated after the peak of i_2 and vice versa as shown in figure 4(d) and zoomed in figure 5. This is because of the uncharging from the parasitic capacitances in Q3 and Q2. The phenomenon brings energy consumption in the transistors and it is actually a side-effect for the self-powered S-SSHI circuit [15]. The fundamental mechanism of self-powered S-SSHI circuit is similar as S-SSHI. Hence the waveshape of V_p is similar as S-SSHI. Self-powered S-SSHI circuit utilizes the diodes and transistors to trigger the switch. The inverse of the V_p therefore is less sharp compared with the conventional S-SSHI because of the delays, which are caused by the charging and discharging in the transistors.

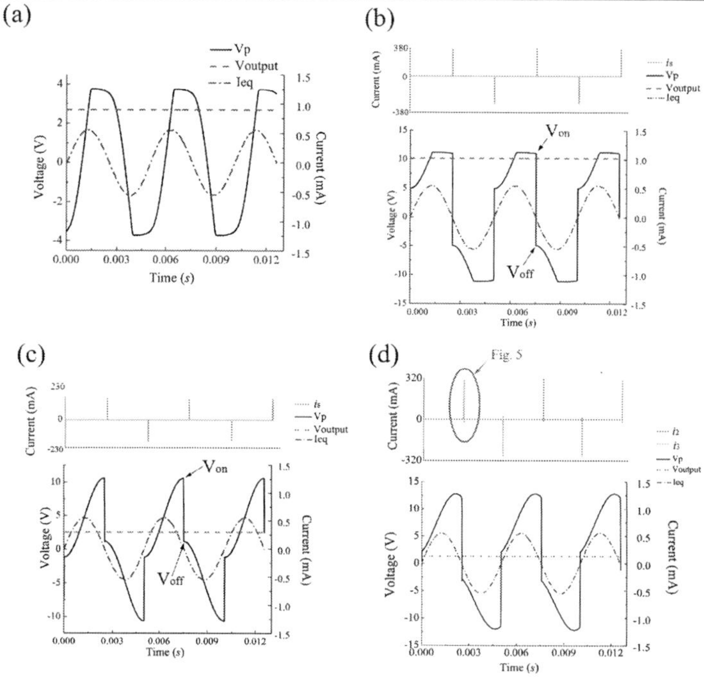

Figure 4. Waveforms of (a) V_p, V_{output} and I_{eq} in standard circuit, V_p, V_{output}, I_{eq} and i_s in (b) conventional P-SSHI and (c) conventional S-SSHI circuit, and (d) V_p, V_{output}, I_{eq}, i_2 and i_3 in self-powered S-SSHI circuit. All circuits are included three identical piezoelectric oscillators.

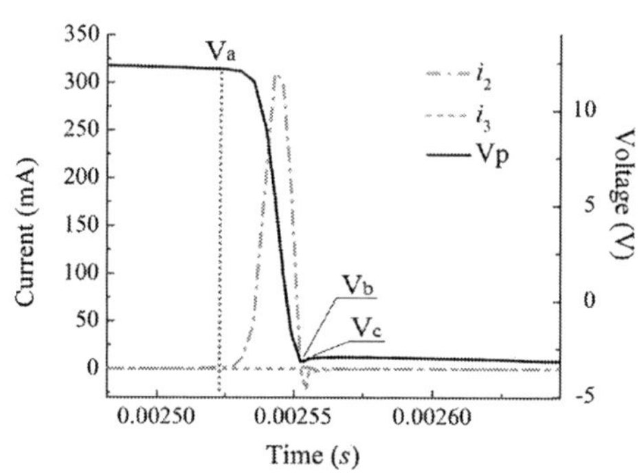

Figure 5. Zoomed in for the voltage reversion and current inversion in the self-powered S-SSHI circuit.

The comparisons of output powers versus loading resistances for standard interface circuit, conventional SSHI and self-powered S-SSHI interface circuits with 1~3 identical piezoelectric oscillators are shown in figure 6. The maximum total output powers of a single piezoelectric oscillator are 0.418 mW at 17 kΩ, 2.232 mW at 70 kΩ, 1.083 mW at 5 kΩ and 0.451 mW at 3 kΩ for the standard, conventional P-

SSHI, conventional S-SSHI and self-powered S-SSHI interface circuits. For three identical piezoelectric oscillators, the maximum total output power of standard circuit increases to 1.196 mW at 6 kΩ. For conventional P-SSHI and S-SSHI, the maximum total output powers are 5.126 mW at 20 kΩ and 2.191 mW at 2 kΩ, which are about 328.6% and 83.2% higher than the standard circuit. The total maximum power of self-powered S-SSHI circuit is 1.685 mW at 0.9 kΩ.

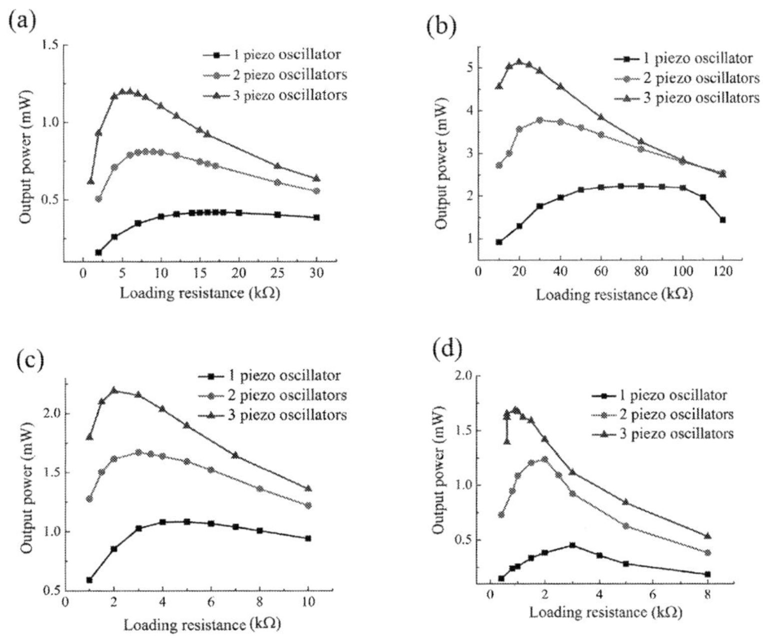

Figure 6. Output powers versus loading resistances of (a) standard circuit, (b) conventional P-SSHI circuit, (c) conventional S-SSHI circuit, and (d) self-powered S-SSHI circuit.

Since there is much more energy loss from the diodes and transistors in the circuit, the harvested power of self-powered S-SSHI circuit is 67.1% and 23.1% lower than P-SSHI and S-SSHI. However, the self-powered S-SSHI circuit's harvested power is 40.9% higher than the standard circuit. Although the harvested powers of conventional P-SSHI and S-SSHI are higher than the self-powered S-SSHI circuit, the requirements for the displacement sensors may make it unrealistic to implement in some energy harvesting systems. In our piezoelectric energy harvester, as shown in figure 1(a) and (b), it is very difficult to install sensors in our tube resonator due to the limited space.

The theoretical values of optimized loading resistances, output voltages and powers of the standard circuit, conventional P-SSHI and conventional S-SSHI were discussed for a single piezoelectric oscillator in previous studies [9-13]. In a very recent study Ref. [4], the optimized loading resistance R_{opt} and output voltage for multiple piezoelectric oscillators were derived by equivalent impedance approach. Here, the maximum output powers of n multiple piezoelectric oscillators are given in table 1 assuming that each oscillator is vibrating in phase and has the same internal capacitance C_{p}.

Table 1. Theoretical optimized loading resistances, output voltages, and Max. output powers of multiple piezoelectric oscillators for standard, conventional P-SSHI, and conventional S-SSHI.

	Standard	**Conventional P-SSHI**	**Conventional S-SSHI**
Ropt	$\frac{1}{2}\frac{\pi}{nC_p\omega}$	$\frac{1}{1-q_I}\frac{\pi}{nC_p\omega}$	$\frac{1-q_I}{2(1+q_I)}\frac{\pi}{nC_p\omega}$

Output voltage	$\frac{R}{\frac{\pi}{2}+nC_pR\omega}\sum_{i=1}^{n}I_{eqi}$	$\frac{2R}{\pi+nC_pR\omega(1-q_I)}\sum_{i=1}^{n}I_{eqi}$	$\frac{2(1+q_I)R}{\pi(1-q_I)+2nC_pR\omega(1+q_I)}\sum_{i=1}^{n}I_{eqi}$
Max. output power	$\frac{1}{2}\frac{1}{nC_p\omega\pi}(\sum_{i=1}^{n}I_{eqi})^2$	$\frac{1}{nC_p\omega\pi}\frac{1}{1-q_I}(\sum_{i=1}^{n}I_{eqi})^2$	$\frac{1}{2}\frac{1}{nC_p\omega\pi}\frac{1+q_I}{1-q_I}(\sum_{i=1}^{n}I_{eqi})^2$

The comparisons of theoretical and simulated powers for standard, conventional P-SSHI and conventional S-SSHI interface circuits are shown in figure 7. For all circuits, the total powers increase linearly with the identical oscillator numbers in simulation and the theoretical calculation. The theoretical powers are 25.7%~31.7% higher than the simulated powers in standard circuit. For the conventional P-SSHI and S-SSHI circuit, the theoretical powers are 4.2%~6.7% and 12.8%~19.7% higher than the simulated powers. The differences mainly derive from the energy loss in the diodes rectified bridges.

The theoretical powers of self-powered S-SSHI circuit calculated based on conventional S-SSHI are obviously higher than the simulated power by 119.3%~212.5%. The energy consumption in self-powered is more significant than the conventional SSHI since there are more diodes and transistors in the self-powered S-SSHI circuit. Thus the theoretical powers for conventional S-SSHI cannot be used to evaluate the output powers of the self-powered S-SSHI interface circuit. Energy flow method [15-16, 18] which considers the energy dissipation in the circuit has been introduced for a single piezoelectric oscillator in self-powered S-SSHI circuit. For n identical oscillators, the harvested power for self-powered S-SSHI interface circuit is:

$$P=\frac{\omega}{\pi}V_{output}(nC_p(V_a+V_c-2V_b)+C_e(V_a-V_b)) \tag{2}$$

where C_e= 5 nF (i.e., C_1 and C_2 in figure 3(c)), V_a, V_b, V_c are corresponding to the piezoelectric voltages when i_2 and i_3 begin to raise, and i_3 returns to zero, as indicated in figure 5. The harvested powers calculated by the Energy flow method are only 10.7~15.8% higher than the simulated powers. It indicates the Energy flow method proving a good evaluation for the output powers of the multiple piezoelectric oscillators.

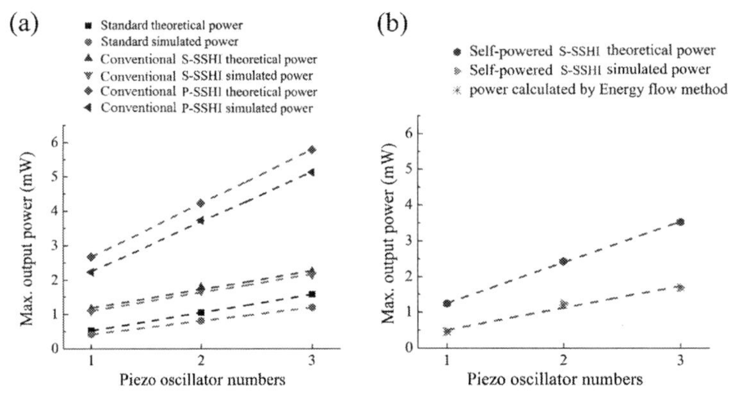

Figure 7. (a) The comparisons of theoretical and simulated powers for standard, conventional P-SSHI and conventional S-SSHI interface circuits with linear fittings. (b) The comparisons of theoretical, simulated powers, and powers calculated by Energy flow method for self-powered S-SSHI circuit with linear fittings.

In figure 8, a comparison of the total maximum output powers when placing three cantilever piezoelectric plates in our piezoelectric energy harvester using standard, conventional SSHI and self-powered S-SSHI interface circuits are shown. The simulated output power of the harvester is 1.041mW by using self-powered S-SSHI circuit, which is 23.9% higher than using standard circuit (0.84mW).

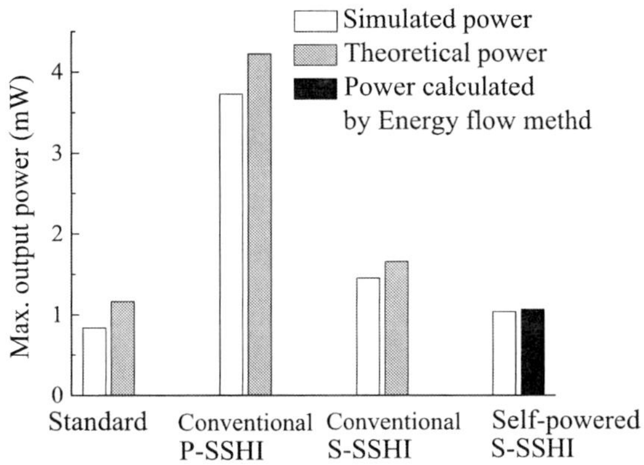

Figure 8. The comparisons of the total output powers of standard, conventional P-SSHI and conventional S-SSHI, and self-powered S-SSHI interface circuits for the piezoelectric energy harvester.

6. CONCLUSIONS

The harvesting efficiency of multiple identical piezoelectric oscillators in piezoelectric energy harvesting is studied using standard interface circuit, conventional P-SSHI, conventional S-SSHI and self-powered S-SSHI circuits. The simulation results indicate that the output power increases linearly as the number of oscillators increases. When three identical piezoelectric oscillators are connected, the total output power of self-powered S-SSHI circuit is 40.9% higher than that from the standard circuit. Although the final harvested power for self-powered S-SSHI circuit is lower than the conventional SSHI, the self-powered S-SSHI circuit is a good candidate of AC/DC conversion for multiple piezoelectric oscillators because it does not require displacement sensors and an external power supply. Compared with the theoretical method, the Energy flow method gives a good evaluation for the harvested power of the self-powered circuit.

References

1. Li, B., Laviage., A. J., You, J. H., and Kim, Y. -J., "Acoustic Energy Harvesting Using Quarter-wavelength Straight-tube Resonator," *ASME 2012 International Mechanical Engineering Congress & Exposition,* 2012, pp. 86989.
2. Li, B., and You, J. H., "Simulation of Acoustic Energy Harvesting Using Piezoelectric Plates in a Quarter-wavelength Straight-tube Resonator," *Proceedings of the 2012 COMSOL Conference in Boston*, 2012.
3. Li, B., You, J. H., and Kim, Y. -J., "Low frequency acoustic energy harvesting using PZT piezoelectric plates in straight-tube resonator," *Smart. Mater. Struct.*, 2013, Vol. 22, No.5, pp. 055013.

4. Lien, I. C., and Shu, Y. C., "Array of piezoelectric energy harvesting by the equivalent impedance approach," *Smart. Mater. Struct.*, 2012, Vol. 21, No.8, pp. 082001.
5. Kim, H. S., Kim, J. -H., and Kim, J., "A review of piezoelectric energy harvesting based on vibration," *Int. J. Precis. Eng. Man.*, 2011, Vol. 12, No. 6, pp. 1129-41.
6. Cook-Chennault, K. A., Thambi, N., and Sastry A. M., "Powering MEMS portable devices—a review of non-regenerative and regenerative power supply systems with special emphasis on piezoelectric energy harvesting systems," *Smart. Mater. Struct*, 2008, Vol. 17, No.4, pp. 043001.
7. Sodano, H. A., Inman, D. J., and Park, G., "Comparison of Piezoelectric Energy Harvesting Devices for Recharging Batteries," *J. Intel. Mat. Syst. Str.*, 2005, Vol. 16, No. 10, 799-807.
8. Lien, I. C., and Shu, Y. C., "Efficiency of energy conversion for a piezoelectric power harvesting system" *J. Micromech. Microeng.*, 2006, Vol. 16, No.11, pp. 2429-2438.
9. Lien, I. C., Shu, Y. C., Wu, W. J., Shiu, S. M., and Lin, H. C., "Revisit of series-SSHI with comparisons to other interfacing circuits in piezoelectric energy harvesting" *Smart. Mater. Struct.*, 2010, Vol. 19, No. 12, pp. 125009.
10. Shu, Y. C., Lien I. C., Wu W. J., "An improved analysis of the SSHI interface in piezoelectric energy harvesting," *Smart. Mater. Struct.*, 2007, Vol. 16, No.6, pp. 2253-2264.
11. Guyomar, D., and Lallart, M., "Recent Progress in Piezoelectric Conversion and Energy Harvesting Using Nonlinear Electronic Interfaces and Issues in Small Scale Implementation" *Micromachines*, 2011,Vol. 2, pp. 274.
12. Lallart, M., Guyomar, D., Richard, C., and Petit, L., "Nonlinear optimization of acoustic energy harvesting using piezoelectric devices," *J. Acoust. Soc. Am.* ,2010, Vol. 128, No. 5, pp. 2739.
13. Guyomar, D., Badel, A., Lefeuvre, E., and Richard, C., "Toward Energy Harvesting Using Active Materials and Conversion Improvement by Nonlinear Processing," *IEEE Trans. Ultrason., Ferroelect., Freq. Contr.*, 2005, Vol. 52, No.4, pp. 584-595.
14. Lallart, M., and Guyomar, D., "An optimized self-powered switching circuit for non-linear energy harvesting with low voltage output" 2008, *Smart Mater. Struct.*, Vol. 17,No.3, pp. 035030.
15. Liang, J., and Liao, W. -H., "Improved Design and Analysis of Self-Powered Synchronized Switch Interface Circuit for Piezoelectric Energy Harvesting Systems," *IEEE T Ind. Electron.*, 2012, Vol. 59, No. 4, pp.1950.
16. Liang, J., and Liao, W. -H., "An Improved Self-Powered Switching Interface for Piezoelectric Energy Harvesting," *Proceedings of the 2009 IEEE International Conference on Information and Automation*, 2009, pp. 945.
17. Tang, L., and Yang ,Y., "Analysis of synchronized charge extraction for piezoelectric energy harvesting," *Smart Mater. Struct.*, 2011, Vol. 20, No.8, pp. 085022.
18. Liang, J., and Liao, W. -H., "Energy flow in piezoelectric energy harvesting system," *Smart. Mater. Struct.*, 2011, Vol. 20, No.1, 015005.

Power Enhancement of Piezoelectric Transformer by Adding Thermal Dissipation Layers

Yu-Hao Su[1,2], Yuan-Ping Liu[3], Dejan Vasic[2,4*], François Costa[2,5], Wen-Jong Wu[1*], Chih-Kung Lee[1,6]

[1] Department of Engineering Science & Ocean Engineering, National Taiwan University, Taipei, Taiwan;
[2] SATIE, ENS Cachan, 94235 Cachan, France;
[3] Miézo Inc., Taiwan;
[4] Université de Cergy-Pontoise, Neuville/Oise, France ;
[5] IUFM, Université Paris Est Créteil (UPEC), Place du 8 mai 1945, 93000 St Denis, France ;

Abstract

Considering the case of large current (>1A), the piezoelectric transformer (PT) easily becomes unstable or even crack with temperature rise because of the excessive internal losses. The output current of PT is also limited by temperature build-up effect owing to the heat generated by mechanical and electrical losses especially when driving at high vibration velocity. Once the PT exceeds a threshold vibration velocity, the temperature rise of PT would significantly increase and lead to instability. PT based power converters are therefore used commonly in high voltage and low current applications with a frequency-tracking mechanism adaptive to loading conditions. However, it is possible to increase the threshold vibration velocity, as well as the limitation of mechanical current, by adding thermal dissipation layers. Once the threshold vibration velocity caused by the temperature rise can be increased, it represents the enhancement of the PT's maximum output current and output power. In this paper, we design a multi-layer PT which possesses high passing current capacity with low profile. A theoretical-phenomenological model is proposed to explain the relationship between the vibration velocity and the temperature rise on the PT. Furthermore, we use flexible graphite tapes and thin aluminum pad on the PT to dissipate the heat generated from internal energy losses. Comparing to the previous research, graphite tapes and the thin copper layers can be directly stick on the PT and can efficiently decrease the temperature rise of the PT without extra power consumption. Our experimental result will confirm that it could satisfy the low profile requirement, while improving the PT output current capacity at the same time. The maximum output current capacity can increase 20% when the condition of PT temperature is kept below 55°C. Moreover, both of the output current and output power of the PT are improved by using the thermal dissipation layers. It implies that this technique allows the piezoelectric transformer to be used in low voltage/high current applications.

1. INTRODUCTION

Piezoelectric materials possess many advantages, such as reversible, easy integration, easy manufacture, low power requirements, abilities of low-profile and low weight, etc. Considering the stiffness and actuation response for different application areas, such as actuators and sensors, piezoelectric materials can be simply classified as follow two groups: piezo-ceramic materials and piezo-polymer materials. Generally speaking, piezoelectric materials can also be classed as hard or soft according to the stability of the domain walls. As to the domain walls in hard compositions, hard compositions are capable of resisting depoling due to high temperatures or high electric fields but possessing lower electromechanical coupling coefficients compare to the soft compositions [1]. In recent years, "hard" piezo-ceramic materials have been used for high power density applications because high coercive field is desirable [2]. However, non-linear effects caused by heat become significant in high power applications. Hysterisis losses caused by cyclic fields tend to intensify under high electric field and thus the thermal stress and heat generation [3]. Thermal problem leads to the result that the material properties no doubt will be changed. Furthermore, high intensification of the temperature gradient may cause rapid crack or thermal shock fracture growth [4]. It should be noted that the fracture behaviors of piezoelectric materials are more complex than that in conventional materials owing to the complicated coupling relationship between mechanical and electric behaviors [5]. Except the fracture problem, thermal depolarization condition cannot be avoided when the component temperature exceeds one-half the Curie temperature [6]. The thermal problems in piezoelectric materials become as the physical limitation and need to be dealt with especially in the high power applications. For instance, excessive temperature rise will decrease the quality factor Q of piezoelectric component during the operational process. Simultaneously the vibration energy cannot be increased even if under higher excitation voltage [7].

Many researchers have made efforts to increase the output current of PT in the aspects of PT's structure and inductive circuits at the PT's secondary terminal. In the structure part, Prieto *et al* presented a design guideline discussed the output power capacity of PT with different thickness of the secondary bulk zone, size, and number of layers in multi-layer PT [8]. As to the inductive circuits part, Yamane *et al* and Ivensky *et al* utilized the step-down transformer and current-doubler rectifier connecting at the secondary terminal of the PT to improve the efficiency and output current (power) in PT-based converters application [9,10]. However, increasing the total volume of PT or adding an inductive circuit sacrifices the inherent advantage of a PT. The two methods above are inadequate for the demands of high power density and low-profile requirement [11]. Furthermore, the root cause of the temperature build-up effect is still an unsolved problem, as is the physical limitation of PTs. This physical limitation is due to the limit value of the vibration velocity of PTs at specific temperature. Once a PT exceeds its vibration velocity threshold, the temperature rise in the PT increases dramatically and the temperature starts to build up, increasing the internal losses and causing instability. Typically, a large output current is directly linked with a high vibration velocity [12, 13]. If the threshold vibration velocity can be improved, it is possible to use PT in high current applications.

In this paper, we present a method to improve the threshold vibration velocity of PTs by using thermal dissipation layers. Comparing to a ring-dot-shaped PT with a central hole to obtain better thermal radiation proposed by Kim *et al* [14], thermal dissipation layers consisted of graphite tapes and thin copper pads can be directly stick on the PT without extra fabrication step. We started by establishing a phenomenological–theoretical model to discuss the thermal balance of PT with or without the thermal layers. From the experimental results including the temperature rise, vibration velocity, and power dissipated in the PT, we can validate that the threshold vibration velocity, as well as the mechanical current, can be improved almost 20% when the condition of PT temperature is kept below 55°C. Moreover, our experimental results revealed that the cooling method is more useful to increase the threshold vibration velocity, rather than to expand the number of layers of multi-layer piezoelectric transformer.

2. THERMAL ANALYSIS OF MULTI-LAYER PIEZOELECTRIC TRANSFORMER

2.1 Heat generation caused by internal losses and thermal balance in PT

PT may enter into unstable state with temperature rises because of the excessive internal losses under high-power conditions [15]. In other words, the temperature rise is a physical limitation for the high power applications of PTs. The excessive temperature rises are typically occurred when carrying high output current especially when output current is larger than 1 A [16]. This fact implies that the PT is only suitable in the case of low output current. To overcome the low-current limitation, the analysis of the heat generation from internal losses and vibration of PT is required. Generally, there are three types of losses in PT: dielectric loss $q_{D\text{-Loss}}$, piezoelectric loss $q_{p\text{-Loss}}$ and mechanical loss $q_{M\text{-Loss}}$ [17, 18]. Dielectric loss is caused by the hysteresis between the electric filed E and electric displacement D; piezoelectric loss is caused by the electromechanical hysteresis between the strain S and electric field E; mechanical loss is caused by the internal elastic hysteresis between the stress T and strain S. Usually the piezoelectric loss is ignored because it is relatively small [19]. Therefore the total losses can by express as:

$$q_{T-\text{Loss}} = q_{D-\text{Loss}} + q_{M-\text{Loss}} \tag{1}$$

When the PT is working, the heat generated by the losses in equation (1) can be dissipated to the ambient environment by heat transfer or can be stored inside the PT.

$$q_{\text{T-Loss}} = q_{\text{E-Loss}} + q_{\text{S-Loss}} \tag{2}$$

Where $q_{\text{E-loss}}$ is the heat flux dissipated to the ambient environment, it is composed of conduction, radiation and convection effects [17], and $q_{\text{S-loss}}$ is the heat flux stored in the PT. It is related to the density ρ, specific heat c_P and total volume V_T of the PT:

$$q_{\text{S-Loss}} = \rho c_p V_T \frac{dT}{dt} \tag{3}$$

Although the PT possesses the heat dissipation capacity $q_{\text{S-Loss,}}$, the excessive heat generation from internal losses leads to the obviously temperature rise. The maximum storage heat capacity $(q_{\text{S-Loss,}})_{\text{Max}}$ of the PT is limited by the piezoelectric material characteristic. Therefore, with the same piezoelectric material, the heat dissipation ability limits the PT power density.

For simplicity, we can use the thermal conductance to model the thermal balance [20]. The relationship between stable temperature rise ($T_s - T_0$) and total power dissipated in the PT is shown in equation (4):

$$q_{\text{T-Loss}} = Y_{\text{PT}}(T_S - T_0) \tag{4}$$

Where Y_{PT} represents the thermal conductance of heat dissipation ability, T_S represents the stable temperature of PT while the energy losses and the generated heat ($q_{\text{T-loss}}$) achieve a balance with the heat dissipation ($q_{\text{E-Loss}}$) to the ambient environment, and T_0 is the ambient temperature. When the PT operates in the low-power transmission, the temperature rise and the heat generation of the PT are limited due to the PT's thermal conductance of heat dissipation ability. On the contrary, both of the PT's working temperature and heat generation increase rapidly in high-power transmission owing to the excessive heat generation. Once the overbalance of heat exceeds the maximum storage heat capacity $(q_{\text{S-Loss,}})_{\text{Max}}$ of the PT, it will decrease the quality factor of the piezoelectric material significantly. Accordingly, the overgreat temperature rise of PT should be regarded as a thermal limitation to prevent PT's susceptibility to crack or depoling. It is the reason why many researchers analyzed or experimentally determined thermally limited maximum power density by restricting the rise in PT temperature above its surroundings to 20°C [22]. In this paper, we measure the maximum mechanical current and output power under the temperature rise constraints 30°C of PT comparing to the ambient temperature, and the 55 °C temperature limitation of PT is determined as the PT's maximum storage heat capacity.

In addition, thermal layers consisted of aluminum pad and graphite tape are applied to dissipate the heat of PT. Based on figure 1, the additional thermal conductance heat dissipation abilities $Y_{\text{Aluminum.-Pad}}$ and $Y_{\text{Graphite -Tape}}$ are contributed from aluminum pad and graphite tape as shown in equation (5).

$$P_{\text{m-Loss}} = (Y_{\text{Total}})(T_S - T_0) = (Y_{\text{PT}} + Y_{\text{Aluminum-Pad}} + Y_{\text{Graphite-Tape}})(T_S - T_0) \tag{5}$$

In equation (5), the increased total thermal conductance Y_{Total} efficiently limited the temperature rise ($T_S - T_0$) of the PT with constant mechanical loss $P_{\text{m-loss}}$.

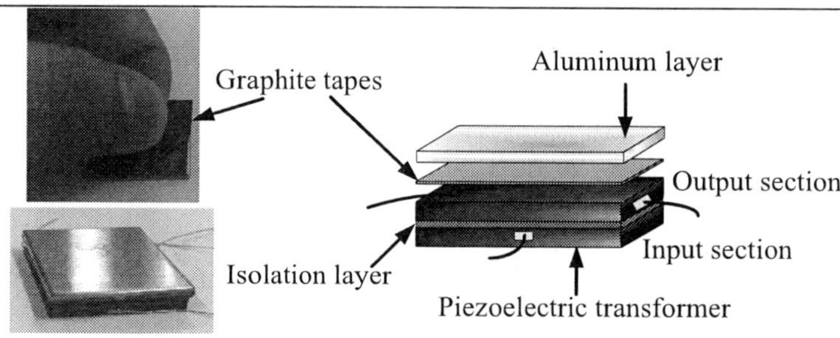

Figure 1. Structure of PT and thermal layers

2.2 Analysis of temperature-dependent nonlinear resistance in PT Equivalent circuit

Equation (4) and (5) shows the balance of the heat flux of the PT. However, it does not mention design parameters of the PT such as the current and the voltage. To link these parameters with heat characteristics of the PT with high vibration level easily, a single-mode non-linear equivalent circuit [23] was adopted as shown in Fig. 2. There are two basic assumptions of this equivalent circuit:

1. The operating frequency is set near the resonant frequency. It should be noted that the resonance frequency is shifted with temperature rise [24], thus we assume that operating frequency don't varies in our model.
2. The piezoelectric losses and the dielectric losses were both neglected since the mechanical losses is the dominated internal loss in the PT. In other words, we assume that the heat is only generated from the mechanical loss:

$$P_{\mathrm{m}-loss} = R_m i_m^2. \tag{6}$$

The typical equivalent circuit of the PT is composed of mechanical resonant branch L_m-C_m-R_m, two clamped capacitors C_1, C_2 and electromechanical transformation ratio N. These parameters are fixed at constant temperature.

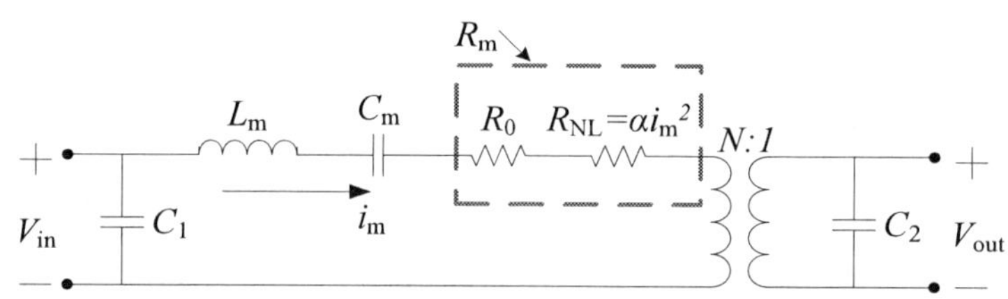

Figure 2. Nonlinear equivalent circuit of piezoelectric transformer.

To take into accounts the temperature effect, a non-linear thermal resistance R_{NL} is introduced in the equivalent circuit of Figure 1. This nonlinear resistance is introduced to take into account the effect of the large vibration amplitude and heat on the quality factor of the piezoelectric material. Thus, the mechanical resistance R_m includes two terms: a low excitation resistance R_0 and a non-linear resistance R_{NL}.

$$R_m = R_0 + R_{NL} \tag{7}$$

According to Albareda's non-linear model of piezoelectric transducers [23], the nonlinear resistance R_{NL} is a function of mechanical current square, such as:

$$R_{NL} = \alpha i_m^2 \tag{8}$$

Where mechanical current i_m can be considered to be equivalent to the vibration velocity passes through the mechanical branch and α characterizes the temperature-dependence of the resistance, *i.e.* $\alpha(T°C)$. In our experimental results, both of the coefficient α and the mechanical resistance R_m were increased with temperature rise. In addition, combining the assumption 2 and equations (4)-(8), the relationship between mechanical current and temperature rise was shown in equation (9):

$$i_{m}^{2}(R_{0}+\alpha i_{m}^{2})=Y_{PT}(T_{S}-T_{0}) \quad (9)$$

In equation (9), R_0 and coefficient α are based on the piezoelectric material properties and piezoelectric structure configuration. Therefore, considering the same PT, the mechanical current is determined by the thermal conductance and acceptable temperature rise. On the other hand, based on circuit analysis, the output current i_{out} of the piezoelectric transformer is positive related to the mechanical current as shown in equation (10):,

$$i_{out} = Ni_{m} + i_{C2}. \quad (10)$$

This is the underlying reason that the PT output current is typically limited by the temperature rise. Fortunately, equation (5) and (10) also imply that the maximum output current of the PT can be increased by improving the heat transfer characteristic of the PT.

3. EXPERIMENTAL RESULTS AND DISCUSSIONS

3.1 Experimental setup

Experimental setup is shown in Figure 3. We incorporated the function generator (Instek SFG-2004) and power amplifier (nF HSA4052) to control the switching frequency and to drive the PT. A contact temperature sensor was used to monitoring the PT temperature in different driving conditions. The contacted aluminum pad and graphite tape were used as the thermal layers to increase the performance of the PT. Furthermore, we also design two types of PTs with the same length and width but different number of layers (thickness) to compare the threshold vibration velocity and thermal characteristic.

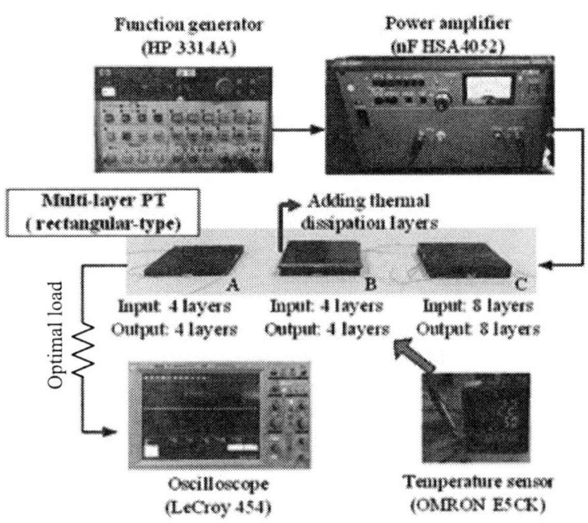

Figure 3. Experimental setup.

Expanding the number of layers of multilayer PT, *i.e.* input and output section, can increase the capacity of passing current, and thus the mechanical current and output current. However, once the effect of temperature build-up becomes heavy in high vibration condition, mechanical current cannot be increased even if under higher excitation voltage. To improve the output current (power) of PT, the accumulated heat must be released rather than the increment of layers or size only. To verify the statements, we present three different experimental cases to analyze the relationships between the input voltage, vibration velocity (mechanical current i_m), non-linear resistance R_{NL}, ability of thermal conductance *Y*, and generated heat of the PT as shown in Table 1. The properties of the PTs specimen and the model parameters measured by the impedance analyzer are given in Table 2.

Table 1 Size and structure of experimental cases

Case A PT size: 25mm*25mm*2.1mm	Input section No. of layers 4 Thickness 0.2mm (0.2*4=0.8mm)	Output section No. of layers 4 Thickness 0.2mm (0.2*4=0.8mm)	Isolation No. of layers 1 Thickness 0.5mm (0.5*1=0.5mm)	Without thermal dissipation layers
Case B PT with thermal layers size: 25mm*25mm*3.7mm	Input section No. of layers 4 Thickness 0.2mm (0.2*4=0.8mm)	Output section No. of layers 4 Thickness 0.2mm (0.2*4=0.8mm)	Isolation No. of layers 1 Thickness 0.5mm (0.5*1=0.5mm)	*Aluminum pad Thickness 1.3mm *Graphite tape Thickness 0.3mm (1.3+0.3=1.6mm)
Case C PT size: 25mm*25mm*3.7mm	Input section No. of layers 8 Thickness 0.2mm (0.2*8=1.6mm)	Output section No. of layers 8 Thickness 0.2mm (0.2*8=1.6mm)	Isolation No. of layers 1 Thickness 0.5mm (0.5*1=0.5mm)	Without thermal dissipation layers

Table 2 Properties of the PTs

Material properties (PZT-QA, ELECERAM TECHNOLOGY Co., Ltd., Taiwan):

K_p= 0.58 is the electromechanical coupling coefficient
d_{33} = 320 × 10^{-12} is the piezoelectric constants (m V^{-1})
Y_{33} = 320 × 10^{-12} is the elastic constants (N m^{-2})
ρ_m = 7.9 is the density (g cm^{-3})
N_p = 2200 is the frequency constants of the plane vibration (kHz mm)
s_{11}^{E} = 1.14 × 10^{-11} is the compliance constant under the constant electric field, i.e. constant *E*

Q_m = 1800 is the mechanical quality factor
d_{31} = -140 × 10^{-12} is the piezoelectric constants (m V^{-1})
Y_{31} = -140 × 10^{-12} is the elastic constants (N m^{-2})
ν = 0.16 is Poisson's ratio
ε_{33}^{S}= 1420 × 8.854 × 10^{-12} is the permittivity at constant strain condition, i.e. constant *S* (F m^{-1})
Unstable temperature of PT: 55 ° C
Operating frequency: 1kHz higher than f_r
Optimal load ($1/\omega C_2$): Case A and B = 5 Ω ;Case C=10 Ω

The experimental result obtained by impedance analyzer:

PT of Case A and Case B

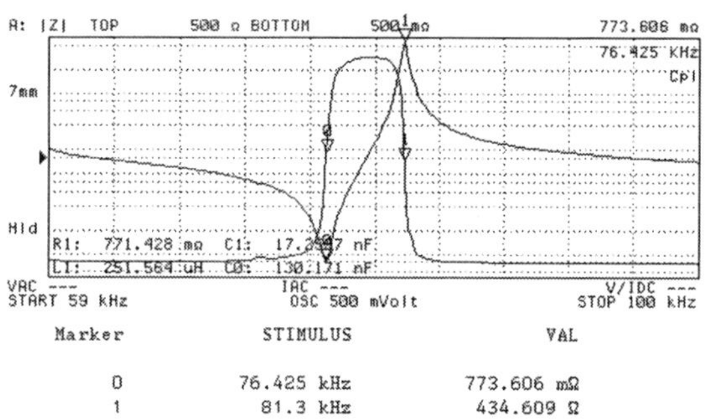

PT of Case C

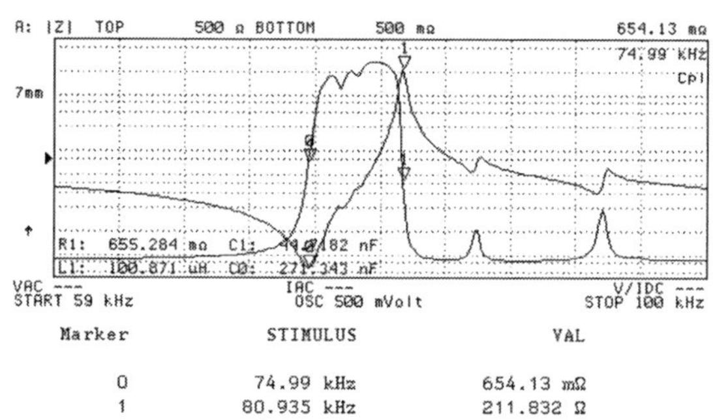

PT of Case A and Case B:

f_r (KHz)	f_a (kHz)	R_0 (Ω)	L_m (mH)	C_m (nF)	C_1 (nF)	C_2 (nF)	N (-)	T °C
76.425	81.3	0.771	0.251	17.25	130.171	129.8	1	25

PT of Case C:

f_r (KHz)	f_a (kHz)	R_0 (Ω)	L_m (mH)	C_m (nF)	C_1 (nF)	C_2 (nF)	N (-)	T °C
74.99	80.935	0.655	0.1	44.01	271.343	272.01	1	25

It is utilized the same PT in case A and case B, but the difference is with or without the thermal dissipation layers. In case C, we expand the number of layers of the multilayer PT without the thermal dissipation layers to compare with other cases. PTs in case A, B, and C are multi-layer rectangular transformer with internal circular electrodes connected in parallel and thickness polarization. The testing specimen was provided by Eleceram Technology Co., Ltd Taiwan. Graphite tape (MS-GTC-025-0.03T) was provided by Miézo Inc. Taiwan. In previous research [25], polypropylene membrane is applied to decrease contact wear of piezoelectric transformer. It should be noted that graphite tape possesses high thermal conductivity to transfer the heat to the heat transfer equipment. Considering the heat transfer capacity, we use the graphite tape to replace the polypropylene membrane as the buffer layer.

3.2 Experimental results of PT with thermal dissipation layers

The relationship between the temperature rise ($T_s - T_0$), total power dissipated in the PT ($P_{m\text{-loss}}$) and thermal conductance heat dissipation ability (Y) is shown in equation (4) and (5). In this paper, we measure the mechanical losses and temperature rise to quantify the Y value (thermal conductance heat dissipation ability) [11]. In Figure 4 below, it is obviously that the Y value can be increased by using the thermal dissipation layers and expanding the layers of PT. Comparing to the case B and case C, the method to improve the Y value by adding thermal layers is more useful than the expansion of PT's layers with the same thickness.

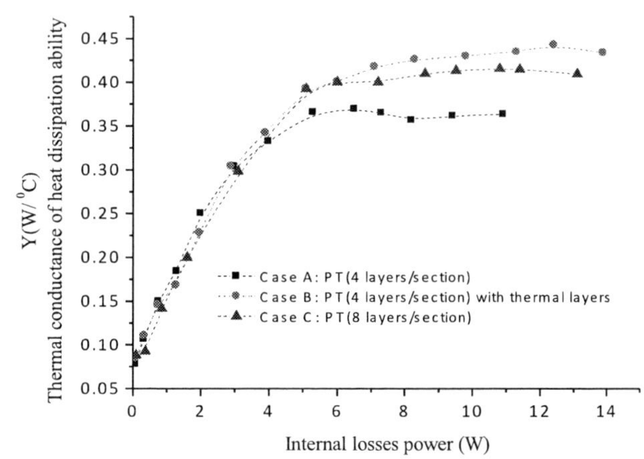

Figure 4. Characteristics between Y (thermal conductance) and temperature rise

According to the Figure 4 and equation (5), the increased thermal conductance (Y value) can efficiently limit the temperature rise ($T_s - T_0$) of the PT owing to the improvement of heat dissipation ability. At the same mechanical current level, the suppression of the temperature rise can be directly decreased the temperature-dependent nonlinear resistance R_{NL} in PT and thus the mechanical resistance R_m as shown in Figure 5. From the equation (6), it is revealed that the mechanical losses $P_{m\text{-loss}}$ are also decreased at the same mechanical current level. In addition, Figure 5 verifies the increment of the mechanical resistance is a function of the square mechanical current i_m^2. In this paper, we determine thermally limited threshold vibration velocity as well as mechanical current by restricting the rise in PT temperature above its surroundings to 30 °C. The 55 °C temperature limitation of PT is determined as the PT's maximum storage heat capacity. The variation of mechanical current and the PT temperature are obtained by applying different input voltages as shown in Figure 6 and Figure 7. According to the experimental results, we know that it is possible to increase the maximum vibration energy of the PT under higher excitation voltage by preventing the overheating problem. The maximum mechanical current as the threshold vibration velocity can be increased from 1.69 A to 2 A at a specific temperature in case B. Comparing the case B and case C in Figure 7, the method utilized the thermal layers (case B) possesses better capacity than case C to improve the mechanical current at each input voltage level. Furthermore, the increment of the threshold vibration velocity in Case B is more obvious than other cases, and the growth is nearly 20%.

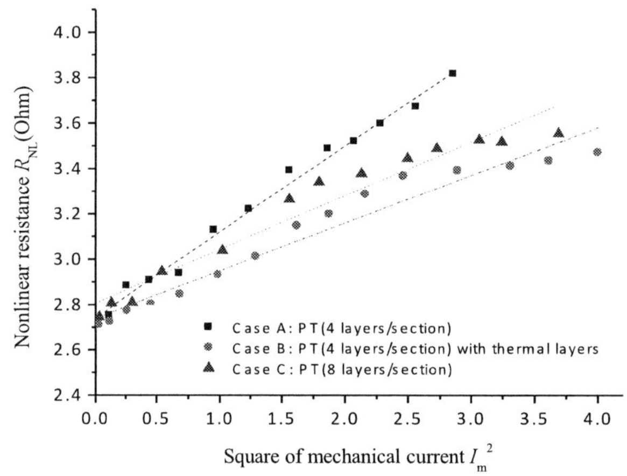

Figure 5. Relationship between the square of mechanical current and the R_{NL} at different temperatures

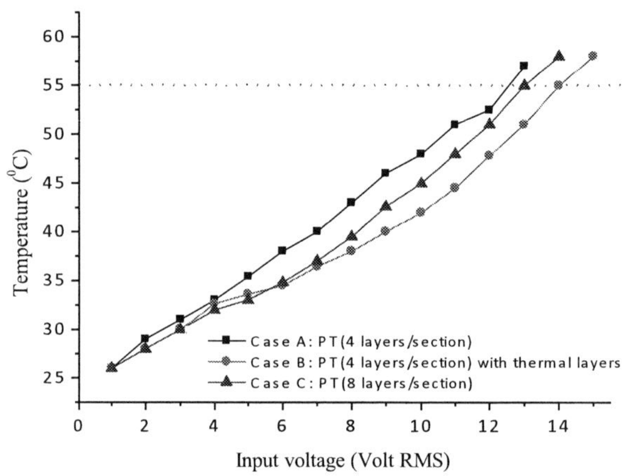

Figure 6. Relationships between temperature and input voltage.

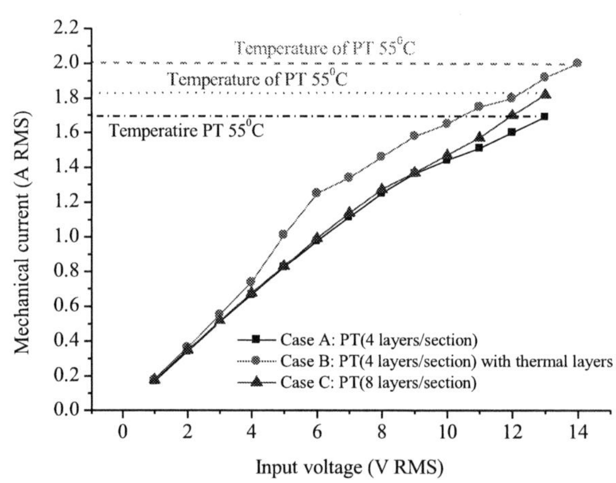

Figure 7. Relationships between mechanical and input voltage.

According to the relationship between output current and mechanical current as shown in equation (10), the maximum output current under optimal load condition can be improved and thus the output power as shown in Figure 8. The measured results show that the highest output power of thermal dissipation layers used for PT can move from 10.9 W to 13.8 W at the temperature of 55°C under optimal load condition (an increase of 26%). Furthermore, all specimens kept a good efficiency (>70%) at the PT temperature limit of 55°C.

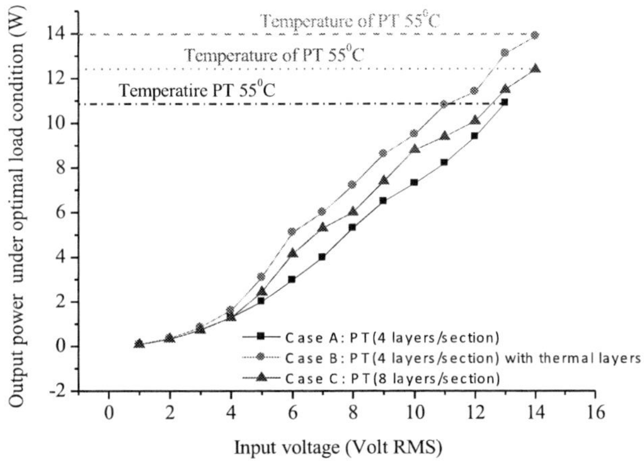

Figure 8. Output power enhancement in different method.

4. CONCLUSION

In this paper, the thermal layers consisted of graphite tape and aluminum pad were used to enhance the output power of a PT. It should be noted that the graphite tape is a perfect layer to decrease contact wear and transfer the heat. According to the experimental results, all specimens retained a satisfactory efficiency even at a temperature of 55°C. Due mainly to the improved ability to dissipate heat by using thermal dissipation layers, the PT passing current can be substantially increased under optimal load conditions instead of just expanding the PT's number of layers. The threshold vibration velocity, as well as the mechanical current, can be improved almost 20% when the condition of PT temperature is kept below 55°C, and thus the highest output power (an increase of 26%). Moreover, we present a theoretical-phenomenological model. The model clearly indicates that it is possible to enhance the performance of the PT by decreasing the temperature-dependent nonlinear resistance R_{NL} value and by increasing the thermal conductance Y value. This paper investigates the temperature build-up problem of PT and presents a useful method allowed the piezoelectric transformer to be used in low-voltage, high-current applications.

Acknowledgements

The authors are grateful to: Eleceram Technology Co. Ltd. and Miézo Inc. for providing us with different types piezoelectric transformers and piezoelectric fans used in the research work. The main funding of this project from National Science Council, Taiwan under Project NSC-101-2917-I-002-014 is gratefully acknowledged.

References

1. R. C. Buchanan, R. W. Schwartz, J. Ballato, and G. H. Haertling, *Ceramic Materials for Electronics*, (chapter 4), 3rd ed: Marcel Dekker Inc., 2004.
2. K. Uchino, J. H. Zheng, Y. H. Chen, X. H. Du, J. Ryu, Y. Gao, S. Ural, S. Priya, and S. Hirose, "Loss mechanisms and high power piezoelectrics," *Journal of Materials Science*, vol. 41, pp. 217-228, 2006.
3. Weisshaar, T. A., & Ehlers, S. M. (1990, April). Static aeroelastic behavior of an adaptive laminated piezoelectric composite wing. In *Proceedings of the 31st Structures, Structural Dynamics, and Materials Conference* (pp. 1611-1623).
4. Wang, B. L., & Mai, Y. W. (2003). Thermal shock fracture of piezoelectric materials. *Philosophical Magazine*, *83*(5), 631-657.
5. Zhang, T. Y., & Gao, C. F. (2004). Fracture behaviors of piezoelectric materials. *Theoretical and applied fracture mechanics*, *41*(1), 339-379.
6. Wang, D., Fotinich, Y., & Carman, G. P. (1998). Influence of temperature on the electromechanical and fatigue behavior of piezoelectric ceramics. *Journal of applied physics*, *83*(10), 5342-5350.
7. M. C. Do, *Piezoelectric Transformer Integration Possibility in High Power Density Applications*: TUDpress Verlag der Wiss., 2008.
8. R. Prieto, M. Sanz, J. Cobos, P. Alou, O. Garcia, and J. Uceda, "Design considerations of multi-layer piezoelectric transformers," in *Applied Power Electronics Conference and Exposition, 2001. APEC 2001. Sixteenth Annual IEEE*, 2001, pp. 1258-1263.
9. Yamane, T., Hamamura, S., Zaitsu, T., Minomiya, T., Shoyama, M., & Fuda, Y. (1998, May). Efficiency improvement of piezoelectric-transformer DC-DC converter. In *Power Electronics Specialists Conference, 1998. PESC 98 Record. 29th Annual IEEE* (Vol. 2, pp. 1255-1261). IEEE.
10. Ivensky, G., Bronstein, S., & Ben-Yaakov, S. (2004). A comparison of piezoelectric transformer AC/DC converters with current doubler and voltage doubler rectifiers. *Power Electronics, IEEE Transactions on*, *19*(6), 1446-1453.
11. Su, Y. H., Liu, Y. P., Vasic, D., Costa, F., Wu, W. J., & Lee, C. K. (2013). Study of a piezoelectric transformer-based DC/DC converter with a cooling system and current-doubler rectifier. *Smart Materials and Structures*, *22*(9), 095005.
12. Hu, J. (2003). Analyses of the temperature field in a bar-shaped piezoelectric transformer operating in longitudinal vibration mode. *Ultrasonics, Ferroelectrics and Frequency Control, IEEE Transactions on*, *50*(6), 594-600.
13. Ebenezer, D. D., Thomas, D., & Sivakumar, S. M. (2007). Non-uniform heat generation in rods with hysteretic damping. *Journal of Sound and Vibration*,*302*(4), 892-902.
14. Kim, I., Jeong, S., Kim, M., Song, J., & Thang, V. V. (2012). Improvement of thermal balance in step-down piezo transformers with ring-dot shapes. *Journal of the Korean Physical Society*, *60*(2), 261-266.
15. Y Sasaki, M Umeda, S Takahashi, M Yamamoto, A Ochi and T Inoue 2001 High-Power Characteristics of Multilayer Piezoelectric Ceramic Transducers *Japanese Journal of Applied Physics* 40 5743-46
16. Y P Liu, D Vasic, W J Wu, F Costa and C K Lee 2009 Design of fixed-frequency controlled radial-mode stacked disk-type piezoelectric transformers for DC/DC converter application *Smart Materials and Structures* 18(8) 085025
17. J Hu 2003 Analyses of the temperature field in a bar-shaped piezoelectric transformer operating in longitudinal vibration mode *IEEE Trans. Ultrason. Ferroelectr. Freq. Control* 50 594-600
18. K Uchino and S Hirose 2001 Loss mechanisms in piezoelectrics: how to measure different losses separately *IEEE Trans. Ultrason. Ferroelectr. Freq. Control* 48 307-321
19. D Thomas, D D Ebenezer and S M Srinivasan 2010 Power dissipation and temperature distribution in piezoelectric ceramic slabs *J. Acoust. Soc. Am.*128(7) 1700–1711
20. W W Shao, L J Chen, C L Pan, Y B Liu and Z H Feng 2012 Power density of piezoelectric transformers improved using a contact heat transfer structure *IEEE Trans. Ultrason. Ferroelectr. Freq. Control* 59 3-81
21. A M Flynn and S R Sanders 2002 Fundamental limits on energy transfer and circuit considerations for piezoelectric transformers *IEEE Trans. on Power Electron.* 17 8-14
22. S Priya, H Kim, S Ural and K Uchino 2006 High power universal piezoelectric transformer *IEEE Trans. Ultrason. Ferroelectr. Freq. Control* 53 810-16
23. A Albareda, P Gonnard, V Perrin, R Briot and D Guyomar 2000 Characterization of the mechanical nonlinear behavior of piezoelectric ceramics *IEEE Trans. Ultrason. Ferroelectr. Freq. Control* 47 844-53

24. Sasaki Y, Umeda M, Takahashi S, Yamamoto M, Ochi A and Inoue T 2001 High-power characteristics of multilayer piezoelectric ceramic transducers Japan. J. Appl. Phys. 40 5743–6

25. Ju, B., Shao, W., Pan, C., & Feng, Z. (2013). Polypropylene membrane decreases contact wear of piezoelectric transformer with heat transfer structure.*Electronics Letters*, *49*(10), 650-651.

ICAST2013: 24th International Conference on
Adaptive Structures and Technologies
October 7-9, 2013, Aruba

A Dynamical Actuated Lip at a Blowing Slot for Active High-Lift

C. Behr[1], P. Wierach[2], M. Sinapius[1]

[1] Institute of Adaptronics and Funktional Integration, Technische Universität Braunschweig, Germany
[2] Institute of Composite Structures and Adaptive Systems, German Aerospace Center (DLR), Braunschweig, Germany

Abstract

This paper presents the design process of an actuation device for novel high-lift devices, based on a very compact oscillating piezo-actuated lip of a blowing slot ahead of a Coanda flap. The actuated lip is used to regulate the mass flow and velocity of a wall near stream over the internally blown Coanda flap. This stream is used to accelerate the boundary layer in order to keep the flow attached to the strongly deflected Coanda flap and to delay the flow separation by the highly energetic Coanda-jet. The combination of MEMS based flow sensors and piezoelectric actuators seem to be a fruitful solution to implement a closed loop control for adaptive airfoils. The goal is the structural coupling of all control systems in one compact module. In a fibre composite lamination process these techniques will be combined in order to manufacture innovative, large-area sensor-actuator-arrays with modular segmentation. These new compact and highly integrated systems provide a real-time monitoring and manipulation of the flow conditions. In a first functional demonstrator the actuators are bonded to a metal substrate as a bending transducer in trimorphic configuration for basic experiments in a water tunnel which allows higher Reynolds-numbers in comparison to a wind tunnel as the flow velocity in water is ten times smaller. This enables much easier time resolved measurements and closed loop control for the research of fundamental physical effects of a blowout lip combined with a Coanda flap. The actuation of the slot-lip is implemented by trapezoid multilayer stack actuators using the d_{33}-effect with high strain rates at low voltages. First measurement results show that the displacement of the actuated lip achieves the required slot height. Another challenging task is the qualification of the adaptive slot-lip for water environment with a waterproof electrical insulation.

1. INTRODUCTION

The collaborative research center (CRC) 880 investigates the fundamentals of high-lift generation for future commercial aircraft, focusing on the fields of aero acoustics, lift generation and flight dynamics.

The motivation for this research lies in the conflict between the public desire towards faster and easier access to air travel on one hand and the resistance to air traffic related, noise and pollutant emission as well as the wasteful use of natural resources. Modern, efficient aircraft with short take-off and landing (STOL) capabilities are one way to approach this conflict.

The underlying research hypothesis of the present work is that further significant increase in lift generation of civil aircraft compared to customary systems is possible using active high-lift devices. The investigated high-lift concept utilizes a combination of internally blown flaps and circulation control to achieve high flow turning over the wing. A flexible leading edge device for the wing without gap or step is designed to reduce noise generation and to increase the efficiency of the active blowing system. The overall objective of the project is to design an active lift system that requires a minimum of additional engine power to generate the required lift.

The necessary technologies and findings will be demonstrated on an oscillating lip of a blowing slot. By the oscillating lip the mass flow in the blowing slot changes dynamically, so that the momentum exchange of the boundary layer over a high lift flap required mass flow can be reduced. Compared to state of the art, an expansion of the frequency range is expected and the benefit due to the compact and highly integrated design is a high aerodynamic efficiency of this configuration.

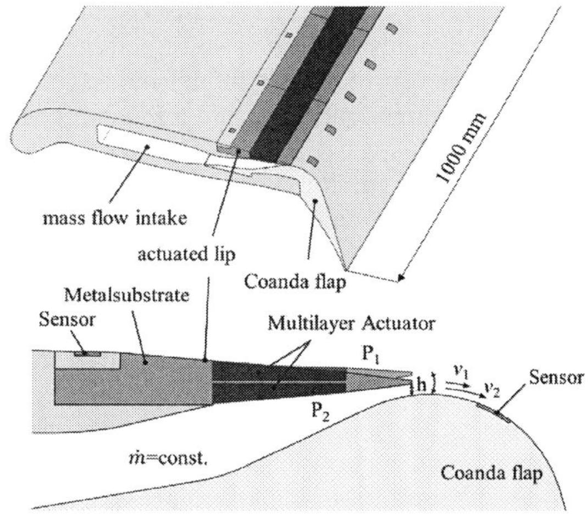

Figure 1. Concept of the actuated slot-lip for water tunnel tests [1]

2. REQUIREMENTS OF THE SLOT-LIP

In the following the main requirements regarding the actuator concept and its design to fulfill the scientific and experimental goals are explained.

The choice of water environment is motivated by the fact that a water tunnel allows higher Reynolds-numbers than achievable in a wind tunnel as the flow velocity in the water tunnel is smaller by a factor of 10. This enables much easier time resolved measurements and closed-loop control for the research of fundamental physical effects of a blowout lip combined with a Coanda flap.

The actuated slot-lip must be operated in static, quasi-static and dynamic modes at frequencies up to 30 Hz with an overall fatigue limit of 10^7 cycles. The slot opening of up to min. 0.2 mm must be warranted at a maximum load level of the differential pressure between inner and outer chamber of the water tunnel model (pressure difference between pressure p_1 and pressure p_2 in Figure 1 while $p_1 \leq p_2$) of 6 bar.

3. DESIGN OF THE SLOT-LIP

The preliminary design of the water tunnel model is based on FEM calculations. A modular design of the slot-lip in span wise direction providing separately controllable digits is aimed for better controllability and adjustability. Figure 2 illustrates the current state of design for one lip segment with its main dimensions and design features.

Different configurations of bending transducers are considered and investigated in order to realize the upwards and downwards movement of the slot-lip tip. Two monomorphic designs and one trimorphic design were investigated. The monomorphic concepts are unable to actuate in two directions except a preload is applied. A preload in this case can be a pre-applied electrical voltage or a pre-applied mechanical deformation. In contrast to this the bending transducer concept featuring a trimorphic configuration enables actuation upwards and downwards without being limited to preload the slot-lip. These considerations and a detailed FEM analysis result in the selection and further development of the trimorphic bending actuator for the slot-lip. This is underlined by Figure 3 as it depicts the higher potential of a possible active tip deflection (Δu) for the trimorphic configuration over the whole differential pressure (Δp) range. At an assumed differential pressure of 6 bar, the overall displacement of the slot-lip is 0.255 mm and meets the required value of 0.2 mm with a buffer of just over 25 %.

The slot-lip comprises of several segments in span wise direction, adding up to an overall length of approx. 1000 mm. It is fixed with screws on the water tunnel model as shown in Figure 1. This design allows later a separate control of each segment and therefore a much higher controllability of flow. The 1 m slot-lip is built up of several segments in span wise direction and fixed with screws to the water tunnel model. Every segment, with its backbone structure (substrate) made of stainless steel holds several trapezoidal actuators on the upper and lower side, bonded with an adhesive to it. The current design comprises 2 segments on each side.

The actuation of the slot lip is realized by trapezoidal multilayer piezo actuators, using the d33-effect. These precisely cut actuators have their origin in the P-888.30 PICMA Multilayer Piezo Stack Actuator of PI Ceramic GmbH, Lederhose, Germany. They are considered to feature a superior lifetime, a large operating temperature range from -40 °C to 150 °C, high humidity resistance, excellent temperature stability, high stiffness, peak currents up to 20 A, UHV compatibility of up to 10^{-9} hPa, sub-millisecond response at sub-nanometer resolution and a superior capability of dynamic operations [3]. The demanded low operating voltage is fulfilled by the use of this actuator type as it operates from -20 V to 120 V. This is a significant advantage over Macro-Fiber Composite (MFC) and Active Fiber Composite (AFC) actuators. As shown in a previous study a maximum free strain of approx. 1400 µm/m at a 10 times lower operating voltage can be achieved for multilayer actuators [4]. Named after their composition of multiple material layers these actuator types are built up of 60 µm thin piezo ceramic layers and alternating interdigital electrode layers connected with an elastic connector electrode on each side of the actuator (in the direction of main expansion). The multilayer actuators used are further more diced from a compact stack that features a ceramic insulation layer as outer surface, and is embedded in a polymer afterwards. This provides a mechanical stabilization on the one hand and an electrical insulation with a diffusion-resistant coating on the other hand.

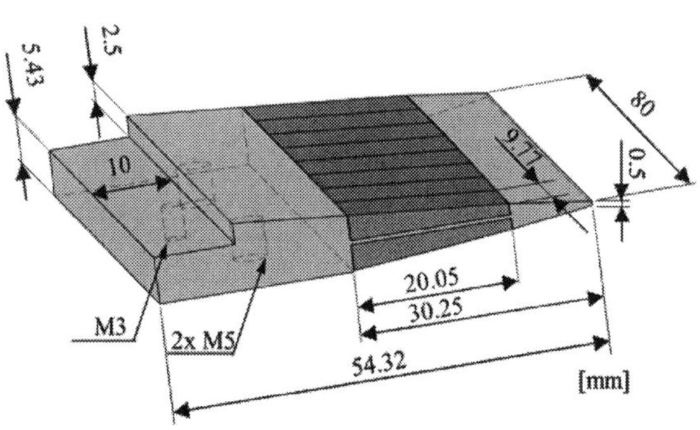

Figure 2. Current slot-lip design with dimensions

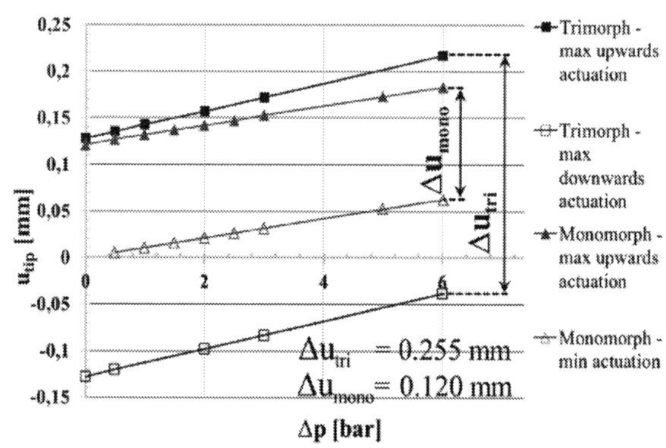

Figure 3. FEM calculation for active tip-deflection of tri- and monomorphic concepts [1]

4. MANUFACTURING OF A FUNCTIONAL DEMONSTRATOR

For the validation of the actuation concept and FE-simulation results a functional demonstrator is required. This demonstrator (Figure 4) is realized as a small lip segment holding two actuators on each side.

The substrate for the functional demonstrator is manufactured from stainless steel (1.4301). The cutouts for the actuators in the lip are achieved by wire-cut EDM. The application of the actuators is realized by an established bonding method with a two-component adhesive under vacuum. The electrical contact is made at the lateral electrodes of the actuators using PTFE insulated copper wires (Ø 0.5 mm). The wires lead to solder terminals which are applied to the passive structure for strain relief during operation.

The piezo ceramics must be molded to the shape of the lip before the actuators are bonded to the structure. The selected PI Ceramic Multilayer Actuators (PICMA®) are manufactured in large batches with tape casting technology. The inner electrode pattern is printed on thin PZT tapes while still unsintered and these are then laminated into a multilayer compound. In the subsequent cofiring process, the ceramic and the inner electrodes are sintered together [3]. After the manufacturing process the ceramics are machined with a wafer saw to bring them into the final shape.

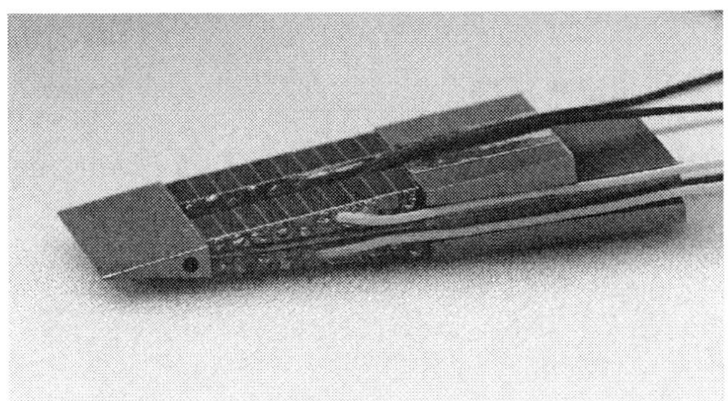

Figure 4. Functional demonstrator for mechanical tests

5. MECHANICAL TESTS OF THE DEMONSTRATOR

The characterization of the electromechanical properties, such as stiffness and blocking force are determined in the context of bending tests on a material testing machine. The measured values are compared with the theoretical data, thus giving more realistic boundary conditions and parameters to optimize and precise the FEM model for further analysis. The scheme of the test set-up for the functional demonstrator is shown in Figure 5. It depicts the clamping device and herein fixed slot-lip. The tip deflection u is measured with laser triangulation in upwards and downwards direction regarding the non-deflection position as neutral (zero) position. The overall possible tip deflection is the sum of upwards and downwards absolute values. In the same test stand the blocking force F_b is measured which is the occurring force value when the slot lip tip is pushed to the neutral (zero) level of deflection at the maximum actuation (max. voltage of 120 V) of the actuators.

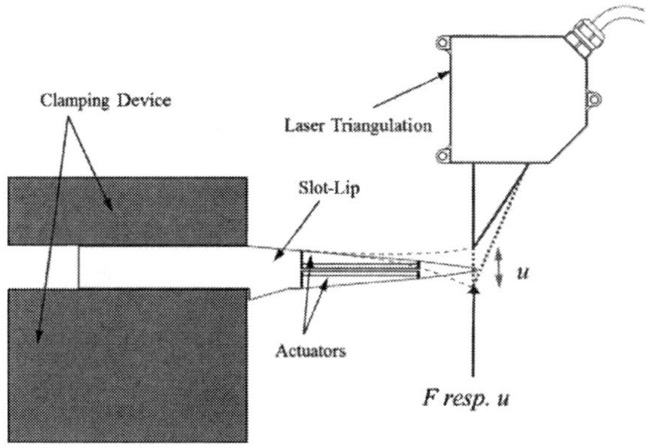

Figure 5. Schematic test setup

5.1 Determination of the Nominal Displacement

For determining the nominal displacement of the functional demonstrator the two actuators on the same side are connected in parallel with each other and are driven by a triangular wave signal and a frequency of 0.1 Hz in a voltage range from 0 V to 120 V. To measure the maximum displacement, top and bottom are driven in phase opposition. By a phase shift of 180 degrees, the elongation of the actuators is converted into an oscillating lifting and lowering of the slot-lip tip, which is measured by laser triangulation.

The deflection of the actuated slot-lip is plotted against the electrical potential in Figure 6. The experimentally determined free deflection of one side without load is 0.13 mm. The measured total deflection of both sides is 0.26 mm which is 30 % higher than the required min. slot height of 0.2 mm. This value is in accordance with the calculated deflection got by FEM-simulations. Indeed the deflection is measured by a constant load and not by pressurization. But we assume a constant pressure in the blowing slot and due to the fact that water is incompressible we also expect a constant load in the water tunnel.

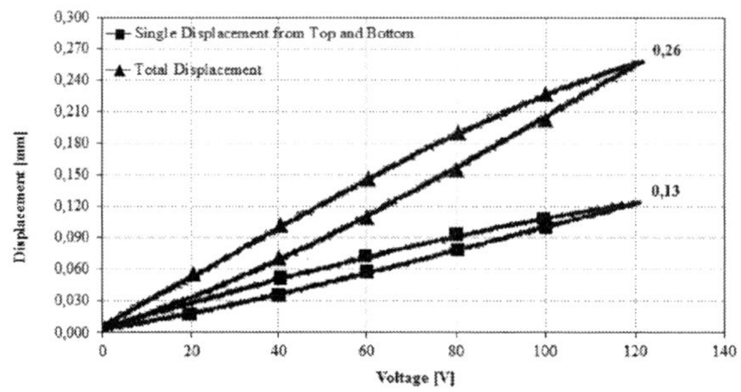

Figure 6. Displacement of the actuated lip

5.2 Determination of the Blocking Force

For the determination of the blocking force the two actuators on the bottom are driven with a static voltage of 120 V. By the elongation of the actuators the leading edge of the lip is displaced upwards. With an anvil a line load is applied to the leading edge which pushes the lip back to its initial position. The required force for this purpose amounts to the blocking force. To exclude an influence of the electrical condition of the passive piezo ceramics on the blocking force, the test is carried out with short-circuited electrodes of both actuators on the topside of the lip.

The curves in Figure 7 show different slopes, which is representative of the stiffness of the lip. Even though the nominal displacements of the individual actuators are near-constant, top and bottom of the slot-lip have different blocking forces. It is assumed that the different thickness of the adhesive layer on top and bottom is responsible for the variation. In addition, the measured blocking force (F_b) of 69 N is only half of the calculated value with FEM (130 N). This can be ascribed to the perfectly bonded actuators to the steel substrate in FEM.

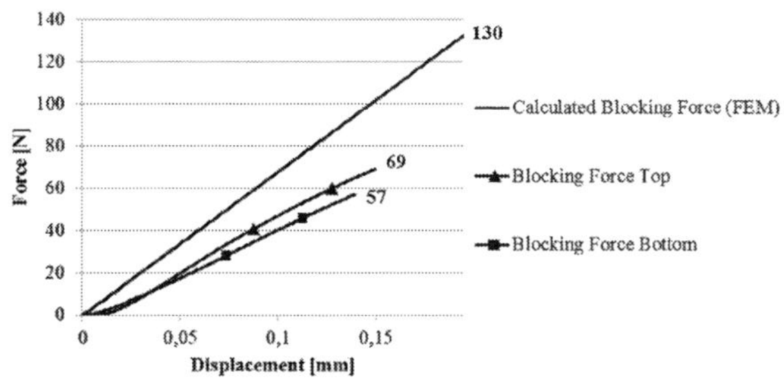

Figure 7. Blocking force of the actuated lip

6. DIFFUSION-RESISTANT COATING

To use water as testing environment brings several advantages. On one hand a water tunnel allows to achieve higher Reynolds-numbers and a ten times smaller flow velocity than a wind tunnel. On the other hand the actuator needs to be completely sealed up against the ingress of moisture. Especially the electric contacts have to be absolutely waterproof because the actuator is operating at a voltage of 120 V. If any moisture diffuses through the coating, it may cause a short circuit, or in the worst case a failure of the actuator.

Currently, it is planned to examine the high-lift system for 12 days in succession in the water tunnel. During this time the actuator has to complete at least 10^7 loading changes at a frequency of 30 Hz. Therefore the coating needs to be as ductile as possible. Otherwise it can break and water can diffuse into the actuator. That is why the yield strength of the coating has to be higher than the actuator which is 0.3 %. After that testing period the high-lift system gets removed from the tunnel to dry out any intruded moisture. Through this redrying procedure during test intermissions an extended underwater operation is expected.

6.1 Coating Materials and Coating Process

The first coating, which is tested, is a two-component tar-epoxy resin. It is used in shipbuilding as a primer to protect hulls against corrosion of sea water. After mixing the two components (base and hardener) in the defined ratio, the primer can be applied with a brush on the specimens. The layer thickness is about 100 μm.

The second coating process, which should protect the actuators against the ingress of water, is a sputter deposition with gold (PVD) with a layer thickness of about 1 μm. An advantage of metallic coatings is the absolute waterproofness but a disadvantage is that they are electrical conductive. That is why the contacts of the actuators have to be insulated by a separate layer before the samples are coated with gold. Therefore, the two-component adhesive, Loctite Hysol 9466 A&B, is used.

Another promising waterproof coating process is the atomic layer deposition (ALD). It is a slightly changed chemical vapor deposition (CVD) and accordingly to the studies of A.I. Abdulagatov et al. an ALD of aluminum oxide (Al_2O_3) and titanium dioxide (TiO_2) can slow the corrosion of static copper specimens in water for ~80 days down [5]. The chosen layer thickness of 250 nm is predetermined by the dielectric strength of Al_2O_3. Because it is about a ceramic coating an separate insulation is not needed.

6.2 Environmental Tests

The functionality of the coated piezoelectric actuators is tested in a water tank. The voltage amplitude is 0 - 120 V and the frequency of the sinus-shaped signal is 30 Hz. With these settings, the actuators are tested until failure. Thus, the different coatings can be compared regarding their ability to protect the actuator from ingress of water.

The results of the dynamic actuator test under water are shown in Figure 8. The hatched bars represent the lifespan of the electrical operated actuators, while the filled bars show the functionality (in days) of the preloaded actuators. As expected, the lifetime of all actuators, which operated under preload, is lower than the life span of the actuators only driven electrically.

The best results have been achieved by the sputtered samples. Only the gold and tar-epoxy coating fulfill the required numbers of 10^7 load cycles but not the minimum service life of 12 days under load. Another coating method, which can be examined in further studies, is a combination of a polymer insulation and a metal coating. The sputtering needs also a separate layer to insulate the electric contacts anyway. Instead of sealing only the contacts with the glue, the whole actuator can be covered with a two-component tar-epoxy resin and afterwards it can be sputtered with the gold. By this way a double coating can be received which could provide more protection against the ingress of water.

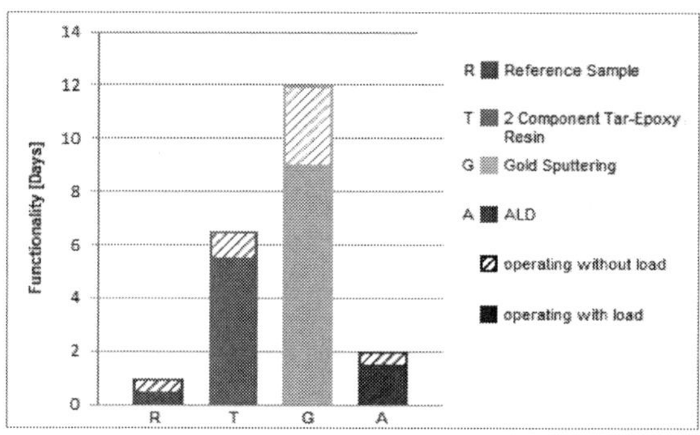

Figure 8. Underwater service life of actuators with and without load

7. CONCLUSION

The feasibility of the proven concept of a piezo-actuated slot-lip intended for flow control of future active high-lift devices is demonstrated by implementing this technology into a functional demonstrator for mechanical tests underlined by FE-simulations. It can be concluded that the current dimensioning of the slot-lip meets the requirements although the determined data where measured with a simplified test setup. Additionally a metallic coating enables underwater operations of piezoelectric actuators for several days. Nevertheless for the future a more sophisticated test bed is needed to apply a pressure load onto the lip which enables realistic test conditions similar to the water tunnel. The enhanced test bed will be a small water tunnel in which a waterproof slot-lip will be actuated statically and dynamically against a constant pressure. These tests will show if the theoretical performance of the slot-lip can be practical implemented. It is assumed that the lip works in a water tunnel against a constant force load, because the differential pressure in the slot should be constant, even if the lip closes. In addition, the medium water is generally regarded as incompressible and does not behave like a "spring". Moreover, it might be possible that due to the narrowing of the slot in direction of "closing" a dynamic pressure is generated and the lip has to work against a dynamic load by an increasing displacement. This has to be confirmed within future experiments and tests.

Acknowledgements

Funding for this work is provided by the German Research Fundation DFG through the collaborative research center (CRC) 880 “Fundamentals of High-Lift for Future Commercial Aircraft”.

References

1. Zander, M. E., Behr, C., Sinapius, M., Wierach, P., “A Piezo-Actuated Slot-Lip for Active High-Lift,” SFB 880 – Fundamentals of high-lift for future commercial aircraft, CFF – Forschungsbericht 2013-03, TU Braunschweig – Campus Forschungsflughafen, Germany, 2013
2. Jensch, C., et al., “Design Aspects of a Gapless High-Lift System with Active Blowing,” DLRK, Aachen, 2009.
3. N.N. (2012). PICMA® Stack Multilayer Piezo Actuators – P-882 – P-888, Technical Datasheet, PI Ceramic GmbH, Lederhose, Germany
4. Wierach, P., “Low Profile Piezo Actuators Based Multilayer Technology,” 17th International Conference on Adaptive Structures and Technologies, Taipei, Taiwan, 2006.
5. Abdulagatov, A. I., et al., “Al2O3 and TiO3 Atomic Layer Deposition on Copper for Water Corrosion Resistance”, ACS Applied Materials & Interfaces, 2011, 3 (12), pp. 4593-4601

Deployment Characteristics of Braided Coated Bi-Convex tape Booms partly used SMA for Deployable Membrane Structures

Nobuhisa Katsumata[1*]

, Masaaki Kaimori[1], Ken Higuchi[1], M.C. Natori[2], Hiroshi Yamakawa[2]

[1] Department of Mechanical, Aerospace, and Materials Engineering, Muroran Institute of Technology, Hokkaido, Japan
[2] Department of Modern Mechanical Engineering, Waseda University, Tokyo, Japan

Abstract

The purpose of this paper is to demonstrate and investigate the concepts of new deployable boom systems, which consist of the Braid Coated Bi-Convex tape (BCON) boom and the Braid Coated Bi-shape memory alloy convex tape (SMA-BCON) boom. Both booms are developed for the deployable membrane structures such as solar sails, thin membrane solar array panels, deorbit mechanisms for small satellites, reflectors of Space Solar Power Satellite, and etc. BCON booms can store around polygonal or cylindrical center hub, and the booms can deploy by the stepwise manner by releasing a constraint mechanism which pin the boons into two or three points for the total length. SMA-BCON booms are mainly developed for a square center body systems, SMA is adapted on the bent points of the booms where stored around each edge of the center hub. Through the deployment experiments of both booms, the stepwise deployment behavior and its tendency are obtained. The design concept of BCON boom and SMA-BCON boom is demonstrated through this study.

1. INTRODUCTION

The combined structures of a deployable membrane and extendible boom has become a key technology for recent and near future gossamer space structures. They can adapt many kinds of space structures, such as light weight high precision space antennas, solar sails, and sun shields. Membrane is a tension elements, centrifugal force due to spinning of spacecraft is also effective to deploy and stabilize it. A spinning solar sail spacecraft "IKAROS" [1] is one of the example and demonstrate such basic structural concept. However a combined structure of membranes and booms is more useful to support compression force during and after deployment, it's depending on deployment manner, and to correspond to various mission requirements. From these view points, there are many kind of researches about how to store and deploy/extend membranes and booms [1-21], and/or how to connect membranes to booms. In the previous studies, various types of extendible boom [8-18] are developed such as the coilable lattice booms, inflatable rigidized booms, shape memory composite booms and convex tape booms. The extendible force of these booms are the stored elastic energy during packing, except for inflatable boom. The extension rate are controlled by a many kinds

* Nobuhisa Katsumata, nobuhisa-k@mmm.muroran-it.ac.jp

of release mechanisms. Although it seems that there are some trade-off design relationship between the mechanical simplex and the extension controllability. In most of the recent studies, some square membrane structures for solar sail spacecraft [1-6] supported by four extendible booms have been well investigated. It is though that a membrane shape may be flexible, its mean pentagonal or hexagonal membranes are also possible, but the square membrane can minimize the number of booms and has more controllability for total spacecraft system operations.

In this paper, from the viewpoints of the simpler extension mechanism and the controllability of boom extension, the newly developed two types of extendible booms are investigated through deployment demonstrations aiming to applying a square membrane structures, one of which is the bi-convex tape boom covered by braid (BCON boom) [18-21] and another of which is the BCON boom partly applied shape memory alloy (SMA-BCON boom). As for the BCON boom, the booms are elastically bended at the corners of the square center body, and they deploy a stepwise manner to control the extension rate just releasing the fixing mechanism without any rotational actuator. As for the SMA-BCON boom, only the convex tape shape like SMA are bended at the corners of the square center body, and they deploy by using shape recovery force of SMA by directory supplying electronic power due to heating themselves. Through these deployment experiment, the

2. BCON BOOM AND SMA-BCON BOOM

2.1 BCON boom and the stepwise deployment device

Figure 1 shows the detail view of the BCON boom and the packed configuration around quadrangular center body. Two independent uni-convex tape booms are covered by braid. Both convex tapes can slide each other toward the longer direction inside the braid, and the braid restrain the deformation toward cross direction to adjust the tightness of braid. This flexible restraint condition toward longer direction help compact packing because the perimeter difference between inner and outer convex tape during rolled-up is well managed. As the result, this boom is easy to be rolled around any polygonal body.

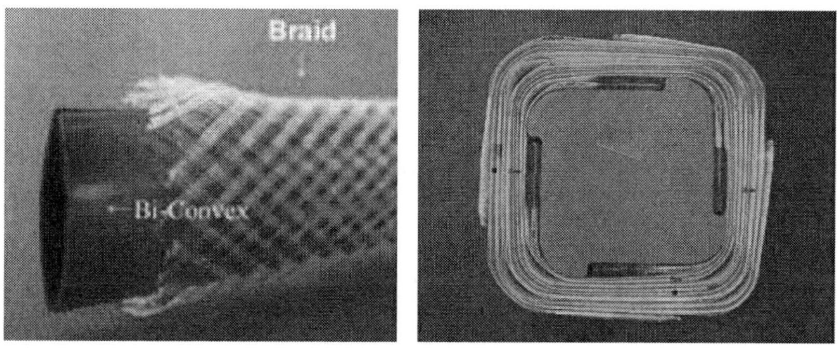

(a) Detail view of BCON boom (b) Packed configuration

Figure 1. The detail view of the BCON boom and the packed configuration around center body

The following Figure 2 shows the stepwise deployment device and the deployment configuration with membrane. The size of center body is 94 mm square. The yellow encircled part in Fig. 2 (a) shows the pin to hold the BCON boom into each length. There are three pin in the figure, it mean this stepwise deployment has three steps. Figure 3 shows the schematic illustration of stepwise deployment. For example, from (a) to (d) configuration shows a 1st deployment step, from (d) to (g) configuration shows a 2nd step, and from (g) to (k) shows a 3rd step. The compartmental location of stepwise deployment depends on the length of the boom held by pins. The springs, shown inside the white circle in Fig. 2 (a), get up the pin in each steps. There has a ball bearing on the bottom side of the device, dark brown column in Fig. 2 (a), it can deploy under free-spin condition and non-spin condition. In this case, the boom length is 500 mm and the membrane thickness is 15μm.

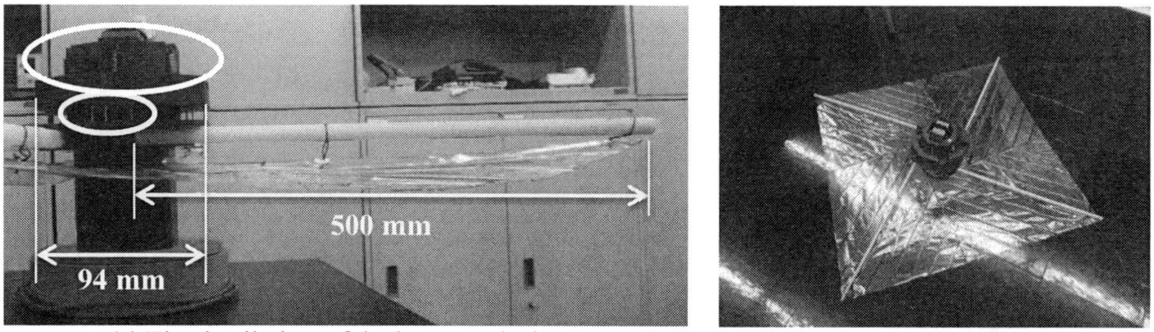

(a) The detail view of deployment device (b) Deployment configuration

Figure 2. The detail view of deployment device and deployment configuration

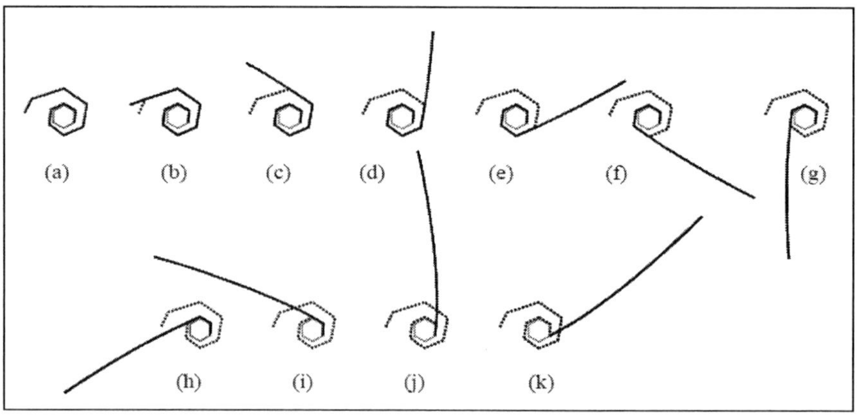

Figure 3. Schematic illustration of the stepwise deployment

2.2 SMA-BCON boom and the stepwise deployment device

The BCON booms is bent only every corner around the quadrangular center body, shown in Fig. 1 (b). In case of the quadrangular center body, about 50 % of the booms are straight (unbent) condition around the each side of tetragon. From the viewpoint of stowed elastic energy and deployment efficiency, the separate design between bent parts and unbent parts through the total booms should be effective. In addition, if we apply SMA in the shape of convex tapes on the bent parts, it is possible to control deployment behavior through temperature control of SMA.

Based on the previously described concept, we design the laboratory level SMA-BCON boom for conceptual study. Figure 4 shows the cross-section view of SMA in the shape of convex tapes. The SMA is made of Ni-Ti alloy and the temperature at the transformation point is about 70 to 80 degrees centigrade. The memorized shape is the straight toward longer direction and the convex shape toward cross-section direction, shown as Fig. 4. The length of the SMA along the longer direction is 150 mm. The unbent parts, set along the each side of quadrangular center body, is made of acrylic plate, shown in Fig. 5. The thickness of the plate is 3mm, which connect to the gap between bi-convex tapes. The joint parts shape and dimensions are designed based on the basic bending and torsion test. Figure 6 shows the covering braid configurations. The material of the braid is fluoroplastic fiver (PFA), and 64 fivers are weaved. The braid is easy to expand just pushing longer direction because of a much stretch properties. The SMAs are inserted inside of braid during expanded condition, shown in Fig. 6 (b), and after inserting, the SMAs are tightly covering by stretching the braid toward longer direction. The assembled configurations of SMA-BCON boom is shown in Figure 7.

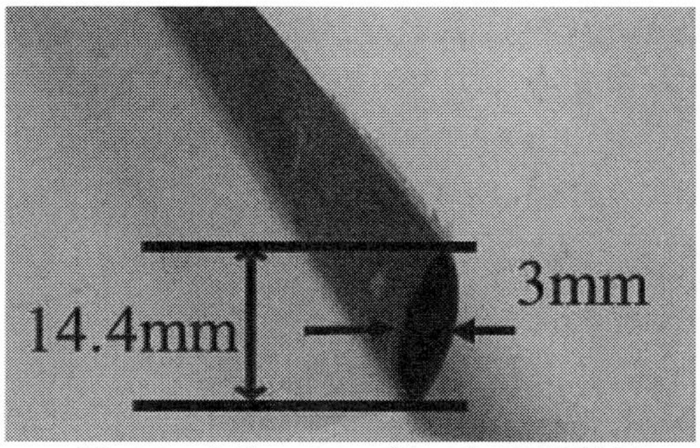

Figure. 4 Cross-section view of SMA

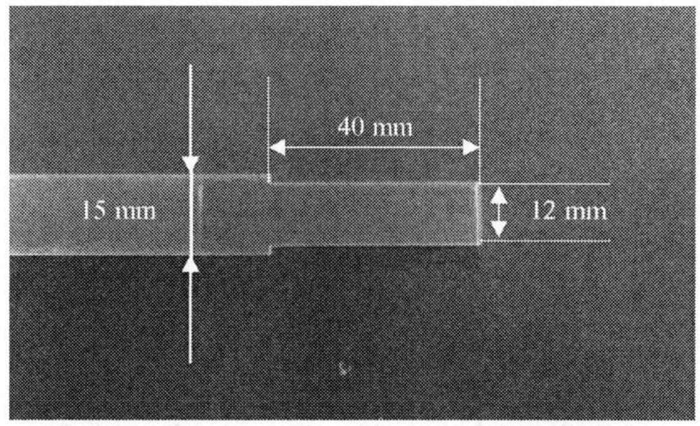

Figure. 5 Joint parts shape and dimensions

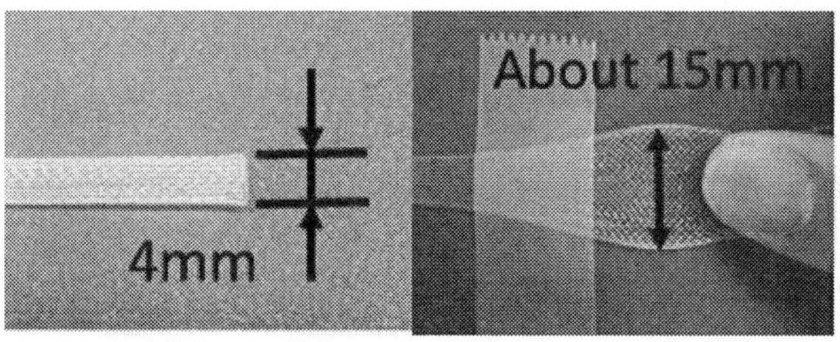

(a) Initial condition (b) Expanded condition

Figure. 6 The covering braid configuration

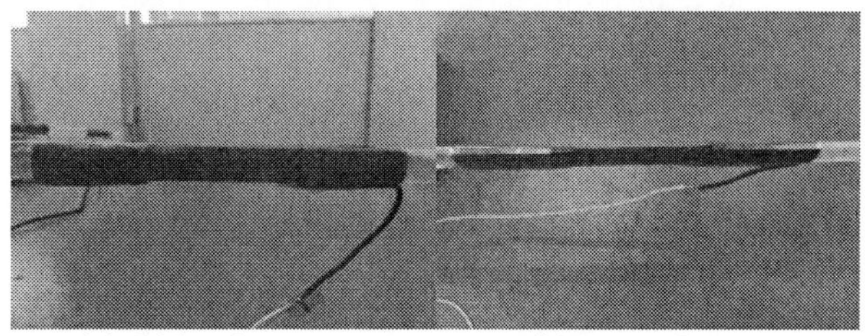
(a) Side view (b) Top view

Figure. 7 Assembly drawing of SMA-BCON boom

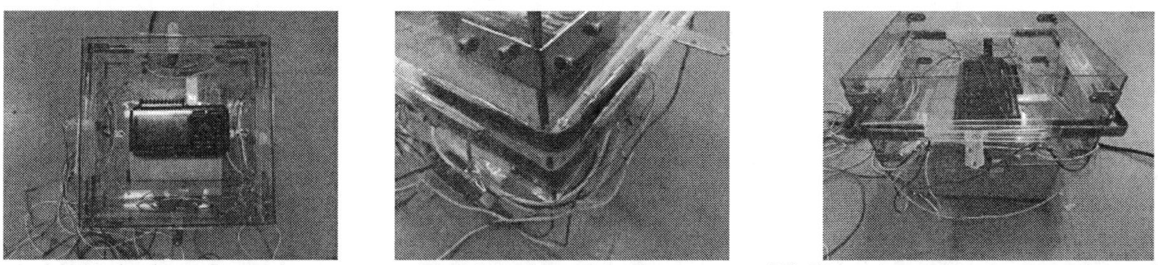
(a) Packed configuration (b) Edge side of the boom (C) Side view of packed configuration

(d) After 1st step deploy (e) Edge side after 1st step deploy (f) Fully deployed configuration

Figure. 8 Packed and deployment configuration of SMA-BCON boom devise

The deployment devise is shown in Figure 8. Based on this SMA-BCON boom concept, the quadrangular center body is designed. Like BCON boom deployment device, there also has thrust bearing on the bottom of the devices to reproduce a rotational free condition. The direct power distribution for SMA is used for heat, the electronic power is supplied by DC power supply located below the devise. The measurement devise is set on the center of the deployment devise.

The size of the square deployment devise is 300 mm on a side. As shown in Fig. 8 (a) and (b), each booms are compactly stored around the center body. The stepwise deployment of this device is two step, the deployed configuration after first step and second step (fully deployed configuration) are shown in Fig. (d), (e) and (f). The total length of the SMA-BCON boom is about 850 mm.

3. DEPLOYMENT BEHAVIOR OF BCON BOOM

The deployment behavior of BCON boom with membrane condition is shown in Figure 9. Three step deployment is achieved under gravity conditions. After the second and third step deployment, the tensional force act to the membrane and the membrane fully deployed after third step. In this case, deployment boom length in the second step is longer than other steps. To investigate the relationship between the deployment length and the deployment steps, three cases deployment experiments are examined. The effect of membrane for deployment behavior is important for this deployable membrane system, the deployment experiments with and without membranes condition is also investigated. The released length of booms in each steps are shown in Table 1. In the case No. 1, the deployment length in the first step is the longest. The second step is the longest in the case No. 2 and the third step is the longest in the case No. 3. The figure 10 shows the average distance of four booms between boom tip and rotational axis in the only boom conditions, and the figure 11 shows them in the booms with membrane condition. The angular velocity during deployment is also measured using gyroscopic sensor, the results are shown in figure 12.

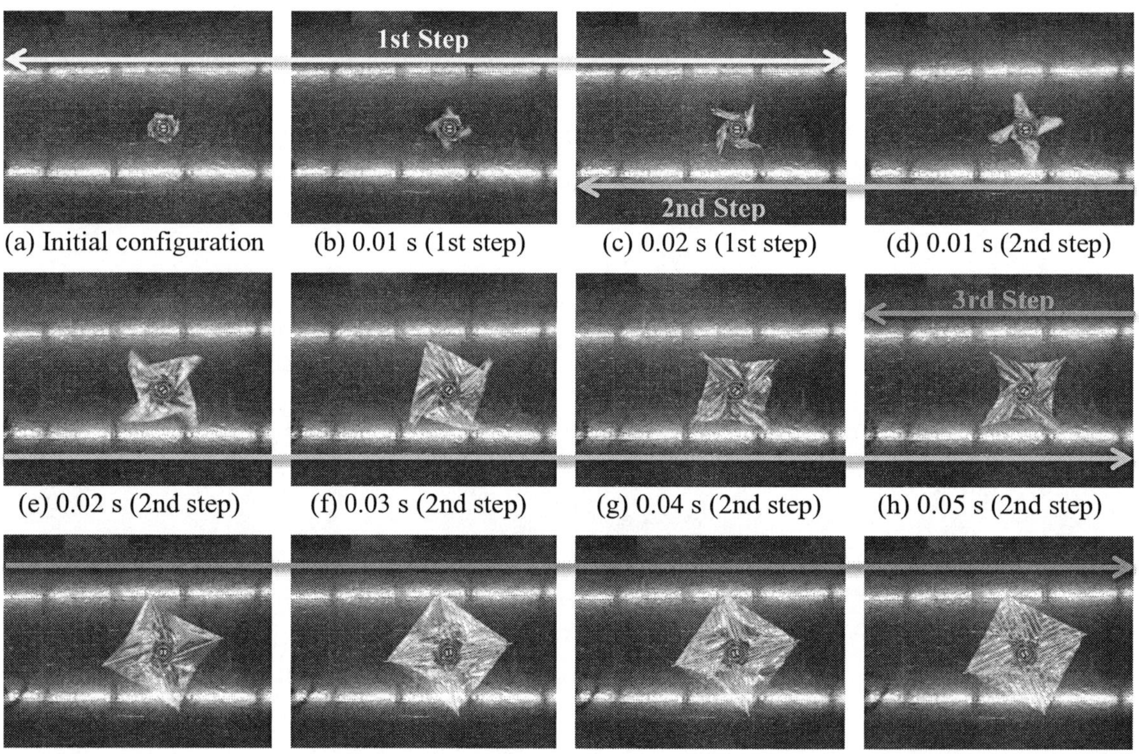

(a) Initial configuration (b) 0.01 s (1st step) (c) 0.02 s (1st step) (d) 0.01 s (2nd step)

(e) 0.02 s (2nd step) (f) 0.03 s (2nd step) (g) 0.04 s (2nd step) (h) 0.05 s (2nd step)

(i) 0.01 s (3rd step) (j) 0.02 s (3rd step) (k) 0.03 s (3rd step) (l) 0.04 s (3rd step)

Figure. 9 Deployment behavior of BCOM boom with membrane (Case 2)

Table 1 Released length of booms in each steps into three cases

Case No.		1	2	3
Initial Distance from rotational axis (mm)		125 (0)	127 (0)	134 (0)
Distance after each steps (mm)	1st step	357 (232)	238 (111)	262 (128)
	2nd step	426 (69)	426 (188)	328 (66)
	3rd step	520 (94)	520 (94)	520 (192)

* The numbers in parentheses mean released length between the steps.

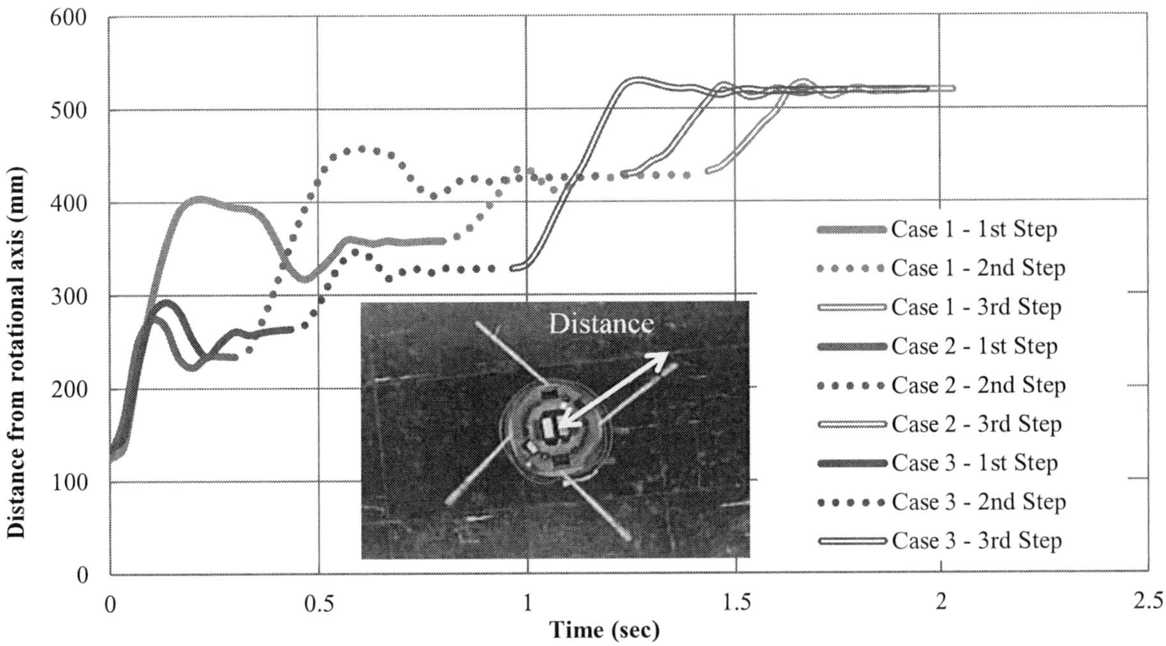

Figure. 10 Average distance between boom tips and rotational axis (Boom only)

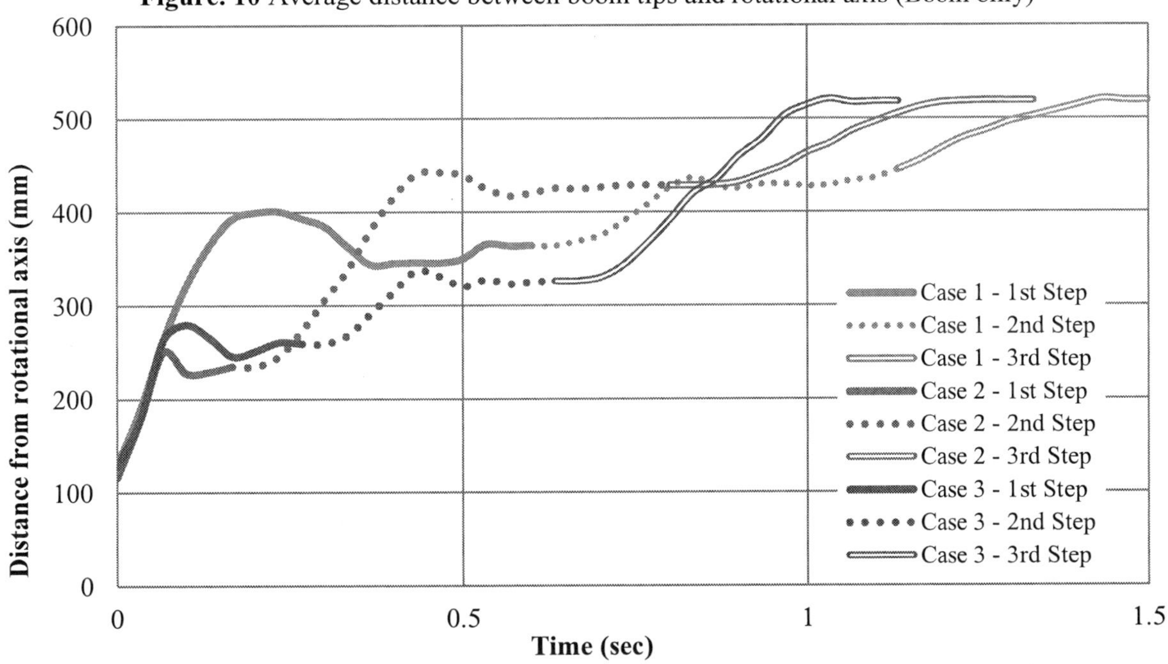

Figure. 11 Average distance between boom tips and rotational axis (Boom with membrane)

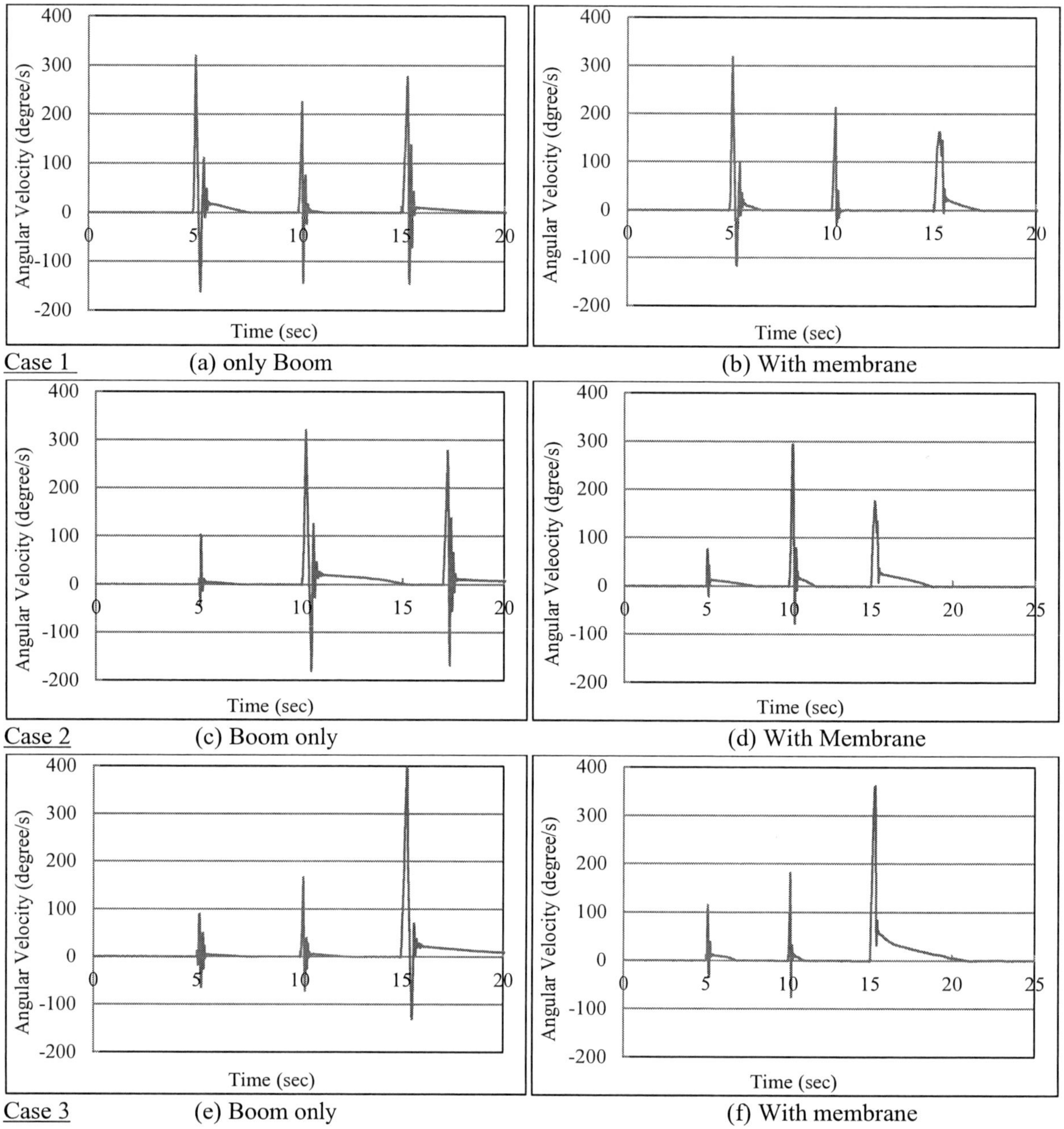

Figure. 12 Angular Velocity during deployment in the three cases.

Shown as Fig. 10 and 11, there are overshoot in each step. In Fig. 9 (f), (g) and (k), the overshoot behavior is clearly tracked. From the angular velocity variations in each steps, shown in Fig. 12, the rotational direction change drastically around the overshoot phenomenon.

However the amount of overshoot decrease according to the higher steps, and at the third steps with membrane condition, there are almost no overshoot, shown in Fig. 10 and 11. This result is notably shown in the angular velocity variations of membrane condition. Due to the overshoot, the only boom condition shows the large opposite angular velocity, shown in Fig. 12 (a), (c) and (e). But in the boom with membrane condition, the amount of opposite angular velocity is decrease, and there are not opposite angular velocity value in the every third step deployment, shown in Fig. (b), (d) and (f).

As for membrane, the total mass and the moment of inertia accretion, and air drag may decrease the overshoot deployment. As for only boom condition, the longer deployment in the higher steps also show less overshoot deployment. It is thought that the friction between the sliding convex tape and the covering braid due to the perimeter difference between outer and inner convex tapes during deployment is one of the cause to decrease overshoot of higher step deployment. But from the view point of the angular velocity, the longer boom deployment lead the large angular velocity shown in Fig. 9 (a) and (b) around 5 s, Fig. 9 (c) and (d) around 10 s, and Fig. 9 (e) and (f) around 15 s.

Through the results, to control the friction force during the deployment become one of the key for stable deployment and the longer deployment toward higher step realize the stable stepwise deployments. We think that the deployment experiments in a large vacuum chamber is a next works to clearly discern the effect of membrane mass, air drag and boom friction. Furthermore, the trade-off design which mean the overshoot and the angular velocity, and/or these optimization using membrane mass, air drag and boom friction adapting to the mission requirements are the future works.

4. DEPLOYMENT BEHAVIOR OF SMA-BCON BOOM

The deployment behavior of SMA BCON booms for the conceptual examination are shown in Figure 13. The supplied electrical power in the first and second steps is 150 W fixed. The each SMA in the same steps are connected series circuit. The fixed power is supplied without any value controlling according to the deployment behavior at the present stage.

The boom are hanged by fishing line from about 3 m high position due to cancelling the gravity force. After the complete deployment and heating condition, this SMA BCON boom can keep the straight shape and can resister gravity force. But during the deployment, the parallel folded part of SMA where located in each corner of center body has less bending stiffness than the convex tape shape, it's difficult to resister the gravity force and the SMA BCON booms are bent down. It is possible to realize unbent down SMA-BCON boom during deployment by increasing the thickness of SMA, but we think those kinds of design may be over stiffness structures in space.

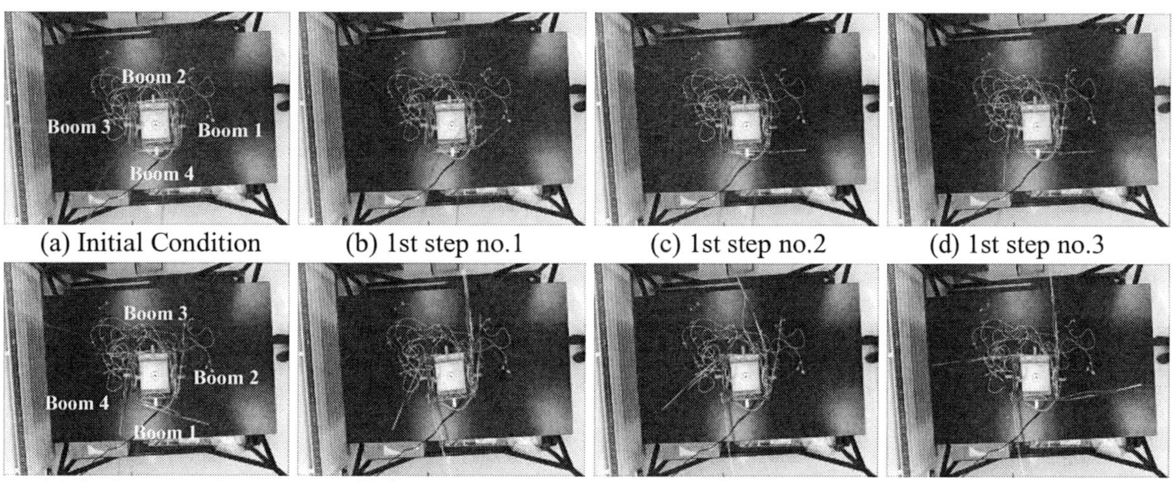

(a) Initial Condition (b) 1st step no.1 (c) 1st step no.2 (d) 1st step no.3

(e) 2nd step no.1 (f) 2nd step no.2 (g) 2nd step no.3 (h) 2nd step no.4

Figure. 13 Deployment behavior of SMA-BCON boom

Table 2 Angular difference between center body side and each booms after 1st step deployment

	Boom 1	Boom 2	Boom 3	Boom 4
Angle (Degree)	2.79	8.13	1.51	7.41

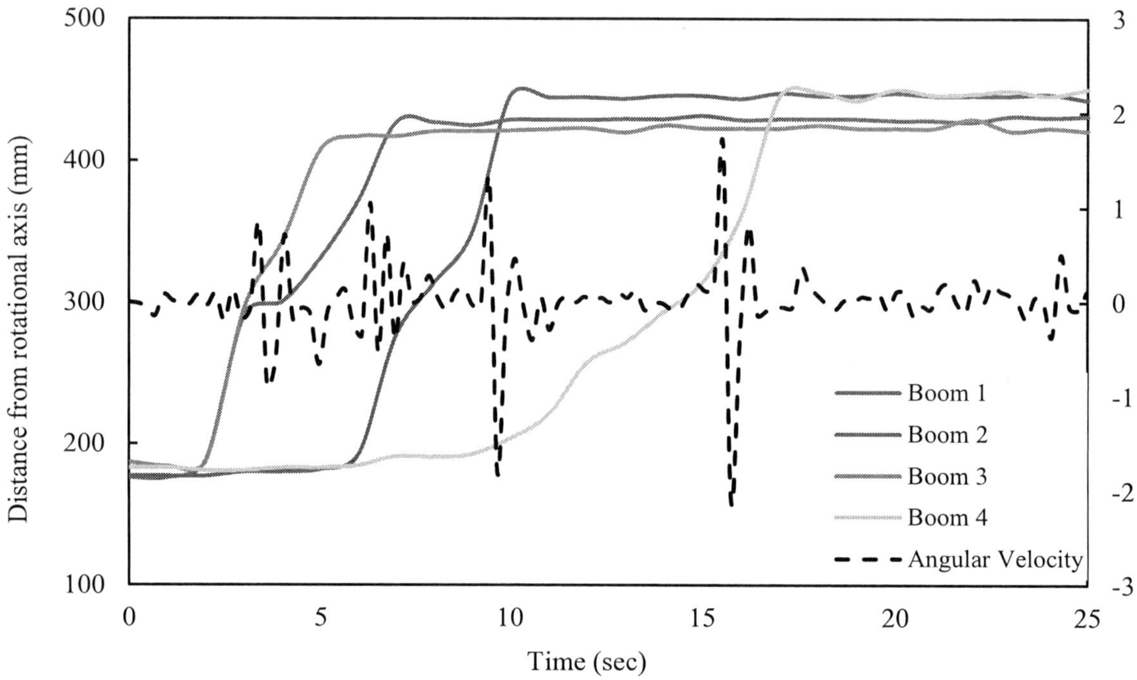

(a) Distance from rotational axis and angular velocity in the 1st Step deployment

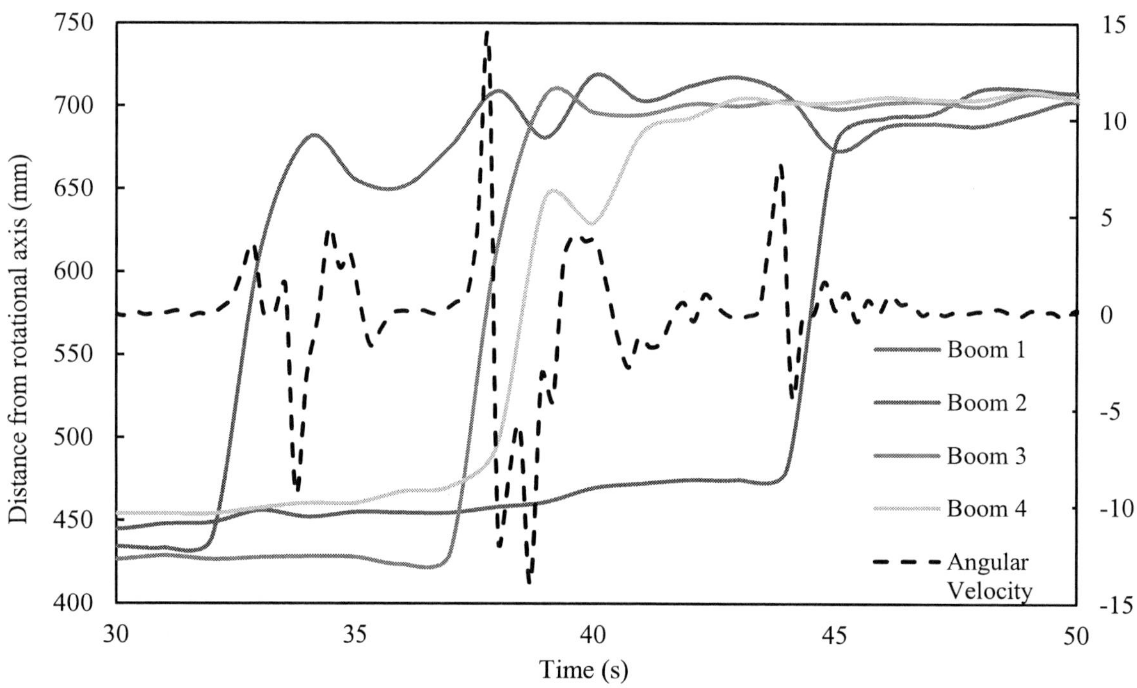

(b) Distance from rotational axis and angular velocity in the 2nd Step deployment

Figure. 14 Distance and Angular Velocity during deployment

In figure 14, the distance between rotational axis and boom tips are shown in the left vertical axis value and the corresponding angular velocity are shown in right vertical axis value. The deployment order and the angular velocity change due to asynchronous deployment are well shown in Figure 14. The table 2 shows the angular difference between center body side and each booms after 1st step deployment.

Through this conceptual study, the synchronous deployment like BCON boom is not achieved. Shown in Fig. 13, each booms deploy independently and the deployment order on the first step is Boom3, Boom 1, Boom2 and Boom 4, and on the second step, the deployment order is Boom 1, Boom 3, Boom 4 and Boom 2. We examined deployment test many times, but it's difficult to find some relation of deployment order. It is thought that the connections between energy supply to SMA, and the hanging tension among each booms may effect asynchronous deployment. However it is not clear at this moment. (現段階では)

Focusing on the overshoot and the angular velocity variations, those of SMA-BCON boom is much less than BCON boom. As for the angular velocity variations, the variations range of BCON boom is more than 20 times bigger than that of SMA- BCON boom. The deployment speed of SMA-BCON boom is too slow comparing to the BCON boom, and it's （明らか） that the slowly deployment decrease the angular velocity variations. Through these comparisons, it will be shown that the deployment controllability of SMA-BCON boom is one of the advantage, and the developments of controlling methods and/or less sensing controlling for stable deployment of SMA-BCON boom is the next step to realize the smart deployable membrane structure systems.

5. CONCLUSION

Deployment characteristics of two kinds of deployable boom system for membrane space structures are investigated through deployment experiments. One is the braid coated bi-convex tape boom (BCON boom), and the other is the braid coated bi-convex SMA boom (SMA-BCON boom). Both booms are newly developed and they can deploy stepwise manner to control the deployment behavior.

As for BCON boom experiments, the effects of the deployable length in each steps are examined and the difference between only boom condition and booms with membrane condition are compared. Through the results, the longer deployment length in the higher deployment steps will decrease the overshoot deployment of the booms due to the total mass and the moment of inertia accretion, air drag and friction force between braid and sliding convex tape. But from the viewpoint of angular velocity variations, the shorter deployment length in the lower deployment steps will be effective. The trade-off design or deployment step optimization will be the next work on this BCON boom.

As for SMA-BCON boom, the laboratory scale conceptual study model is developed, and the deployment is demonstrated. Comparing the deployment behavior of BCON boom system and SMA-BCON boom system, it is shown that the advantage of the SMA-BCON boom is the controllability of the deployment speed or order. The control methods and the sensing methods are future works of SMA-BCON boom systems.

Acknowledgements

The Authors appreciate Messrs. Naoya Fujiwara (Muroran Institute of Technology), Masanobu Yamasaki (Waseda University), Akihito Watanabe, Toshiyuki Hori (Sakase Adtech Co.,Ltd) for their cooperation on this study.

References

1. Sawada, H., et al., "Mission Report on The Solar Power Sail Deployment Demonstration of IKAROS," 52nd AIAA/ ASME/ASCE/AHS/ASC Structures, Structural Dynamics and Materials Conference, April 2011, AIAA 2011-1887.
2. Greschik, G. and Mikulas, M.M., "Design Study of a Square Solar Sail Architecture," J. Spacecraft and Rockets, 2002, Vol. 39, pp. 653-661.
3. Lichodziejewski, D., Derbès, B., Slade, K. and Mann, T., "Vacuum Deployment and Testing of a 4-Quadrant Scalable Inflatable Rigidizable Solar Sail System," 46th AIAA/ASME/ASCE/AHS/ASC Structures, Structural Dynamics and Materials Conf., April 2005, AIAA 2005-2122.
4. Murphy, D.M., McEachen, M.E., Macy, B.D. and Gaspar, J.L., "Demonstration of a 20-m Solar Sail System," 46th AIAA/ASME/ASCE/AHS/ASC Structures, Structural Dynamics and Materials Conf., April 2005, AIAA 2005-2126.
5. Banik, J.A. and Murphey, T.W., "Synchronous Deployed Solar Sail Concept Demonstration," 49th AIAA/ASME/ASCE/AHS/ASC Structures, Structural Dynamics, and Materials Conf., April 2008, AIAA 2008-2213.
6. Fernandez, J.M., Lappas, V.J. and Daton-Lovett, A.J., "The Completely Stripped Solar Sail Concept," 53rd AIAA/ASME/ASCE/AHS/ASC Structures, Structural Dynamics and Materials Conf., AIAA 2012-1745.
7. Greschik, G. and Palisoc, A., "CubeSat-Deployable Photon Sieve Design for Strength and a High Degree of Deployment Control," 53rd AIAA/ASME/ASCE/AHS/ASC Structures, Structural Dynamics and Materials Conf., April 2012, AIAA 2012-1671.
8. Schultz, M., Francis, W., Campbell, D. and Lake, M., "Deployment Accuracy and Mechanics of Elastic Memory Composites," 48th AIAA/ASME/ASCE/AHS/ASC Structures, Structural Dynamics, and Materials Conf., April, 2007, AIAA-2007-2401.
9. Seffen, K.A. and Pellegrino, S., "Deployment Dynamics of Tape Springs," Proc. R. Soc. Lond. A, 1999, 455, pp. 1003-1048.
10. Jeon S.K. and Murphey, T.W., "Design and Analysis of a Meter-class CubeSat Boom with a Motor-less Deployment by Bi-stable Tape Springs," 52nd AIAA/ASME/ASCE/AHS/ASC Structures, Structural Dynamics and Materials Conf., April 2011, AIAA 2011-1731.
11. Kwok, K. and Pellegrino, S., "Viscoelastic Effects in Tape-Springs," 52nd AIAA / ASME / ASCE / AHS / ASC Structures, Structural Dynamics and Materials Conference, April 2011, AIAA 2011-2022.
12. Mallikarachchi, H.M.Y.C. and Pellegrino S., "Design and Validation of Thin-Walled Composite Deployable Booms with Tape-Spring Hinges," 52nd AIAA/ASME/ASCE/AHS/ASC Structures, Structural Dynamics and Materials Conference, April 2011, AIAA 2011-2019.
13. Straubel, M., Block, J., Sinapius, M. and Huehne, C., "Deployable Composite Booms for Various Gossamer Space Structures," 52nd AIAA/ASME/ASCE/AHS/ASC Structures, Structural Dynamics and Materials Conf., April 2011, AIAA 2011-2023.
14. Picault, E., Bourgeois, S., Cochelin, B. and Guinot, F., "On the Folding and Deployment of Tape Springs: A Large Displacements and Large Rotations Rod Model with Highly Flexible Thin-Walled Cross-Section," 53rd AIAA/ASME/ASCE/AHS/ASC Structures, Structural Dynamics and Materials Conf., April 2012, AIAA 2012-1956.
15. Lichodziejewski, D., Veal, G. and Derbes, B., "Spiral Wrapped Aluminum Laminate Rigidization Technology," 43rd AIAA/ASME/ASCE/AHS/ASC Structures, Structural Dynamics, and Materials Conf., April 2002, AIAA-2002-1701.
16. Fang, H., Lou, M. and Har, J., "Deployment Study of a Self- Rigidizable Inflatable Boom," J. Spacecraft and Rockets, 2006, Vol. 43, No.1, pp.25-30.
17. Higuchi, K., Watanabe, K., Watanabe, A., Tsunoda, H. and Yamakawa, H., "Design and Evaluation of an Ultra-light Extendible Mast as an Inflatable Structure," 47th AIAA/ASME/ASCE/AHS/ASC Structures, Structural Dynamics, and Materials Conf., May 2006, AIAA 2006-1809.

18. Watanabe, A., Ito, H. and Hori, T., "Study of the Extensible Structure Which Coated a Braid," Japanese 56th United Conf. Space Science and Technology (Ukaren), Beppu, Japan, Nov. 2012, JSASS-2012-4496.
19. Natori, M.C., Hori, H., Sawai, K., Okuizumi, N. and Yamakawa, H., "Stepwise Deployment of Membrane Space Structures Rolled-Up together with Support Booms," 63rd Int'l. Astronautical Congress, Oct. 2012, IAC-12-C2.2.9.
20. Ito, Y., Okuizumi, N., Natori, M.C., Yamakawa, H., Katsumata, N., "Deployment Analysis of Membrane Space Structures Using Multi-Particle Approximation Method," Japanese 56th United Conf. Space Science and Technology (UKaren), Beppu, Japan, Nov. 2012.
21. Katsumata, N., Tashiro, M., Tomomatsu, H., Yamasaki, M., Natori, M.C., Yamakawa, H., Watanabe, A. and Kawabata, N., "Experimental Evaluation on the Boom for Stepwise Deployable Membrane Structures,"Japanese 56th United Conf. Space Science and Technology (Ukaren), Beppu, Japan, Nov. 2012.

ICAST2013 000000

Shape Memory Alloy Column Buckling: An Experimental Study

Ryan Watkins* and John Shaw

Department of Aerospace Engineering, University of Michigan, Ann Arbor, MI, USA

Abstract

The frame-like structure of many shape memory alloy devices motivates a better understanding of their buckling characteristics. The present work experimentally examines the buckling behavior of superelastic NiTi and 2024 Al columns under clamped-clamped boundary conditions. For the first time, through the use of digital image correlation, full field displacements and strains are measured, quantifying lateral buckling displacements, "no-slip" axial displacements, and local strain fields. The experimental critical buckling load is measured for a range of column aspect ratios, showing good agreement with inelastic buckling theory. The mechanical characteristics of each material are also compared, demonstrating the effects of phase transformation on post-buckling. For the case of the lowest aspect ratio NiTi column, it is shown that superelastic columns exhibit "unbuckling" behavior, or the straightening of a buckled column under continued loading (which is similar to the "Twinning Deformation" effect found by [12]). This behavior is then modeled using a modified Shanley column, illuminating the geometric conditions under which NiTi columns "unbuckle".

1. INTRODUCTION

A prominent group of active materials, shape memory alloys (SMAs) exhibit the unique thermomechanical properties of the shape memory effect and superelasticty. The more well known shape memory effect is the process by which the deformed material returns to a unique "memorized" configuration upon heating above the austenite transformation temperature, whereas superelasticity refers to the material's ability to recoverably deform strains on the order of 10% when functioning at temperatures somewhat above the austenite transformation temperature. The large deformation kinematics and high work density of shape memory alloys motivated their use in the biomedical sector as vascular stents, arch wire in orthodontic braces, highly resilient eyeglass frames and more generically as couplers, actuators, deployable devices, and vibration suppressors.

Beam buckling is an inherent concern for many SMA devices, such as stents, due to their slender frame-like structures. In the case of traditional metals, buckling often results in plastic deformation (either at the initiation of buckling or in the post-buckling response) due to yield strains on the order of 0.1%; however, in the case of NiTi SMAs, in which the material can recover deformation on the order of 10% strain, a much wider range of loading conditions exists in which the structure can buckle recoverably. It is therefore conceivable that an SMA structure could be designed to functionally buckle while maintaining recoverable deformation, presenting the possibility for novel applications.

Buckling of SMA beams presents a complex interaction between material nonlinearity and structural instability. Although expected to behave similarly to elasto-plastic materials at the onset of instability, a small literature exists with respect to SMA beam buckling. Rahman et al. were of the first to examine the buckling behavior of SMAs, comparing the buckling response of superelastic NiTi with aluminum and stainless steel[8]. Later work examined the modeling of this behavior through finite element analysis[9]. Urushiyama et al. discovered the "Twinning Deformation" effect for shape memory NiTi (i.e., in the

*Corresponding author: watkinrt@umich.edu

martensite phase at room temperature), which is the process by which a buckled SMA column straightens ("unbuckles") upon further loading given certain geometric and material conditions[12]. Finally, Movchan et. al. have modeled a variety of SMA buckling configurations, examining both columns and circular plates[1–7].

The present work examines the buckling behavior of NiTi columns under clamped-clamped boundary conditions. Section 2 experimentally quantifies the buckling behavior of NiTi and Al rods of varying aspect ratios, comparing the similarities and differences between the two materials. Although similar to the work by [8], the full field displacements and strains were additionally measured through the use of stereo digital image correlation (DIC). This allowed for the measurement of the critical buckling load based on the column lateral displacement, characterization of local strain fields, and quantification of the "unbuckling" behavior found in low aspect ratio NiTi columns, along with an accurate measurement of axial displacement free from adverse effects such as grip slippage. Section 3 addresses the "unbuckling' behavior originally discovered by [12] and verified here in Section 2. A modified Shanley column is examined, illustrating the mechanism behind the "unbuckling" behavior.

2. EXPERIMENTS

Experiments were performed on 3.175 mm (1/8 in) diameter (d) *superelastic* NiTi and 2024 Al solid rods. The room temperature uniaxial tension and compression behaviors for each material are shown in Figure 1a. A single experiment was conducted to measure the tensile properties of the Al rod. Assuming symmetric behavior, the compressive response of the Al material was extrapolated as denoted by the dotted line in Figure 1a. In the case of the NiTi rod uniaxial characterization, two separate experiments were necessary to fully characterized the tension-compression asymmetry found in the material. Furthermore, measuring the compressive response of a beam-like structure, especially one which softens during phase transformation, is challenging. Great efforts were made to minimize loading misalignment so as to compress the material while avoiding premature buckling. For more details on the methods used to measure the compressive response, see [10].

2.1 Experimental Setup

The experimental setup for the column buckling experiments is shown in Figure 2. Each rod specimen was painted with a matte white paint (Golden Airbrush Titanium White – #8380) using an airbrush, followed by black speckles (Golden Airbrush Carbon Black – #8040) for use with digital image correlation (DIC). The specimen was then loaded into an Instron 5585 electromechanical load frame and gripped using ER collets, resulting in a clamped-clamped column with free length of L. Special care was taken to align the top and bottom grips to minimize the effects of geometric and loading imperfections on the buckling characteristics. This alignment was undisturbed throughout the different experiments, resulting in consistent boundary conditions for all experiments. Two Point Grey GRAS-50S5M/C Grasshopper CCD cameras were focussed on the specimen at a stereo angle ranging from 8° to 15°, depending on the column aspect ratio, and used to image the specimen for DIC analysis. In the case of the NiTi experiments, ambient and specimen thermocouples were used to verify isothermal loading. The specimen thermocouple was attached to the specimen using OmegaTherm 201 thermal paste to ensure good thermal contact. Each specimen was compressively loaded under displacement control at a slow constant strain rate of $\dot{\delta}/L = 10^{-4}\,\mathrm{s}^{-1}$ to a maximum grip displacement $\delta_{\max}$. The specimen was then unloaded at a constant strain rate of $\dot{\delta}/L = -10^{-4}\,\mathrm{s}^{-1}$ to a state of zero load. In the case of the NiTi experiments, an additional 10 cycles were performed to gain a sense of the shakedown characteristics during buckling. During the loading procedure, the axial load (P) and grip displacement (δ) where measured along with the DIC imaging. Using the DIC analysis, the full-field strain profiles, along with axial displacement (δ_e), lateral buckling displacement (δ_r) and axial strain (δ_e/L_e where L_e is the DIC gauge length), were measured.

Due to issues with grip slippage, a specialized analysis of the DIC data was used to measure the axial displacement of the column. Traditionally, mechanical or laser extensometers are used to accurately mea-

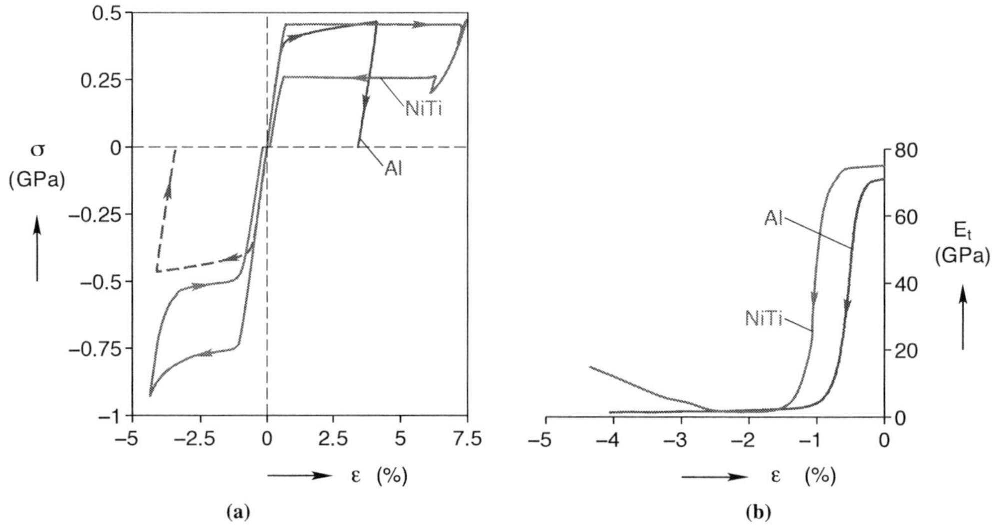

Figure 1. Material data for the NiTi and Al rods. (a) Uniaxial responses of 3.175 mm diameter NiTi and Al rods, where σ and ε are engineering stress and strain, respectively. For the case of the Al characterization, the dotted line of the compressive response signifies the assumed tension-compression symmetry of the material behavior. (b) Loading tangent modulus (E_t) of the uniaxial compressive response for the NiTi and Al rods.

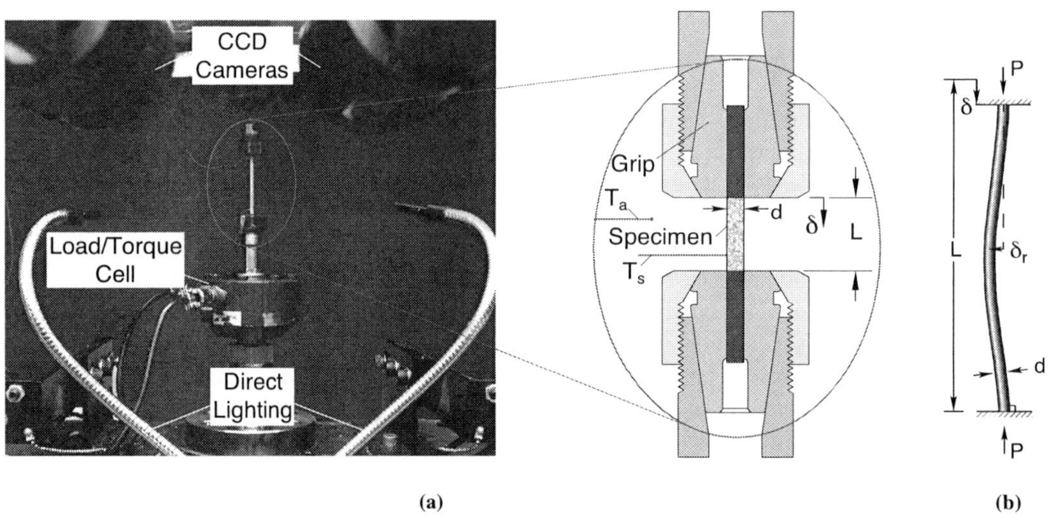

Figure 2. Column buckling experimental setup

sure axial displacement over a specified extensometer gauge length; however, the lateral displacement that occurs during buckling can result in inaccurate measurements. With respect to the DIC measurements, the axial displacements directly measured by DIC are with respect to the column surface, not the column centerline. In the case of bending, the surface displacements under/over predicts the axial displacements. Therefore, to accurately measure the axial displacement, the full 3D surface data from the DIC analysis were used to reconstruct and measure the deformation of the column centerline. An

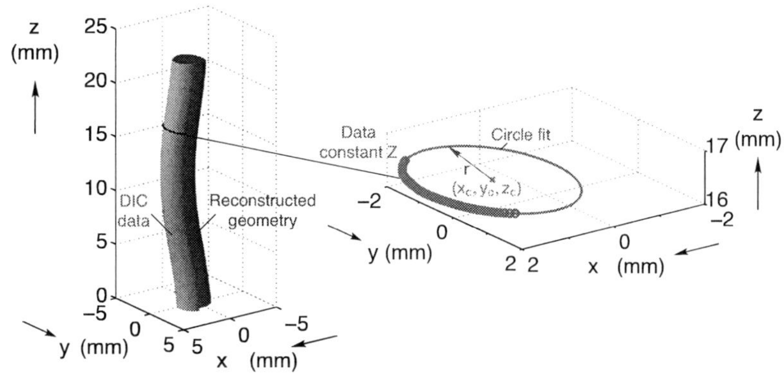

Figure 3. Schematic of the column centerline reconstruction. For each bin of material points with the same Z coordinate in the reference frame (black line in the column surface data), a circle is fit to the data and the corresponding center point (x_c, y_c, z_c) and radius r are determined.

Table 1. Column aspect ratios

Target L/d	Actual L/d	
	NiTi	Al
5	4.89	4.61
7.5	7.95	7.58
10	9.55	9.94
20	19.18	18.56

illustration of the centerline reconstruction is shown in Figure 3. A discrete number of material points with coordinates (X, Y, Z) were established in the reference configuration of the structure (the first stereo image pair). These material points were then tracked during the experiment, measuring the (x, y, z) coordinates for every material point at each time step (each successive stereo image pair). This results in a Lagrangian measurement where all displacements and strains are with respect to the reference configuration. To reconstruct the column centerline, all of the material points were first grouped together according to their Z location along the column length, resulting in bins of material points that all have the same Z coordinate in the reference configuration. For a given instant in time, a circle was fit to each bin of material points, determining the column center point (x_c, y_c, z_c) and radius r at each location along the length of the column. This procedure was conducted for each frame of the DIC data, resulting in a reconstruction of the column centerline through the course of an experiment. The end displacement of the reconstructed center line was then used to measure the axial displacement (δ_e) with respect to the reference centerline length L_e. It should be noted that this approach assumes that the column bends such that cross-sectional planes remain plane and normal to the deformed centerline. It has recently been shown that phase transformation localization results in distortion of the cross-sectional planes during bending of superelastic tubes[10]; however, phase transformation localization was not seen in the DIC full field strain measurements for any of the experiments on solid rods presented here.

2.2 Experimental Results

A range of column aspect ratios (L/d) were tested for both the NiTi and Al materials, examining the effects of column length on the critical buckling load and post-buckling behavior. Aspect ratios of 5, 7.5, 10 and 20 were chosen to capture the buckling response in both the elastic (high aspect ratio) and phase transformation/plasticity (low aspect ratio) regimes of the structural response. The exact aspect ratios of each column are shown in Table 1. The results for the experiments are shown in Figures 4 to 9.

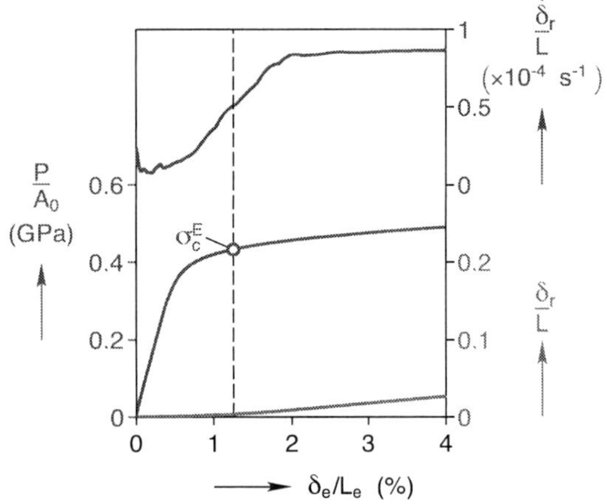

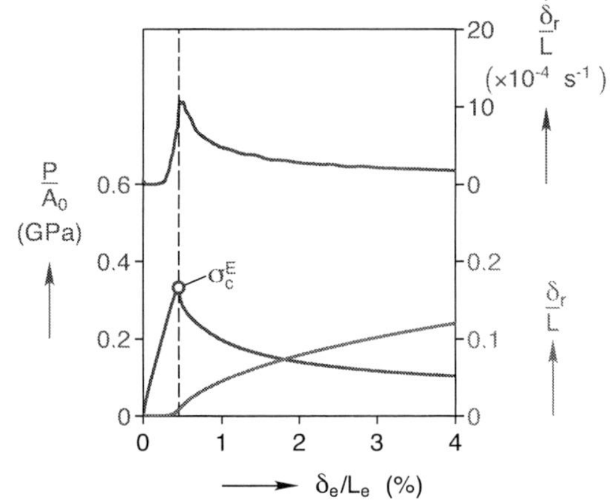

Figure 4. Measuring the critical buckling stress σ_c^E of the Al data for (a) L/d = 5 and (b) L/d = 20

For each experiment, the experimental critical buckling stress (σ_c^E) was measured and the theoretical critical buckling stress (σ_c^T) was calculated using the tangent modulus theory for inelastic buckling of clamped-clamped columns:

$$\sigma_c^T = \frac{4\pi^2 E_t I}{A_0 L^2} \tag{1}$$

where E_t is the loading tangent modulus of the uniaxial compression response as calculated using the data in Figure 1a and shown in Figure 1b, $A_0 = \pi d^2/4$ is the reference area of the column cross-section, and $I = \pi d^4/64$ is the reference area moment of inertia of the cross-section. The measurement method for σ_c^E is shown in Figure 4, using the Al columns with aspect ratios 5 and 20 as examples. In the lower portions of the figures, the engineering stress and normalized lateral displacement are plotted with respect to engineering strain, while the upper portions show the corresponding time derivative of the lateral displacement. Since buckling is defined as the bifurcation from the principle solution to a secondary path, a theoretical step should exist in the derivative of the lateral displacement ($\dot{\delta}_r$) at the point of bifurcation. In reality, the change is smooth, so the halfway point between the initial lateral displacement derivative and the buckled lateral displacement was taken. This method was found to be more accurate in determining σ_c^E than using the lateral displacement directly. The results for σ_c^E and σ_c^T are shown in Figure 5. The theory and experiments match within a few percent except for the NiTi column with L/d = 20. The source of the discrepancy in the calculation and measurement of the critical buckling load for this high aspect ratio NiTi column is unclear and is being investigated.

The effects of column aspect ratio on the buckling mechanical responses are shown in Figure 6. As expected, the critical buckling load of the columns for each material decreased with increasing aspect ratio (which is also illustrated in Figure 5). After buckling, the Al columns continued to respond in a smooth monotonic fashion. Contrary to elastic buckling, where the tangent modulus of the mechanical response approaches zero, the Al column with an aspect ratio of 5 buckled under increasing load, maintaining a positive tangent modulus deep into the post-buckling response. In the case of the NiTi columns, the post-buckling response is more complicated. The post-buckling response is non-monotonic with changes in the tangent modulus, likely due to the interplay between portions of the column changing phase at different times during the loading. Furthermore, as the column aspect ratio is increased, the effects of phase transformation on the NiTi post-buckling response decrease, which is likely due to the transition from axial to bending dominated deformation.

A comparison between the mechanical response for the Al and NiTi columns, for each aspect ratio,

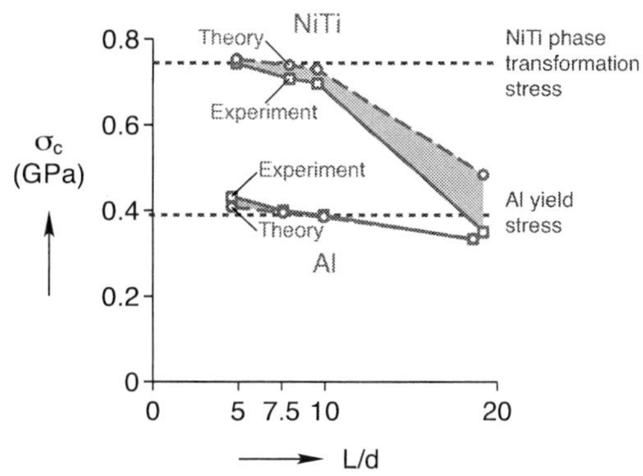

Figure 5. Comparison of critical buckling stress measurements and theoretical predictions.

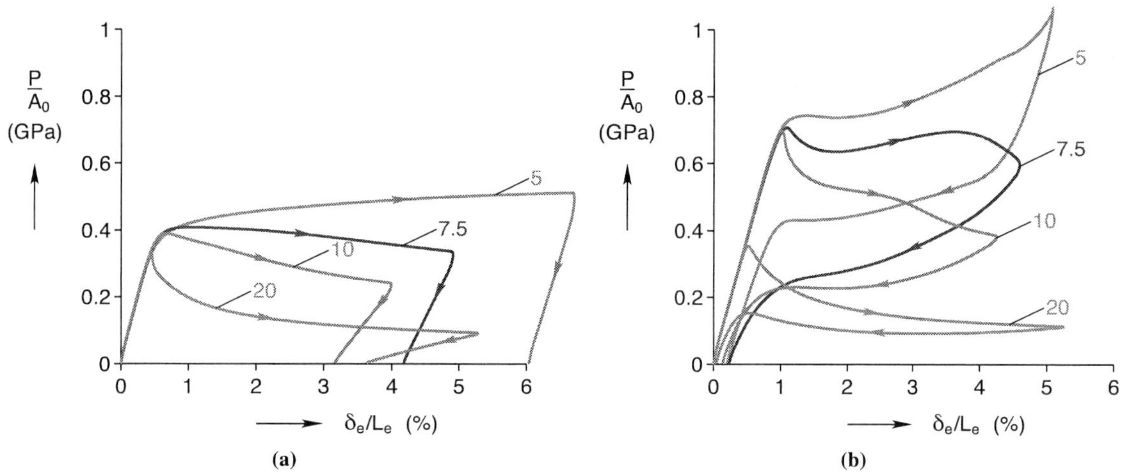

Figure 6. Comparison of buckling responses with respect to the aspect ratios 5 (green), 7.5 (purple),10 (teal), and 20 (orange) for the (a) Al and (b) NiTi columns.

are shown in Figure 7. The bottom portion of the plots report the engineering stress and strain, while the top portion reports the normalized lateral displacement. To help delimit loading and unloading, the loading portion of each plot is denoted by a solid line while the unloading portion is denoted by a dash–dot line. For all aspect ratios, the enhanced recoverable deformation of the NiTi columns, for both axial and lateral displacement, are obvious. The critical buckling stresses for the NiTi columns are roughly twice that of the Al columns, corresponding to the two–fold difference between the NiTi phase transformation stress and the Al yield stress. The only exception to this is for an aspect ratio of 20 in which both materials buckle at approximately the same stress. Since NiTi and Al have similar elastic moduli (of the order of 70 GPa), this case corresponds to elastic buckling of both materials. Interestingly, the effects of phase transformation on the NiTi response are less noticeable in the case of elastic buckling, resulting in similar post-buckling behavior to the Al material during loading. Focussing on the lateral displacements, the loading behavior is similar between the two materials except for the the columns with an aspect ratio of 5. In Figure 7a, it can be seen that the lateral displacement of both columns increases upon buckling; however, at a critical load, the lateral displacement of the NiTi column begins to decrease under continued loading. In other words, the column reaches a point in which further loading causes the column to "unbuckle." This is similar behavior to the phenomenon termed the "Twinning Deformation"

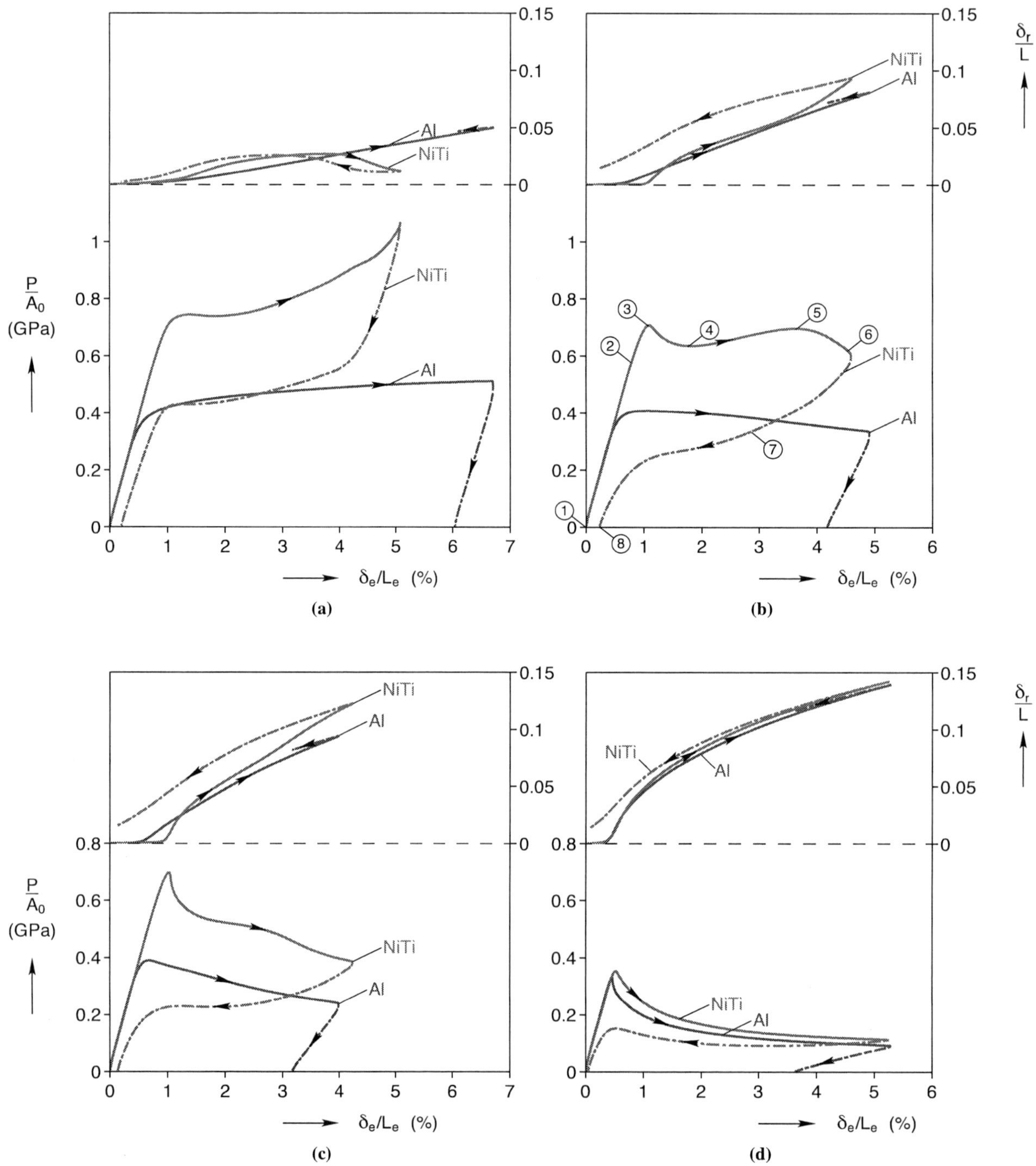

Figure 7. Comparison of Al and NiTi column buckling responses for column aspect ratios (L/d) of (a) 5, (b) 7.5, (c) 10 and (d) 20

effect by Urushiyama et al. [12].

A sequence of axial strain fields for the NiTi column with $L/d = 7.5$ are shown in Figure 8, where the labels correspond to Figure 7b. It should be noted that, since the rod has no preference with respect

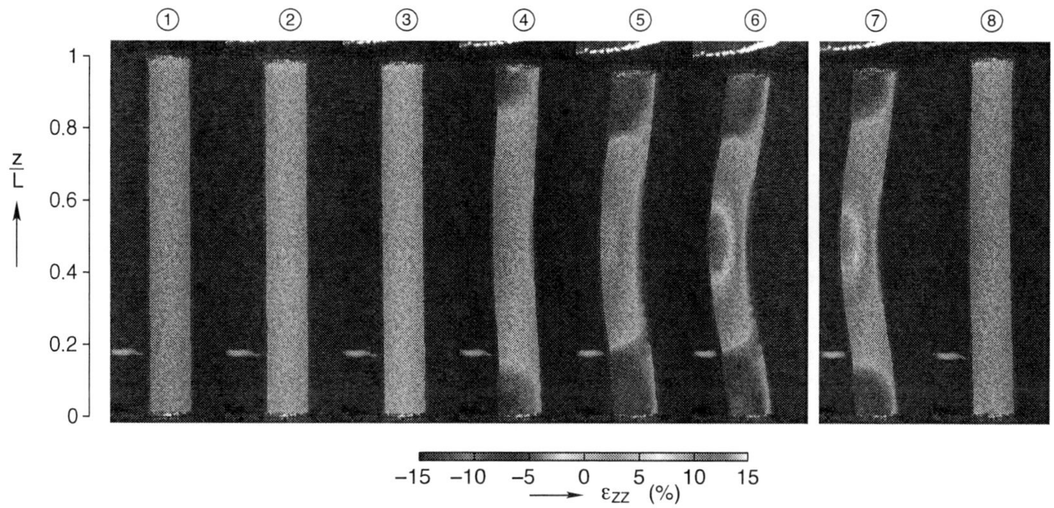

Figure 8. DIC axial strain (ε_{zz}) field images for the NiTi column with L/d = 7.5 which correspond to the labels in Figure 7b.

to the buckling direction ($I = I_{xx} = I_{yy} = \pi d^4/64$), the column did not buckle perfectly in the plane of the image. Starting in the reference configuration, ①, the column deformed uniformly up through ③, after which it buckled. In the post-buckling portion of the loading, ④–⑥, the strain fields were smooth, suggesting that no localization phenomena occurred (i.e., no propagating fronts). This was consistently seen for all of the NiTi experiments and is contrary to what was seen in [10] for the case of a NiTi tube under pure bending. It is postulated that the restrictions on kinematic compatibility for macroscopic phase boundaries due to the solid cross-section of the columns prevented the macroscopic formation of phase boundaries during buckling.

The cyclic behavior of the NiTi column buckling is shown in Figure 9. For each aspect ratio, the column was cycled an additional 10 times after the first cycle. The engineering stress and strain are presented, demonstrating the shakedown characteristics of each column. For the applied deformations, the behavior for each column aspect ratio converges to a repeatable response within the cycling procedure. Interestingly, the complex post-buckling behavior simplified with increasing cycles, resulting in nearly monotonic loading paths. Furthermore, the post-buckling behavior transitioned from unstable (negative tangent modulus) to stable in all cases except L/d = 20, where a slightly negative tangent modulus persisted.

3. SHANLEY COLUMN ANALYSIS

Although Euler's solution for elastic buckling was developed in 1744, there was controversy regarding inelastic buckling until the 1950s. It wasn't until Shanley's seminal work in 1947 that the debate was settled[11]. To illustrate the complexities and mechanisms of elastoplastic buckling, Shanley examined a simple 2-degree of freedom model, now commonly referred to as the Shanley column. Although Shanley focused on elastoplastic buckling, the idea was later expanded to general inelastic structures. A slight modification of the Shanley column is used here to illustrate the mechanisms behind the "unbuckling" behavior found in SMA columns.

A schematic of Shanley's column is shown in Figure 10. The column consists of a rigid inverted "T," with height L and half width d, which rests on two nonlinear springs (labeled 1 and 2) located at the outer ends of the "T." Although the uniaxial behavior of SMAs is asymmetric and hysteretic in general, a symmetric trilinear constitutive law for the spring was found to be sufficient to capture the "unbuckling" response demonstrated in Figure 7a for the NiTi column. An illustration of the constitutive law is shown in the center of Figure 10, where i is an integer specifying either spring 1 or 2, F_i is the

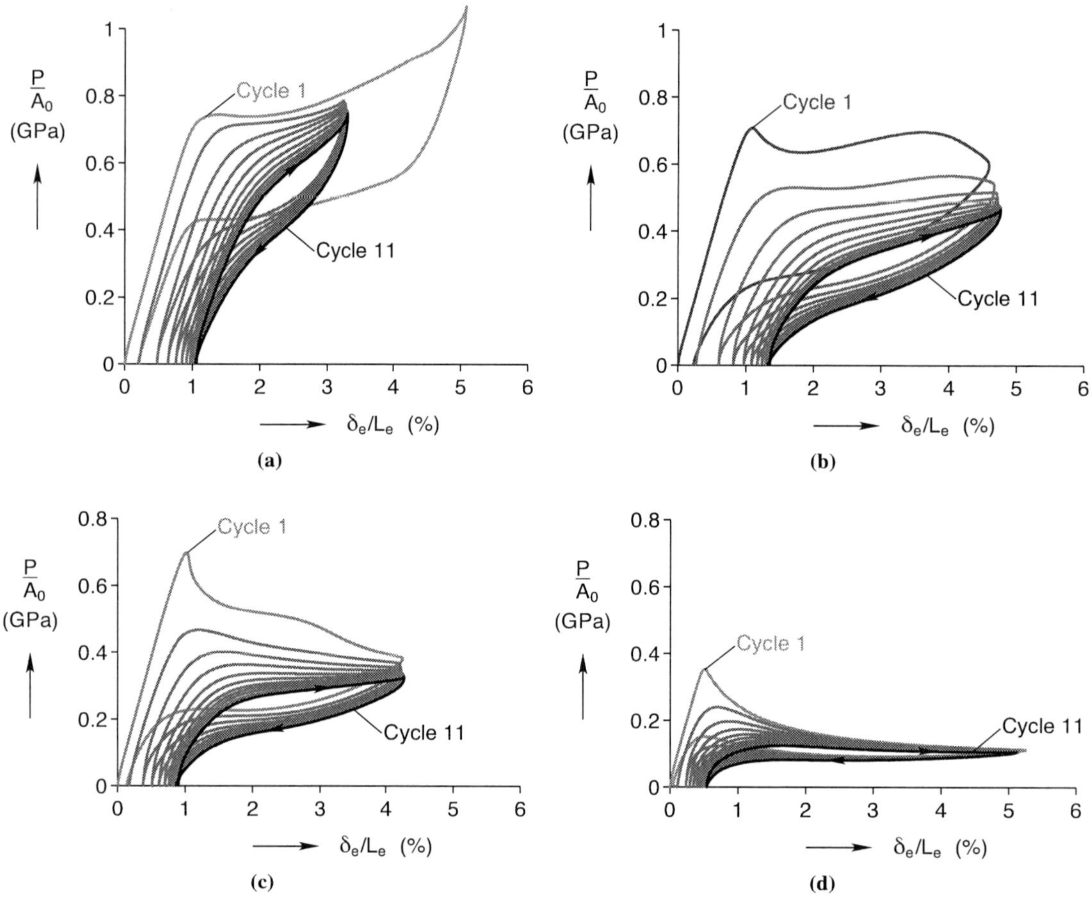

Figure 9. Cycling response of NiTi column buckling for column aspect ratios (a) 5, (b) 7.5, (c) 10 and (d) 20

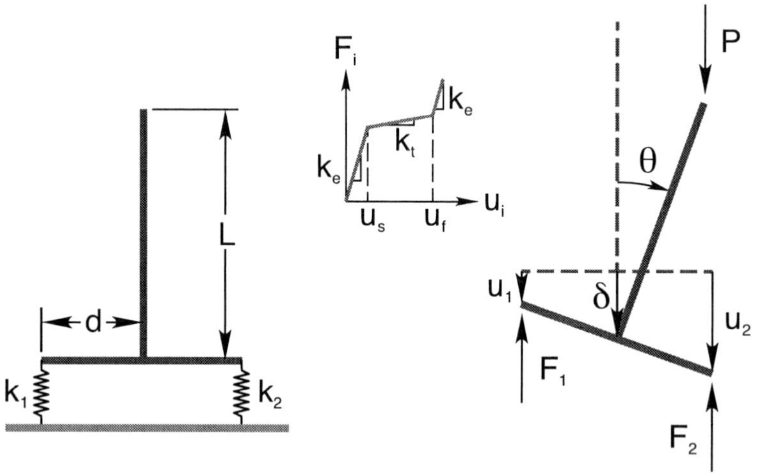

Figure 10. Schematic of the Shanley column

compressive spring force of spring i, u_i is the compressive displacement of spring i, and u_s and u_f are the displacements at the start and finish of phase transformation, respectively. The first and third linear sections of the constitutive law have linear spring stiffness k_e, while the second linear section has linear spring stiffness k_t. Mathematically, the constitutive law is defined as

$$F_i = \begin{cases} -k_e u_s - k_t(u_f - u_s) + k_e(u_i + u_f) & : u_i < -u_f \\ -k_e u_s + k_t(u_i + u_s) & : -u_f \leq u_i \leq -u_s \\ k_e u_i & : |u_i| < u_s \\ k_e u_s + k_t(u_i - u_s) & : u_s \leq u_i \leq u_f \\ k_e u_s + k_t(u_f - u_s) + k_e(u_i - u_f) & : u_i > u_f \end{cases} . \tag{2}$$

Initially in the upright position, the column is loaded with an axial displacement δ and corresponding compressive force P. Upon buckling, the column rotates by angle θ. According to the rigid body kinematics of the column, the spring displacements are defined by

$$u_1 = \delta - d\sin\theta \quad \text{and} \quad u_2 = \delta + d\sin\theta. \tag{3}$$

The nonlinear equilibrium equations for the described system are

$$\sum F = F_1 + F_2 - P = 0 \tag{4a}$$

and

$$\sum M = (F_2 - F_1)d\cos\theta - PL\sin\theta = 0, \tag{4b}$$

balancing forces and moments, respectively. Solving for P in Eq. (4a) and substituting it into Eq. (4b), the governing equation for quasi-static loading of the Shanley Column is

$$(F_2 - F_1)d\cos\theta - (F_1 + F_2)L\sin\theta = 0. \tag{5}$$

As the self-straightening behavior only occurs for low aspect ratio columns, in which displacements and rotations are small, Eqs. (3) and (5) are linearized to the forms

$$(F_2 - F_1)d - (F_1 + F_2)L\theta = 0 \tag{6}$$

and

$$u_1 = \delta - \theta d \quad \text{and} \quad u_2 = \delta + \theta d, \tag{7}$$

respectively. It should be emphasized here that, although the constitutive law is nonlinear, it is still *elastic*. This is contrary to the inelastic problem originally examined by Shanley.

Depending on the geometric and constitutive parameters, a variety of solution paths exist for the described Shanley column. Of interest here is the case in which the column buckles due to the initiation of phase transformation. This mode of deformation can be broken into the three sections shown in Figure 11, which are denoted *Stage a* (Figure 11a), *Stage b* (Figure 11b) and *Stage c* (Figure 11c). During *Stage a*, the system deforms along the principle branch (i.e., $\theta = 0$, $F_1 = F_2$, $u_1 = u_2$ and $k_1 = k_2 = k_e$). For the range of aspect ratios of interest here the structure does not buckle elastically; instead, the first bifurcation point occurs when the springs begin to first transform phase (according to the tangent modulus theory), or when

$$P_c = F_1 + F_2 = 2k_e u_s. \tag{8}$$

In *Stage b*, the solution follows the secondary branch, so $\theta \neq 0$. Due to the symmetry of the system, it is assumed that $\theta > 0$ without loss in generality. There are three hypothetical loading paths for *Stage b*: 1) springs 1 and 2 begin to unload, 2) springs 1 and 2 continue to load, or 3) spring 1 unloads while spring 2 continues to load. The case in which springs 1 and 2 begin to unload is physically impossible for continued loading and can therefore be discarded. Likewise, it can be shown that the only case in

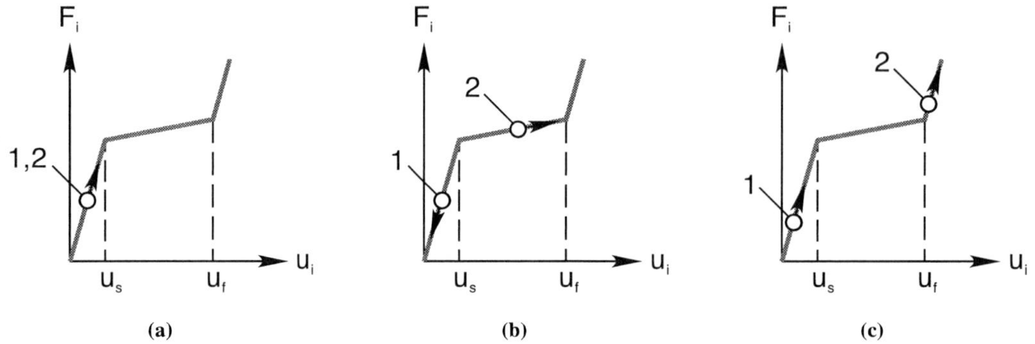

Figure 11. Stages of the Shanley column buckling for the case of trilinear elastic springs. The numbers indicate springs 1 and 2, respectively, along the constitutive relationship.

which both springs 1 and 2 continue to compress corresponds to $\theta = 0$, which is not of interest here as it does not follow the secondary branch. The only remaining case is option (3) where spring 1 begins to unload while spring 2 continues to load as illustrated in Figure 11b. This phase of loading progresses until phase transformation is completed in spring 2 ($u_2 = u_f$).

Stage c of loading is defined by both springs loading again with stiffness k_e as depicted in Figure 11c. The constitutive relationships for this portion of the loading are defined by

$$F_1 = k_e u_1 \quad \text{and} \quad F_2 = k_e u_s + k_t(u_f - u_s) + k_e(u_2 - u_f). \tag{9}$$

The corresponding governing equation reduces to

$$-d(k_e - k_t)\left(u_f - u_s\right) + \theta\left(k_e\left[2d^2 + L\left(-2\delta + u_f - u_s\right)\right] + Lk_t\left[-u_f + u_s\right]\right) = 0 \tag{10}$$

with the solution

$$\theta = \frac{d(k_e - k_t)\left(u_f - u_s\right)}{k_e\left(2d^2 + L\left[-2\delta + u_f - u_s\right]\right) + Lk_t\left(-u_f + u_s\right)}. \tag{11}$$

Depending upon the geometric conditions and material behavior, there are two possible phenomena which can occur during this stage of the loading: secondary buckling or unbuckling. Secondary buckling occurs when another bifurcation point is met, resulting in further unstable post-buckling behavior. Unbuckling occurs when the material response of spring 1 softens due to the initiation of phase transformation, resulting in a restorative moment towards the principle path. The axial displacement at which spring 1 begins to change phase ($u_1 = u_s$) is defined by

$$\delta_{pt} = \frac{1}{4Lk_e}\left[Lk_t\left(-u_f + u_s\right) + k_e\left(2d^2 + L\left[u_f + u_s\right]\right) \pm \zeta^{1/2}\right] \tag{12}$$

where

$$\zeta = L^2(-3k_e + k_t)^2 u_s^2 + \left[2d^2 k_e + L(-k_e + k_t)u_f\right]^2 - 2L\left[2d^2 k_e(k_e + k_t) + L\left(3k_e^2 - 4k_e k_t + k_t^2\right)u_f\right]u_s. \tag{13}$$

As δ_{pt} is in general a complex number, unbuckling only occurs when δ_{pt} is a real number ($\zeta \geq 0$); otherwise, the structural response is defined by secondary buckling. Therefore, the critical point dividing "unbuckling" and secondary buckling behavior is defined by $\zeta = 0$, or when

$$\Lambda_c = \frac{2\left[-2\sqrt{2}\sqrt{k_e^3(k_e - k_t)\left(u_f - u_s\right)u_s} + k_e k_t\left(-u_f + u_s\right) + k_e^2\left(u_f + u_s\right)\right]}{\left[k_e\left(u_f - 3u_s\right) + k_t\left(-u_f + u_s\right)\right]^2}. \tag{14}$$

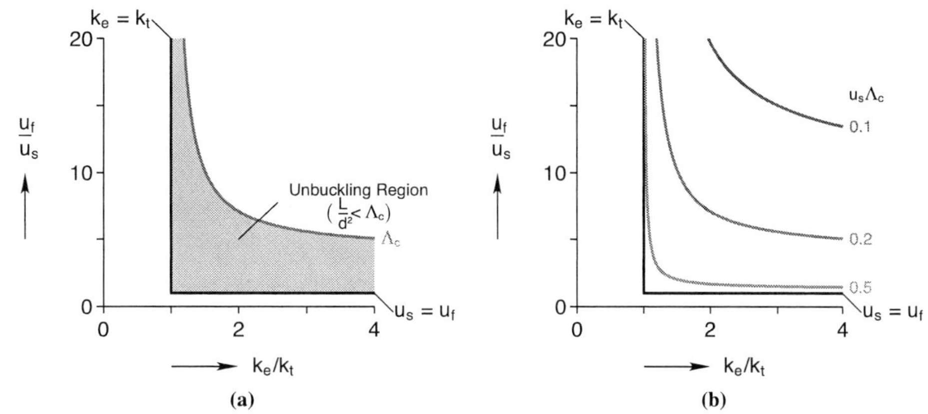

Figure 12. Relationship between constitutive parameters (u_s,u_f,k_e,k_t) and geometric parameters (d,L), illustrating (a) the region in which "unbuckling" behavior occurs for a given Λ_c and (b) contours of constant Λ_c.

Such that $L/d^2 < \Lambda_c$, as shown by the red region in Figure 12a, the Shanley column will exhibit "unbuckling" behavior. The relationship between Λ_c and the constitutive parameters is shown in Figure 12b, illustrating the decrease in the feasible unbuckling region as Λ_c increases. Thus, "unbuckling" can only occur for sufficiently stout columns (small aspect ratio), which is consistent with the experimental results where only the smallest aspect ratio NiTi specimen exhibited unbuckling.

4. CONCLUSIONS

The buckling behavior of NiTi and Al columns was experimentally investigated. A series of clamped-clamped buckling experiments were performed on NiTi and Al solid cylindrical rods of varying aspect ratios, comparing and contrasting the buckling behavior of superelastic and elastoplastic materials. Through the use of digital image correlation, full field displacements and strains were measured, quantifying lateral buckling displacements, axial displacements, and local strain profiles. It was found that the critical buckling stress for both materials across all aspect ratios matched well with inelastic buckling theory (except one NiTi case in which geometric imperfections caused premature buckling). In the post-buckling region of the mechanical response, the NiTi columns exhibited a non-monotonic behavior in contrast to the monotonic behavior of the Al columns. This behavior is attributed to the interplay of localized phase transformation with column buckling throughout the loading process. For the NiTi column with the smallest aspect ratio of 5, "unbuckling" behavior (similar to that discovered in [12]) was observed. Examining a modified Shanley column, in which superelastic SMA behavior was modeled through the use of a trilinear material law for the foundation springs, a relationship for the appropriate geometric and material conditions for "unbuckling" behavior was developed, qualitatively explaining why this phenomenon can occur for SMA columns of low aspect ratio.

References

1. Movchan, A., Movchan, I., and Silchenko, L. "Effect of Structural Transformation and Deformation Non-linearity on the Stability of a Shape Memory Alloy Rod." *Mechanics of Solids*, 2010. Vol. 45, No. 6, pp. 876–884. doi:10.3103/S0025654410060117.
2. Movchan, A., Movchan, I., and Silchenko, L. "Stability of an Annular Plate of a Shape Memory Alloy." *Journal of applied mechanics and technical physics*, 2011. Vol. 52, No. 2, pp. 279–287. doi: 10.1134/S0021894411020167.
3. Movchan, A. and Sil'chenko, L. "Buckling of a Rod Undergoing Direct or Reverse Martensite Transformation Under Compressive Stresses." *Journal of applied mechanics and technical physics*, 2003. Vol. 44, No. 3, pp. 442–449.
4. Movchan, A. and Sil'chenko, L. "Analysis of Buckling Induced by the Direct Thermoelastic Transformation Under the Action of Compression Stresses." *Mechanics of solids*, 2004. Vol. 39, No. 2, pp. 104–114.
5. Movchan, A. and Sil'chenko, L. "The Stability of a Plate of Shape-memory Alloy in a Direct Thermoelastic Phase Transition." *Journal of applied mathematics and mechanics*, 2004. Vol. 68, No. 1, pp. 53–64.
6. Movchan, A. and Silchenko, L. "The Stability of a Circular Plate of Shape Memory Alloy During a Direct Martensite Transformation." *Journal of applied mathematics and mechanics*, 2006. Vol. 70, No. 5, pp. 785–795. doi:10.1016/j.jappmathmech.2006.11.013.
7. Movchan, A. and Silchenko, L. "Buckling of a Circular Plate Made of a Shape Memory Alloy Due to a Reverse Thermoelastic Martensite Transformation." *Mechanics of Solids*, 2008. Vol. 43, No. 1, pp. 100–111. doi:10.3103/S0025654408010093.
8. Rahman, M., Qiu, J., and Tani, J. "Buckling and Postbuckling Characteristics of the Superelastic SMA Columns." *International journal of solids and structures*, 2001. Vol. 38, No. 50, pp. 9253–9265.
9. Rahman, M., Qiu, J., and Tani, J. "Buckling and Postbuckling Characteristics of the Superelastic SMA Columns–Numerical Simulation." *Journal of intelligent material systems and structures*, 2005. Vol. 16, No. 9, pp. 691–702. doi:10.1177/1045389X05054022.
10. Reedlunn, B., Churchill, C. B., Nelson, E., Daly, S., and Shaw, J. A. "Tension, Compression, and Bending of Superelastic Shape Memory Alloy Tubes." *Journal of the Mechanics and Physics of Solids*, 2013. doi: 10.1016/j.jmps.2012.12.012.
11. Shanley, F. "Inelastic Column Theory." *Journal of the Aeronautical Sciences (Institute of the Aeronautical Sciences)*, 1947. Vol. 14, No. 5. doi:10.2514/8.1346.
12. Urushiyama, Y., Lewinnek, D., Qiu, J., and Tani, J. "Buckling of Shape Memory Alloy Columns (Buckling of Curved Column and Twinning Deformation)." In "The Japan Society of Mechanical Engineering International Journal," Vol. 46, 2003 pp. 60–67.

Knitted Electromagnetic Textile Surfaces

A Tennant[1], W Hurley[2] and T Dias[2]

[1] Department of Electronic and Electrical Engineering, The University of Sheffield, UK
[2] Advanced Textiles Research Group, School of Art and Design, Nottingham Trent University, UK

Abstract

We report a commercially attractive approach to manufacturing conducting textiles which is based on computerised flat-bed knitting technology using conducting yarns. We examine how flat-bed knitting can be used to manufacture large area samples of functional electromagnetic structures such as frequency selective surfaces (FSS). In addition we show how the knitting process can be adapted to allow the integration of conducting vias into a 3-D knitted spacer structure to form an electromagnetic high impedance surface (HIS).

1. INTRODUCTION

Passive and active electromagnetic components are usually manufactured from sheet metal or using lithographic techniques for printed circuit boards. However, over the past decade new materials and manufacturing techniques have been developed that permit the creation of electromagnetic structures, such as antennas and transmission-lines using conducting textiles [1,2].

Electromagnetic structures based on conducting textiles have certain advantages over equivalent structures constructed from conventional materials. Textile structures offer advantages such as light weight, flexibility, softness, and high-strength-to-weight ratios. A unique advantage of electromagnetic structures based on textiles is that they can be integrated into clothing to provide functions such as on-body and off-body communications. Functional conducting textiles may also be integrated into composite structures where they provide a light weight and flexible alternative to conventional structures based on circuit-board technology. Advanced textile manufacturing techniques, such as computerised flat-bed knitting, can also be used to create a variety of 3-D and conformal structures to prescribed geometries [3]

Much of the previously published work on conducting textiles at RF frequencies considers structures constructed from combinations of traditional (non-conducting textiles) with commercially available conducting textiles, such as Zelt to form hybrid structures [4]. The disadvantage of this approach, particularly when applied to complex structures such as Frequency selective surfaces (FSS), is that the conducting textile needs to be "cut-to- shape" and then attached to the non-conducting textile substrate using adhesive or some other means. This is obviously a time consuming process that is not amenable to mass manufacturing and alternatives are sought. One alternative technique that has been investigated is the use of computerised embroidery in which a conducting yarn is used to embroider conducting features, such as a patch antenna, onto a non-conducting textile base, or substrate, layer [5]. Although,

computerised embroidery is a step forward with regard to efficient manufacturing, it is still a relatively slow process and the conducting surfaces created often suffer from poor continuous conductivity.
In this contribution we investigate the use of another textile manufacturing process – knitting – to fabricate conducting textile structures. Knitting is a very old and traditional process that has been revolutionised over the past few decades to allow the efficient, commercial manufacture of a wide variety of textile products using advanced, large-scale, flat-bed knitting machines under computer control and driven by advanced CAD software interfaces [6]. The computerised flat-bed knitting process can easily be used to manufacture of large area textile structures with repeated patterns and is therefore ideally suited to the manufacture of periodic structures such as FSS and meta-material surfaces. FSS typically consist of a grid of periodically arranged conducting patches on a dielectric substrate. At microwave frequencies, FSS can act as transmission or reflection filter elements and can be configured for low-pass, high-pass, band-pass or band-stop operation. Applications of FSS include dichroic sub reflectors for multi-band reflector antennas, radomes, electromagnetic shielding, and as components in radar absorbers and electromagnetic bangap (EBG) structures.

2. TEXTILE MANUFACTURING TECHNOLOGY

Textiles have a major role to play in strategically important areas and offer many advantages over traditional materials such as support for technologically advanced products, flexibility, softness, and high-strength-to-weight ratios. The highly developed mass production techniques that characterise the textile sector facilitate the path for the development of new products and their cost effective manufacture for aerospace applications. In particular modern computerised flat-bed knitting technology is one of the most technologically advanced textile production methods.

Flat-bed knitting machines are precisely engineered with two primary needle beds of hardened steel that are arranged in an inverted V-form. In the needle beds, latch or compound needles are placed inside needle tricks. This arrangement facilitates the movement of needles individually and linearly during the knitting process. It is this unique combination of needle tricks and latch needles that has paved the way for the creation of complex 3D knitted structures on these machines.

On current machines the technology used to select needles is based on electro-mechanical principles and microprocessors control. This has resulted in the development of CAD/CAM systems for generating pattern information for needle selection and storing it on workstations. The development of holding-down sinkers has complemented the existing methods of knitted loop control (yarn loop in the needle hook), including fabric take-down and presser foot technology. These developments allowed for the creation of previously un-knittable structures and shapes. On a modern computerised flat-bed knitting machine needles of the two needle beds can also be selected individually to form stitches, tuck loops and floats, which are the three primary elements of knitted structures. The technology for transferring knitted loops between the needle beds has also been perfected.

2.1 **Conductive knitted spacer structures**

A knitted spacer fabric (3D textile structure) is a double layer fabric with both of its layers separated by a defined distance due to the use of a stiff spacer yarn, usually a monofilament, knitted in between the two outer layers. The gap between the two outer layers of the fabric is defined by the stiffness of the spacer yarn and its length between the two layers (Figure 1).

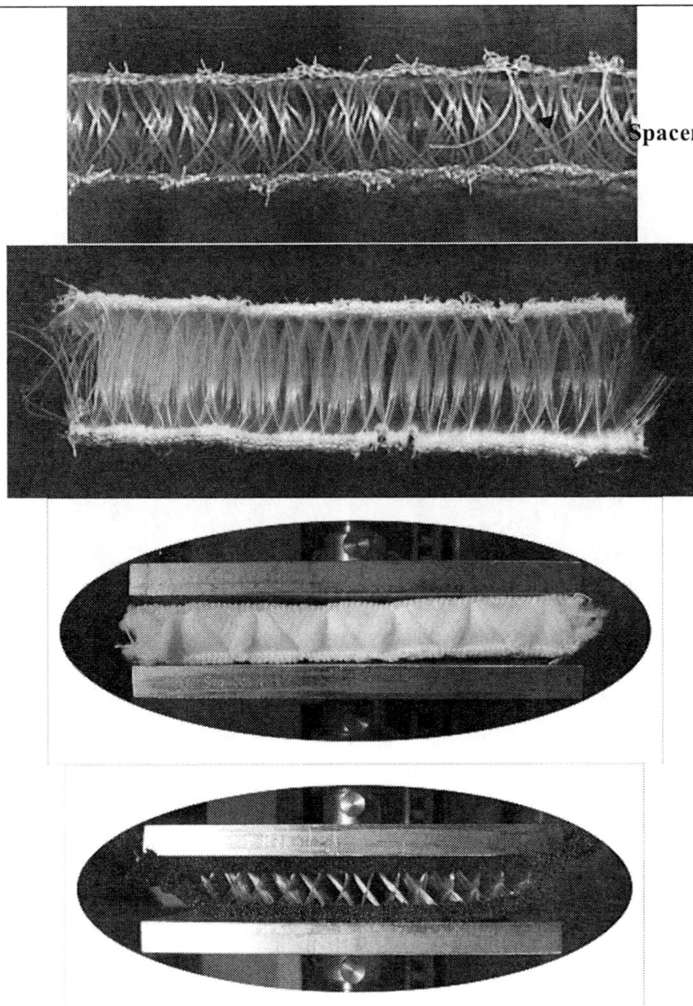

Figure 1. Examples of knitted spacer fabric

During the last few years Nottingham Trent University (NTU) has developed the technique of producing spacer structures on computerised flat-bed knitting machines. The spacer yarn is anchored into the fabric layers using the different combinations of the binding elements (stitches, tuck loops and floats). Structural modifications are also possible using loop transfer and racking techniques.

3. SINGLE LAYER KNITTED FSS

Initially NTU investigated techniques for manufacturing single layer patterned FSS textiles using a combination of non-conducting and conducting yarns to knit textiles with an insulating base-layer and a conducting surface pattern. As an example, Fig. 2 shows samples of two types of textiles: one containing a conducting diamond grid surface pattern and the second containing a triangular conducting patch pattern.

Figure 2a and 2b. Examples of single layer knitted textiles with a conducting mesh pattern

The single layer structures, knitted on one needle bed, were developed on a Shima Seiki SWG 15g 091. The textiles were knitted using both silver and polyester yarns. The silver yarn provided the conductive grid and the high bulk polyester the base structure. When considering the grid mesh design, the grid would normally consist of vertical and horizontal elements which would then intersect at predetermined points along the horizontal and vertical axes. However, to manufacture a square grid would be difficult and could only be realised by a knitting method called intarsia. This is when multiply yarn carriers, which deliver the yarn, are positioned across the length of the fabric being produced to effectively become the vertical elements of the grid. The major limitation of this method is that it is limited by the number of carriers available on a single machine, with dedicated intarsia machines having up to 45 carriers. This technique would therefore become quite limiting in terms of fabric widths that could be produced and also because of challenging and complicated nature of producing intarsia structures.

3.1 Measurements

The knitted textile FSS are inherently flexible structures, but to characterise them they were arranged and measured as flat planar surfaces. To facilitate the measurements the textile FSS were placed over a 30cm by 30cm square, low-loss, polystyrene slab (3cm deep) and gently stretched in both linear dimensions to equalise any distortion and to produce a flat surface with a symmetrical surface pattern. The normal incidence, free-space reflectivity (S11) characteristics of the textile FSS were measured in a fully calibrated NRL arch using an automated measurement system employing an Agilent 8510B Vector Network Analyser. The reference, calibration standard was a flat aluminium plate of equivalent size (30cm by 30cm). The measurements were performed in a dedicated anechoic chamber and a standard time-domain gating procedure was used to remove any spurious returns. Fig. 3 shows the reflectivity response of the single layer knitted textile with the conducting diamond grid pattern. The response is typical of an inductive FSS and is strongly reflective at lower frequencies and non-reflecting (indicating transmission) at higher frequencies. Fig. 4 shows the reflectivity response of the single layer knitted textile with the small triangular conducting patch pattern. The response is typical of a capacitive FSS and is transmissive at lower frequencies and reflecting at higher frequencies.

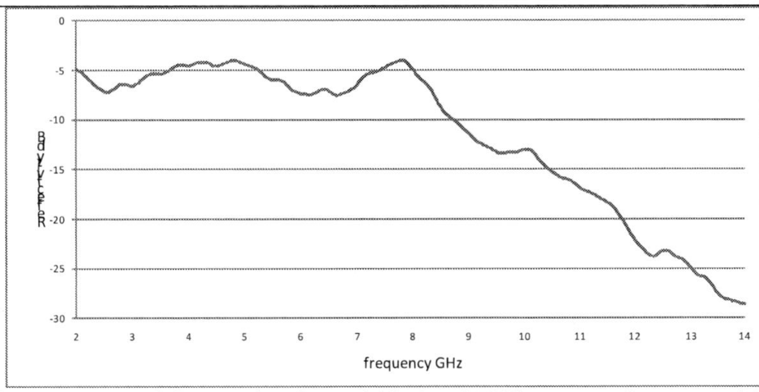

Figure 3. Reflectivity response of the conducting grid knitted FSS measured at normal incidence.

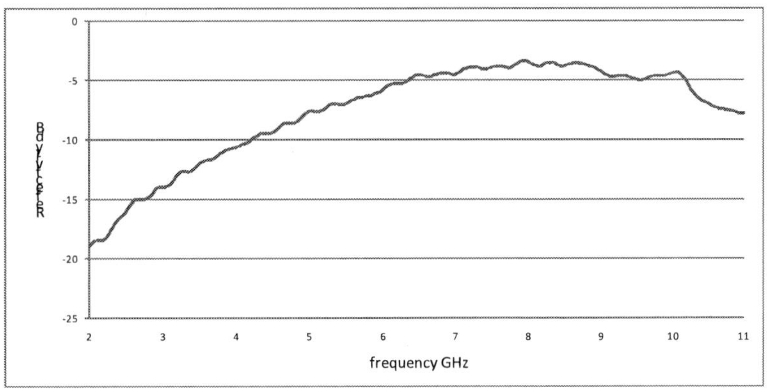

Figure 4. Reflectivity response of the conducting patch knitted FSS measured at normal incidence.

4. DOUBLE LAYER KNITTED FSS

The double layer, 3-D knit structure, knitted on two needle beds, was developed on a Shima Seiki SWG 15g 091. The textiles were knitted using both silver and polyester yarns. The silver yarn provided the conductive triangular patches and the high bulk polyester the base structure and spacer material. The technically challenging and more complicated double layer structure also included vias connecting the conductive ground plate back layer to the triangular patches arranged within a base layer of high bulk polyester. Spacer yarns are normally monofilament and provide a semi rigid separation layer between a double layer structure however these monofilament spacers have many open spaces through which the conductive layers can connect . To minimise this risk and to provide a more robust isolation layer it was decided to use high bulk polyester which would fill the space between layers. The partial knit techniques used in the previous knit sample were again utilised to produce the triangular conductive patches. An example of a double layer FSS (also known as a high impedance surface (HIS)) structure containing vias is shown in Figure 5.

4.1 Measurements

To characterise the properties of the knitted textile HIS a free-pace reflectivity phase measurement was made in an anechoic chamber using a fully calibrated NRL arch system employing an Agilent 8510B

Vector Network Analyser. The reference calibration standard was a flat aluminium plate of equivalent size to that of the HIS (36cm by 36cm). The measurement was made for illumination at normal incidence and a standard time-domain gating procedure was used to remove any spurious returns. The free-space reflection phase measured from the knitted HIS for normally incident illumination is shown in Figure 6. The surface exhibits the classic band-gap reflection phase response with zero phase reversal at approximately 4.7GHz and with a +/-90° band-gap of approximately 750MHz

Figure 5. Knitted triangular patch , 3-D sandwich structure with integrated knitted conducting back-plane. Overall sample size approximately 30mm by 30mm. Thickness approximately 3.5mm. This sample contains conducting vias located at the base corners of each triangular patch

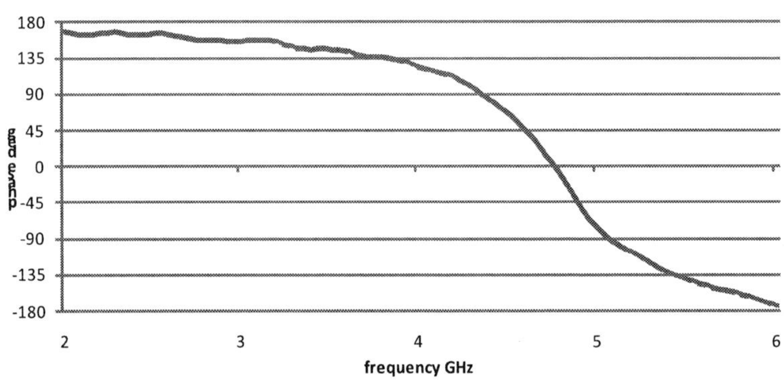

Figure 6. Measured reflection phase response of the knitted HIS at normal incidence.

5. CONCLUSIONS

Several experimental frequency selective surfaces, knitted from conducting yarn on a computerised flat-bed knitting machine have been presented. Two knitted FSS were configured for low-pass and high-pass operation respectively. Measured data shows that the knitted textile FSS exhibit the fundamental characteristics of capacitive and inductive FSS designs and that the HIS exhibits a band-gap reflection phase response. No simulated data is presented in this paper as, of yet, we do not have any accurate models to describe the highly complex interactions and contact processes between the individual conducting yarns in the knitted structure (particularly at microwave frequencies) and this is an area for future research. Other microwave components have been fabricated using the flat-bed knitting process, including 3-D waveguide structures, and these will be reported at a later date.

References

1. Cottet D, Grzyb J, Kirstein T and Troster G, "Electrical characterization of textile transmission lines," Advanced Packaging, IEEE Transactions on, vol. 26, pp. 182-190, 2003
2. Kennedy, T.F.; Fink, P.W.; Chu, A.W.; Champagne, N.J.; Lin, G.Y.; Khayat, M.A.; , "Body-Worn E-Textile Antennas: The Good, the Low-Mass, and the Conformal," Antennas and Propagation, IEEE Transactions on , vol.57, no.4, pp.910-918, April 2009
3. Dias T., Hurley W., Monaralgala R. and Wijeyesiriwardana R., "Development of Electrically active knitted textiles", Advances in Science & Technology, vol. 60, pp 74-84, 2008
4. Locher I, Klemm M, Kirstein T, and Tröster G, "Design and characterization of purely textile patch antennas, IEEE Transactions on advanced packaging, vol. 29, No. 4, pp 777-788 2006
5. T. Acti, S. Zhang, A. Chauraya, W. G. Whittow, R. Seager, T. Dias, and Y. Vardaxoglou, "High Performance Flexible Fabric Electronics for Megahertz Frequency Communications," presented at 2011 Loughborough Antennas & Propagation Conference (LAPC), Loughborough, UK, 2011
6. Dias, T., Knitting seamless 3D shell structures, Textile Technology International 2001, ISBN 185938, pp 57-61, 2002
7. Dias, T., Wijesiriwardana, R. and Mukhopadhyay, S., Resistive Fibre-Mesh Transducers, "7th IEEE Conference of wearable computers", October 2003
8. P130 Technical data sheet, Statex Produktions & Vertriebs GmbH, Bremen, Germany

ICAST2013: 24th International Conference on Adaptive Structures and Technologies October 7-9, 2013, Aruba

Actuator placement for shunt damping of panel structures – numerical simulation and experimental validation

Dipl.-Ing. Martin Pohl[*1], Dr. Michael Rose[2], Dipl.-Ing. Thomas Haase[3]

[1-3] German Aerospace Center (DLR)
Lilienthalplatz 7, 38108 Braunschweig, Germany

Abstract

For the reason of ecologic sustainability and decreasing reserves of fossil energy sources efficiency is a great concern. A very attractive solution is the use of lightweight fiber reinforced structures. But due to high stiffness and low mass they are much more prone to vibration and noise emission.

Besides passive damping or active structural acoustic control so called shunt damping offers an opportunity for noise and vibration attenuation. For broadband damping performance negative capacitance shunt networks show promising results. Actuator placement is a great concern for this concept because a good coupling is required for sufficient damping efficiency.

For active noise and vibration damping a pre-design tool was developed at DLR to obtain the positions of a certain number of actuators. This algorithm has been improved by a coupled model of electric networks to obtain the closed loop transfer function of the whole structure equipped with shunted actuators to simulate network damping performance.

In order to prove the simulated damping result, an experimental validation for a steel panel is presented in this paper. An actuator placement for this panel is derived with the predesign tool and a test panel is equipped with actuators according to the placement. It is investigated in a panel test stand, where vibration amplitudes can be measured with a laser vibrometer and sound power by a sound intensity probe. Finally, numerical and experimental results are compared and discussed.

1. INTRODUCTION

In modern transportation, fuel efficiency becomes more and more important due to ecologic sustainability, limited reserves of fossil energy sources and increasing costs. A major concept consists in the usage of lightweight structures, which is addressed in passenger aircraft or for electrically driven cars. To reduce structural weight, fabric materials as carbon fiber reinforced plastics are increasingly used instead of steel or aluminum. But the high stiffness and low density and damping of these materials is a disadvantage for their noise and vibration behavior. To ensure cabin comfort, e.g. in aircraft passenger compartments, noise and vibration attenuation has to be considered.

* martin.pohl@dlr.de

In most cases passive damping is not an acceptable solution, because additional mass is needed, especially at lower frequencies. On the other hand, active damping is investigated, such as Active Structural Acoustics Control (ASAC). This comes along with the need for sensors, controllers, amplifiers and the need for adaptive system and controller identification.

A third opportunity is given by shunt damping, where multifunctional materials combining actuator- and sensor effects are bonded to a vibrating structure. These actuators are equipped with appropriate electric (shunt) networks which enhance damping performance. Therefore, this concept is also known as "Piezoelectric shunt damping" because in most cases piezoelectric transducers are used.

For maximum shunt damping performance, actuator placement has to be considered. Therefore, an ASAC predesign tool, developed by DLR, has been improved by shunt damping prediction to generate optimized actuator placements. In this paper, an actuator placement for a panel structure is derived by the design tool and then damping prediction is compared with experimental results.

2. Piezoelectric Shunt Damping

"Piezoelectric shunt damping" subsumes all concepts, where an oscillating structure is damped with applied piezoelectric actuators connected to electric shunt networks. The basic principle can be seen in Figure 1.

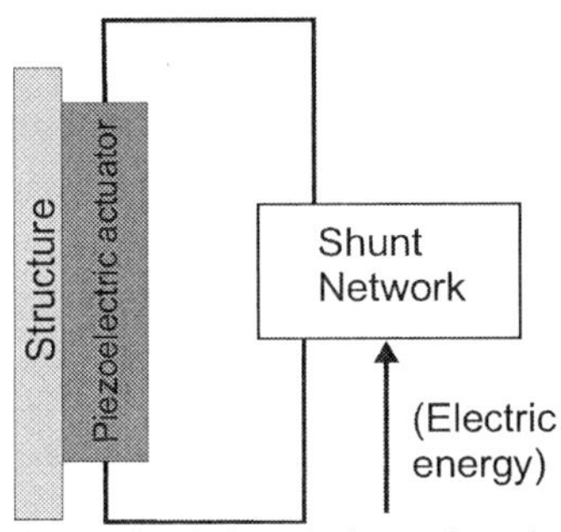

Figure 1: Working principle of piezoelectric shunt damping

Generally the piezoelectric actuator is used as energy transducer with the aim to transfer vibration energy into electrical energy. By this, vibrations are damped. Therefore the electrical network has to fulfill two characteristics: First it has to receive and dissipate the vibration energy from the piezoelectric patch. Second, the electric network ensures proper impedance matching between structure, piezoelectric transducer and the dissipation network. This is required for a sufficient energy transportation rate between mechanical and electrical system.

Therefore, several concepts were investigated. Most of them are addressing the inherent capacitance of the piezoelectric transducer. If the capacitive reactance is eliminated, the energy transfer can be enhanced. Hagood and Flotow [1] use an inductor together with a resistor creating a damped resonant system with the capacitance of the piezoelectric transducer. By this, a behavior comparable to a tuned mass vibration absorber is achieved, so that one single mode can be damped. For multiple modes, this concept can be expanded by multiple inductors, where the modes are separated with filters. Wu [2] and Behrens et al. [3] introduced several filter concepts where multiple modes can be damped using a single piezoelectric transducer. All of these shunts can be built from passive electrical components as inductors, capacitors and resistors. As advantage, no external power supply is required. The disadvantage lies in the lack of adaptability. Changes in the resonant frequencies of the structure cannot be regarded. In addition, due to the large amount of electrical components, this is only practical for a few modes.

For applications with a lot of varying eigenfrequencies the negative capacitance shunt is better suited. In this case, the capacitive reactance of the piezoelectric transducer is compensated over a wide frequency range in comparison to resonant shunts. This is achieved by an active circuit based on negative impedance converters. In [4] and [5] good damping performance was demonstrated for the first couple of modes of a clamped beam with a negative capacitance shunt on a single actuator.

Compared to the beam, panel structures show a lot more modes due to the additional lateral dimension. Therefore multiple piezoelectric transducers with individual negative capacitance shunts are required to achieve sufficient damping. By this, equaling currents within the actuator patches can be prevented as described in [6].

To cover most modes a custom actuator placement has to be derived. It is dependent from the mode shapes of the structure, the present excitation and from the goal function. Minimizing sound radiation or structural vibration yields in different actuator configurations why an optimization with respect to the mentioned characteristics is oblique.

In [7] an optimization method is presented which is based on a genetic algorithm capable of optimizing the positions of multiple actuators on a structure. There, several actuator placements were derived for a test panel for minimizing sound and vibration amplitude.

Subsequently, the theoretical results will be experimentally verified in the following paper. First a finite element model of the test panel according to [7] is created with the use of a model updating tool. Then an actuator placement is derived to build a test specimen. This panel is experimentally investigated with a laser vibrometer and a sound intensity probe.

3. Modeling of test panel

3.1. Test Stand and finite element setup

For the experimental investigation a panel test bed is used, where a panel of 800mmx600mm can be mounted in a wooden box. Four bolts are used to fix the panel in the corners. A photograph of the test stand is shown in Figure 2.

Because the actuator placement is based on modal data from the structure, these data have to be provided. This can be done by numerical and experimental modal analysis. Experimental modal analysis delivers exact values for the eigenfrequencies, however in most cases the eigenvectors show significant noise and not all eigenmodes can be identified with the limited number of excitation points. Therefore a numerical modal analysis is used for generating the required data, based on a finite element model of the test panel.

A model based on shell elements is used in the finite element tool ANSYS. In [7] the fixation in the corners was realized by clamping the nearest node to the fixation point. Mode shape comparisons with measurements show that strain amplitudes are getting too high due to discretization. Therefore, all nine nodes around the fixation point are blocked in their translational degrees of freedom. This better represents the bolt fixation and smoothes strain amplitudes at the mounts. An image of the finite element model is displayed in Figure 3.

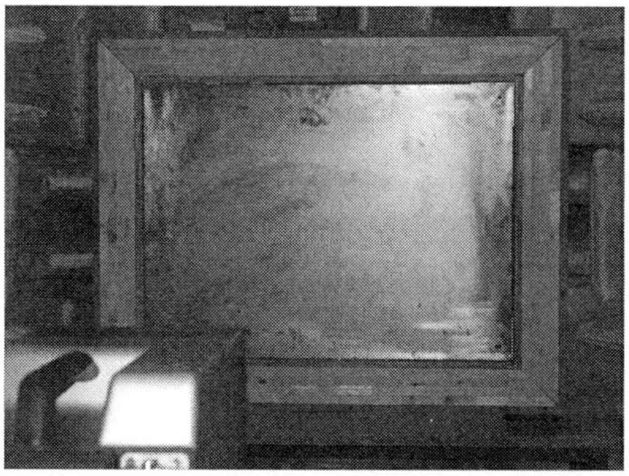

Figure 2: Photo of Panel Test Stand

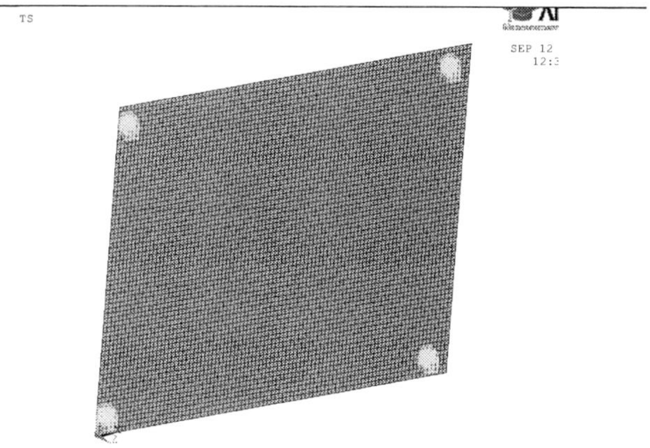

Figure 3: Finite element model

3.2. Finite element model updating

Due to deviations between model and experiment, especially for material and structure parameters, model updating helps to fit the eigenfrequencies. Therefore, four parameters of the finite element model are varied in narrow areas: young's modulus, poisson ratio, material density and the panel thickness. Especially the thickness has major influence on the eigenfrequencies due to the quadratic influence on panel stiffness.

As goal function the difference of the eigenfrequency of congruent eigenmodes in the finite element model and experimentally obtained modal data is used. For a 2mm steel panel the parameters listed in Table 1 were generated with the modal upating tool.

Table 1: Model parameters after updating

Panel thickness	Young's Modulus	Poisson ratio	Density
2mm	207GPa	0.277	7830kg/m³

In Figure 4 the frequency response functions (FRF) are shown for the experimental reference and the numerically obtained modal model based on the updated finite element data. A modal damping of 0.2% is used for all modes in the numerical data.

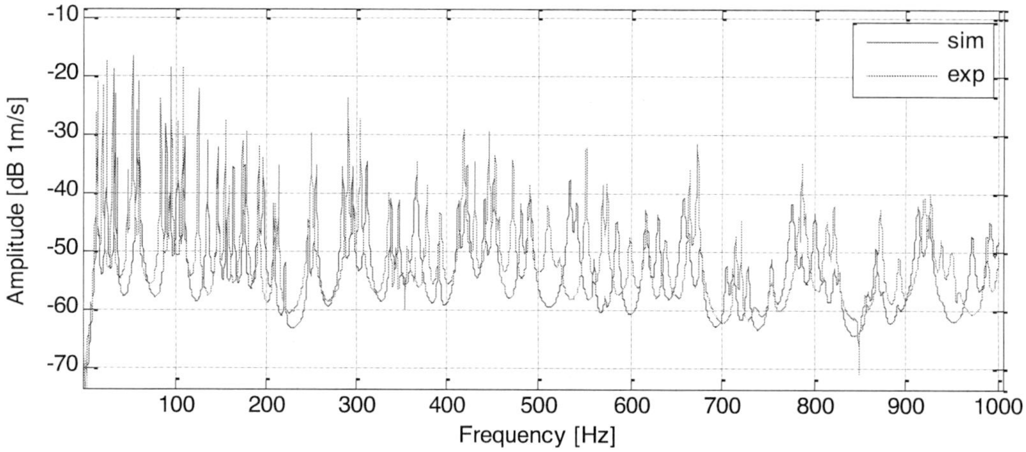

Figure 4: FRF comparison between simulation and reference experiment

Regarding the FRF plot in Figure 4 it can be seen, that the fundamental behavior of the panel is mirrored in the numerical domain. Up to 500Hz, the mode groups are nearly congruent. Because even around 800Hz the FRFs are comparable, the updated model is sufficient for an actuator placement. Deviations in the height of resonance peaks are likely to be caused by the assumption of constant damping in the numerical domain. Because this does only affect a fraction of the total modes, it will be tolerated for model simplicity.

4. Generating an actuator placement

4.1. Piezoelectric patch actuators

As actuators, piezoelectric patches with a thickness of 0.2mm are used. For improved applicability and individual electric connection, the ceramic foils are bed in an epoxy matrix containing the electric contact planes. Figure 5 shows a photo and the setup of the actuators. The parameters of the actuators are taken from [8].

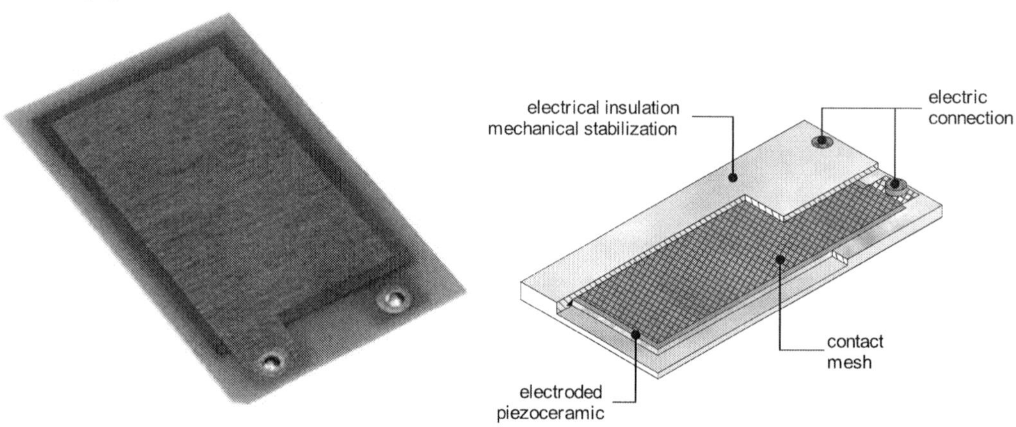

Figure 5: DuraAct® piezoelectric patch actuator photo and schematic [8]

4.2. Optimization of actuator placement

With the numerical modal data from 3.2 actuator placements were calculated for minimized sound and vibration amplitude with the improved design tool as shown in [7]. Therefore ten locations with four standard patches of 30mmx50mm are used.

Actuator placement simulations showed, that the goal function is supposed to be relatively flat because visually different actuator settings produce nearly the same result. To achieve a sufficient damping performance it is useful to look for regularities. Therefore multiple individuals of a population are superposed to reveal regions where actuators are often placed [9]. Figure 6 shows a superposition of the best individuals for minimized sound (left) and vibration amplitude (right). Excitation was chosen to be a constant point force at 200mm from left and 100mm from the lower edge of the panel.

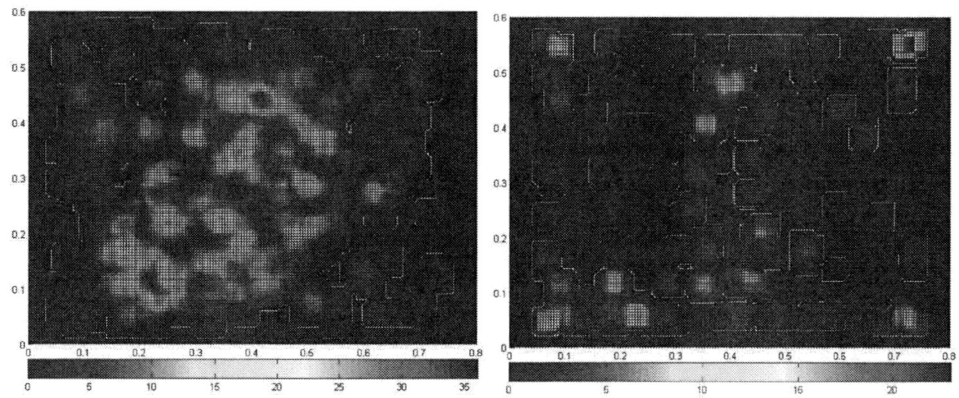

Figure 6: Superposed best individuals for sound (left) and vibration (right) reduction [9]

If the optimization is made for vibration reduction the actuators are mainly placed in the corners of the panel and in the middle as shown on the right in Figure 6. This is so far consistent, as piezoelectric patches show best controllability in regions with high strain. In fact this occurs near the fixations in the corners why many actuators are placed there. When minimizing sound radiation, the actuator placement is more variable and shows a concentration in the middle of the panel.

For the experimental validation the actuator placement a superposition of the images in Figure 6 is used which is displayed in Figure 7. A good performance of sound power reduction and vibration reduction is expected by this. For maximum flexibility, the actuator groups are transversally divided in their electrical connection, so that each group can be connected to two separate negative capacitance shunts. This also minimizes equaling currents within the actuator group and is expected to produce higher damping.

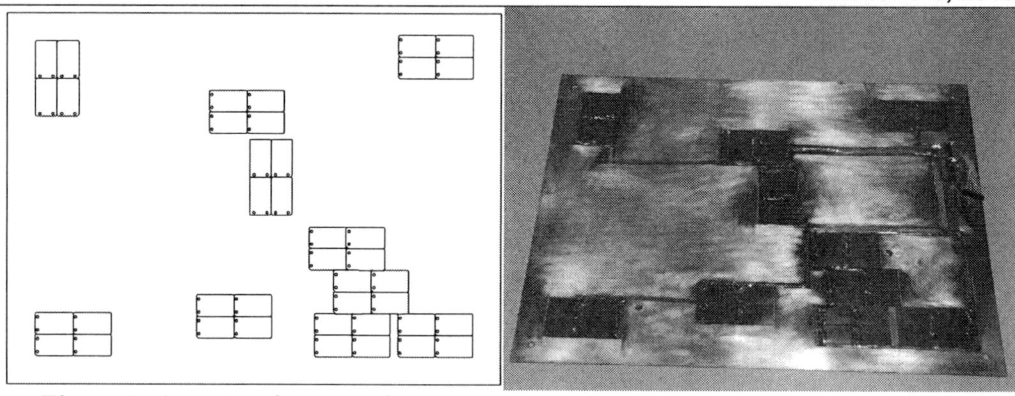

Figure 7: Actuator placement for test panel, sketch and photograph of test specimen

5. Experimental validation

5.1. Vibrational FRF Comparison for simulation and experiment

In order to validate the design tool used to calculate actuator placements, the simulated damping is faced to measurements. Therefore the panel prototype in Figure 7 is also mounted into the test stand. Vibration amplitude is obtained by a scanning vibrometer in the undamped case with open actuator electrodes and with negative capacitance shunts connected. Comparable to the simulation, series negative capacitance shunts were used with C_{shunt}/C_{PZT} =-0.87 and a series resistance with a time constant with the capacitance of the piezoelectric patch of R_sC_{PZT}=1.4E-7s. The excitation point is also chosen at the same place for simulation and experiment.

With these conditions, the undamped and damped FRFs for simulation and experiment are shown in Figure 8 and Figure 9.

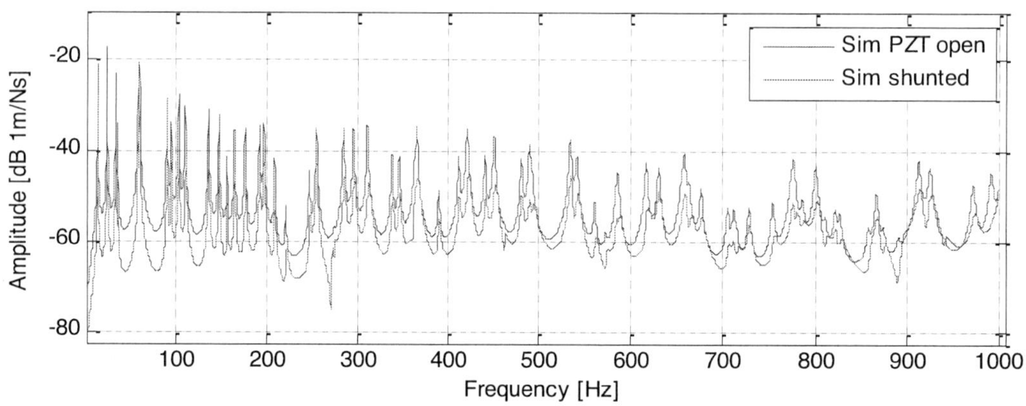

Figure 8: Damped and undamped FRF from simulation

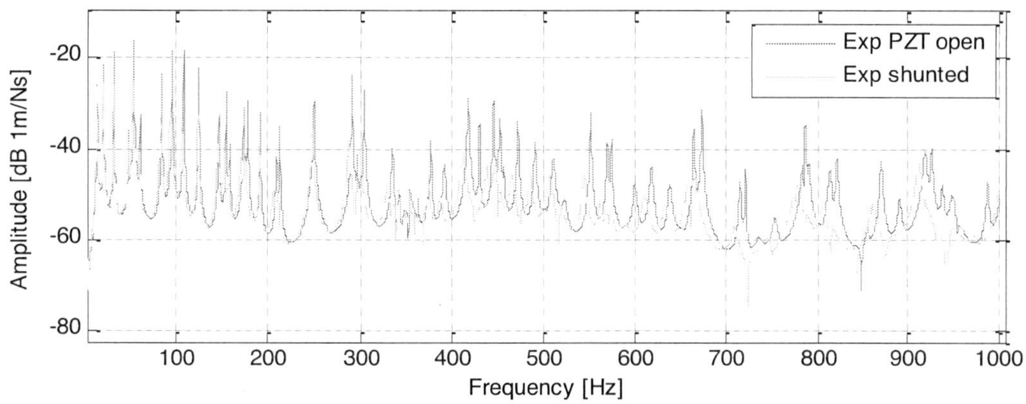

Figure 9: Damped and undamped FRF from experiment

Despite deviations at some modes, damped and undamped FRFs show a comparable behavior in simulation and experiment. A remarkable difference can be noticed in the lower frequency domain in the regions between the eigenfrequencies, where the damped FRF is calculated lower than the experiment. This is supposed to be a consequence from modal modeling where the influence of the negative capacitance is described as additional modal damping. In chapter 5.2 this is investigated in detail.

5.2. Impact of the shunt to modal damping

For precise comparison of the simulated and experimental damping, the FRF plots are not clear enough, because the curves arise from a superposition of all modes. Therefore, damping of single modes cannot be compared.

A better way is offered by directly comparing modal damping values. For the experimental data, they can be derived from an experimental modal analysis of the damped panel with negative capacitance shunts applied. For the simulation, the transfer function of the coupled system with a number of nact actuators can be used.

$$G_{sys} = \frac{G_{qF}}{1+\sum_{j=1}^{nact} \frac{sG_{QUj}}{sC_{PZT}+\frac{1}{Z}}} \quad (1)$$

There G_{qF} stands for the transfer function of the undamped system. G_{QUj} represents the reaction charge of an actuator j if an external voltage is applied. Z stands for the impedance of the electric network, which is a combination of negative capacitance and resistance in series. The "s" is the Laplace operator in frequency domain. The exact derivation of equation 1 can be found in [7].

Due to the low damping of the steel panel, it can be assumed, that resonance peaks in the FRF are dominated by one single mode. Therefore, a change in the modal damping parameter for each mode can be calculated, when the negative capacitance shunt is connected.

For one mode i, G_{qF} and G_{QUj} can be rewritten in equation 1. The parameter fp_{ij} represents the coupling factor for the mode i of the transducer j.

$$G_{sysi} = \frac{\frac{s}{s^2M_i+sD_i+K_i}}{1+\sum_{j=1}^{nact}\frac{sfp_{ij}^2}{(s^2M_i+sD_i+K_i)\left(sC_{PZTj}+\frac{1}{Z_j}\right)}} \tag{2}$$

This can be rearranged using the impedance of the series negative capacitance.

$$G_{sysi} = \frac{s}{s^2M_i+sD_i+K_i+s\sum_{j=1}^{nact}\frac{fp_{ij}^2}{\left(sC_{PZTj}+\frac{1}{R_j+\frac{1}{sC_j}}\right)}} \tag{3}$$

Obviously the sum in equation 3 adds to the damping term D because both are multiplied with s. Therefore it is regarded as complex modal network damping DN.

$$\mathrm{DN}_i = \sum_{j=1}^{nact}\frac{fp_{ij}^2}{\left(sC_{PZTj}+\frac{1}{R_j+\frac{1}{sC_j}}\right)}$$

In contradiction to the constant structural damping factor D, the network damping DN shows first a frequency dependence and is second a complex value due to the present laplace operator within the sum. Hereby the real part of DN acts as additional damping to D. The imaginary part of DN will lightly change the eigenfrequency.

Due to the assumption of the resonance peaks being dominated by a single mode, values for DN for each mode can be calculated using the eigenfrequency of the corresponding mode for the parameter s. To compare damping results with the experiment, simulated and experimental modes were assigned to each other by the modal assurance criterion (MAC). The result for 21 congruent modes is shown in Figure 10. Here the difference of modal damping with shunts applied or with actuator electrodes left open is displayed. Table 2 lists the eigenfrequencies of the mapped modes, which nearly cover the whole frequency range of the experimental measurement between 0 and 1kHz.

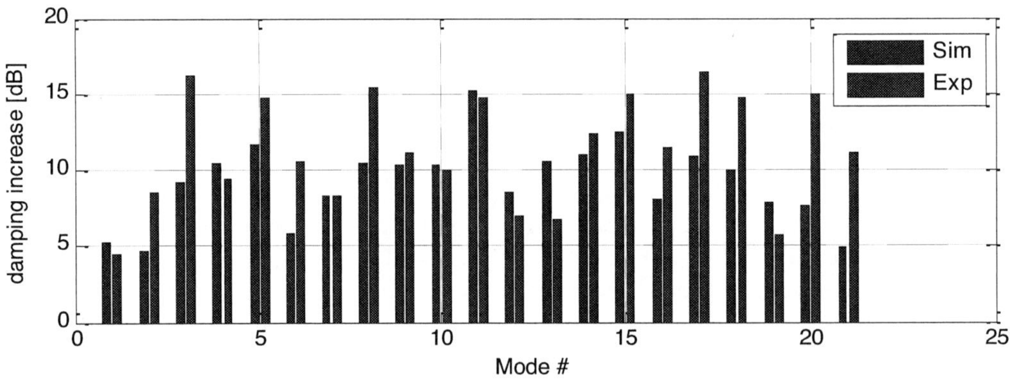

Figure 10: Simulated and measured damping increase for congruent modes

Table 2: Eigenfrequencies of mapped modes

Mode #	1	2	3	4	5	6	7	8	9	10	11	12	13	14	15	16	17	18	19	20	21
Sim [Hz]	15	25	58	90	103	136	147	164	156	191	175	196	220	208	285	422	421	451	481	658	867
Exp [Hz]	15	21	61	84	95	125	146	154	158	174	178	191	210	213	291	418	430	453	490	673	869

For the majority of the modes damping prediction of the simulation shown in Figure 10 is within a few dB of deviation. Therefore the simulation shows an adequate result. The network-induced damping of several modes (like number 2,3,8 and the last modes at high frequencies) is estimated too low.

5.3. Sound power reduction

Besides vibration measurement also sound power was measured using a sound intensity probe in an anechoic room to be compared with the simulation. Figure 11 shows the sound power reduction in 3rd octave bands for simulation in blue and experiment in green. Although both curves have comparable characteristics, the simulation is too optimistic especially at low frequencies. When looking at the simulated FRF for sound reduction in Figure 12, this is expected to be an effect of the high damping between the resonances. Referring to vibration reduction, this behavior has also been experienced. The measurements do not show this effect.

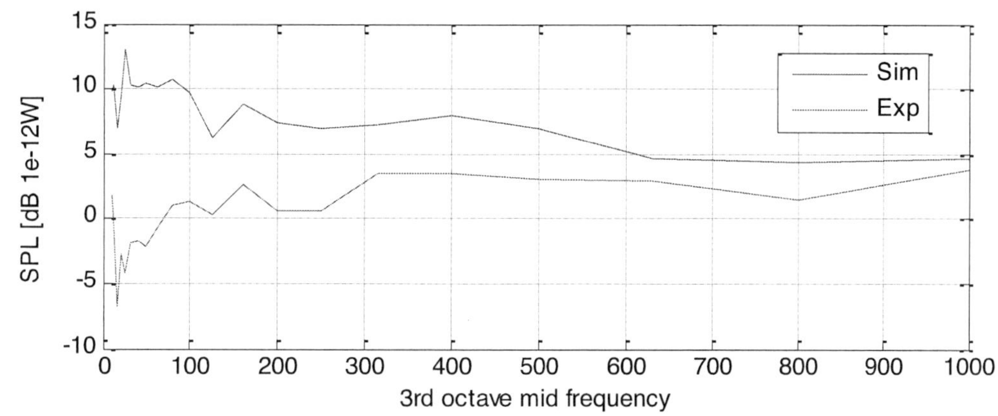

Figure 11: Comparison of sound power reduction

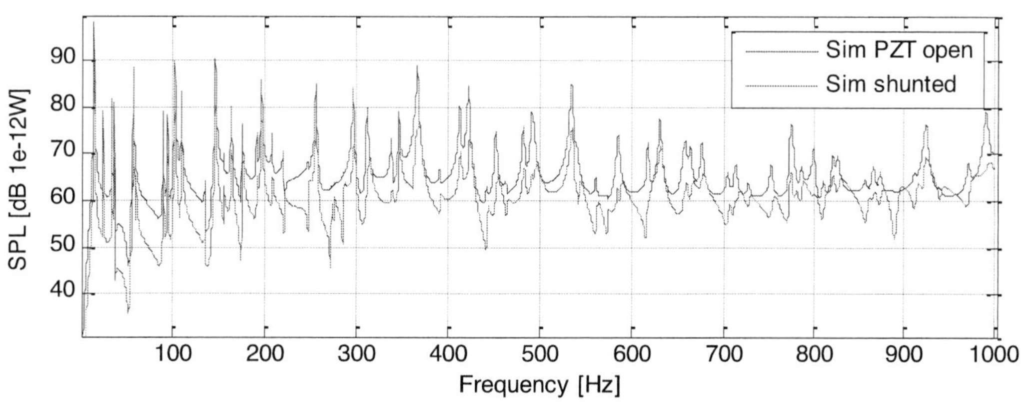

Figure 12: Simulated FRF for sound reduction

5.4. Discussion

In both cases, for sound power and vibration reduction, deviations between simulated and experimentally measured amplitude reductions are present. Especially at lower frequencies the simulation produces high amplitude reductions between the resonances, which is not reproduced in the measurements. Therefore multiple reasons are probable.

One difference from the simulation to the experiment is caused by modeling of the actuators. In the simulation, every patch consists only of the active region. In the experiments, the actuators are embedded into a matrix, which laterally separates the four actuators one patch consists of. The separation lines can be seen in Figure 5 circumfering one actuator and in Figure 7 for the test panel. This may be a reason for an altered behavior at high frequencies, when the structural wavelengths are small.
A second reason comes from the parameter estimation of the electric networks where the adjustment of the negative capacitance is done manually for every actuator with the help of an oscilloscope. This will cause deviations in the adjustment which are not present in the simulation. There the network parameters are put in dimensionless in relation to the piezoelectric capacitance. In reality, this capacitance is also subjected to a statistic variation.
The third source for deviations arises from structural modeling of the test panel. In the simulation, a fixation of all translational degrees of freedom at nine nodes around the fixation point is used. Practically such a clamp shows a flexibility influencing the mode shapes and by this damping efficiency of the shunted actuators.

6. Conclusion and outlook

In the paper, a modal design tool for piezoelectric shunt damping has been experimentally investigated. Therefore, an actuator placement was derived for negative capacitance shunt damping of a 600mmx800mmx2mm steel panel. A finite element model was created with the help of measurements and a genetic model updating tool to get a sufficient reproduction of the experimental modal behavior in the simulation.
These data were introduced into the design tool to produce an actuator placement for negative capacitance shunt damping. In parallel, a test panel with the same actuator configuration was built for experimental measurement. The additional damping was obtained either by simulation or by experiment to be compared to each other. Therefore modes with congruent eigenvectors were mapped together to be compared. It could be shown, that damping prediction is in medium frequencies quite accurate while it is too optimistic for low frequencies. A set of probable causes was mentioned.

For future work, these deviations are subject of further investigations. Therefore the boundary conditions in the corners should be integrated into the model updating tool. Another potential for improvements is situated in the modeling of the shunt networks. Here a method of precisely measuring the exact parameters of the network and the piezoelectric transducers will help to increase the accuracy of the damping prediction.

References

1. N.W. Hagood, A.H. von Flotow, "Damping of structural vibrations with piezoelectric materials and passive electrical networks", *Journal of Sound and Vibration* 146 (2) (1991) 243–268.
2. S.Y. Wu, "Method for Multiple Mode Piezoelectric Shunting with Single PZT Transducer for Vibration Control", *Journal of Intelligent Material Systems and Structures* (1998) 9; 991
3. S. Behrens, S.O.R. Moheimani, A.J. Fleming, "Multiple mode current flowing passive piezoelectric shunt controller", *Journal of Sound and Vibration* 266 (2003) 929–942
4. C. H. Park, H. C. Park "Multiple-Mode Structural Vibration Control Using negative Capacitance Shunt Damping", KSME International Journal Vol. 17 No. 11 (2003), pp 1650-1658
5. C. H. Park, A. Baz, "Vibration Control of Beams with Negative Capacitive Shunting of Interdigital Electrode Piezoceramics", *Journal of Vibration and Control* (2005); 11; 331
6. M. Pohl, "Noise and vibration attenuation of a circular saw blade with applied piezoceramic patches and negative capacitance shunt networks", *ISMA Leuven*, Belgium, (2010), 411-424
7. M. Pohl, T. Haase, Dr. M. Rose, „Actuator Placement for Shunt Damping of Panel Structures", *EURONOISE* Prague 2012, 421-426
8. DuraAct Datasheet
http://www.physikinstrumente.de/de/produkte/prdetail.php?sortnr=101790
9. T. Nieland, M. Pohl, „Untersuchung und Simulation der Schall- und Schwingungsreduktion an einer Plattenstruktur mittels Piezokeramiken und elektrischen Netzwerken", Master Thesis, Hochschule Osnabrück University of Applied Sciences, 2013

Experimental study of composite fibre concrete floor slabs in fire condition

Jan Bednář[1], Shota Urushadze[2*], Michal Kloiber[2]

[1] Czech Technical University in Prague, Thákurova 7, Prague CZ16629, Czech Republic
[2] Institute of Theoretical and Applied Mechanics, v.v.i., Academy of Sciences of the Czech Republic, Prosecka 76, CZ 190 00 Prague 9, Czech Republic

Abstract

The paper describes behaviour of the composite floor slabs with fibre-concrete at ambient and at elevated temperature. At elevated temperature, the floor was only partially fire protected. Intermediate beams and a fibre-concrete slab in steel sheeting were not fire resistant. Concrete slabs were reinforced by steel fibre but steel bars were not added. The aim of the tests was to demonstrate the suitable properties of a steel fibre-concrete slab in fire. For fire resistant floor slabs, it is important that the material has sufficient ductility and adequate tensile and shear strength. These material properties of the fibre-concrete allow the slab to create a different load-bearing mechanism, which increases fire resistance of the floor slabs. The fibre-concrete was tested at ambient and elevated temperatures with a focusing on ductility and tensile strength of the fibre-concrete. Fire resistance of the composite floor slabs was calculated normally as resistance of the simply construction elements, but fire tests showed that floor slabs had much better fire resistance. This phenomenon, related to a joint acting of simple elements and the membrane action of the slabs. Due to a description of this phenomenon, slabs 4.5 x 3m were tested. The experiments with fibre-concrete slabs at ambient temperature and the fibre-concrete behaviour at elevated temperature showed it was feasible to apply the ductility of the fibre-concrete.

1. INTRODUCTION

Resistance of the composite floor structures came from the resistance of the concrete slab and composite beams. This provides double protection against fire and was demonstrated by a number of fire tests and investigation of fires in existing buildings. The tests detected that weakened construction can create a new load-bearing system that more easily transferred the load to the surrounding construction. Lightly reinforced concrete slabs can create a new load-bearing system (see [1] and [2]). Owing to new technology, concrete constructions are reinforced by steel fibres without adding steel bars. With steel fibres, the material created better properties in tensions and in compression compared to traditional reinforcement (see [4] and [5]).

The principal objective of this paper was to show that concrete reinforced only by steel fibres can attain ductility and tensile strength which ensures fire resistance during an actual fire. Tests were conducted in 2010 and 2011 (see [3]) with four slabs at ambient temperature and two slabs at elevated

* Corresponding author, urushadze@itam.cas.cz

temperature. In 2010, two slabs were tested in ambient temperature (AMB-PL-1 and AMB-BE-1) and one at elevated temperature (ELE-BE-1). In 2011, two slabs were tested in ambient temperature (AMB-PL-2 and AMB-PL-3) and one at elevated temperature (ELE-BE-2) (see Fig. 1). The slabs 3.0 m to 4.5 m were bonded steel frame by headed studs of diameter 12.5 and height 70 mm.

The slabs tested in ambient temperature had no internal beams. The slab at elevated temperature was supported by internal beams. Concrete trapezoidal sheets TR40/160/0.75 were used as lost formwork. The concrete mixtures were the same but fibres were added with different sizes and strength (see Tab. 1). Fibres DE 60/0.8N were 60 mm in length with diameter 0.8 mm and strength 1200 MPa. Fibres DE 60/0.8M had the same dimensions but strength 1550 MPa. Fibres HE 75/50 were 50 mm length, diameter 0.75 mm and with strength 1100 MPa.

Figure 1. Bottom view the of slab ELE-BE-2 during fire test at 63rd min

Table 1 Composition of the tested steel fibre reinforced concrete

Composition, kg/m^3	Mixture for tests 2010	Mixture for tests 2011
	Plant H	**Plant H**
CEM I 42,5 R Mokrá	330	350
Aggregate 0/4 mm	948	948
Aggregate 8/16 mm	669	669
Addition of fly ash	140	140
Addition ViscoCrete	3.96	3.96
Water	190	190
Steel fibre HE 75/50 Arcelor	70	-
Steel fibre Krampe Harex DE 60/0.8N	-	50
Polypropylene fibre	1.5	-

2. TESTS AMBIENT TEMPERATURE

The tests were performed in a laboratory of the Institute of Theoretical and Applied Mechanics (ITAM) in Prague. The slabs were loaded by four hydraulic jacks through triangular steel plate(see Fig.2). During the test, were possible development of clearly visible yield lines responded to the calculated pattern. The pattern of yield lines was calculated by the form (see [2]). The behaviour of the fibre-concrete slab was recorded by 11 deflectometers on each slab. Five deflectrometers recorded a vertical deflection and six recorded the horizontal deflections in lateral and longitudinal directions. A slab without internal beams in 2010 reached the deflection of 160.7 mm and the load-bearing capacity was 16.25 kN/m^2. Slabs in 2011 attained the deflection 179.5 mm and load-bearing capacity was 11.9 kN/m^2 (DE 60/0.8N. The second slab achieved deflection 157.5 mm with load-bearing capacity 13.5kN/m^2 (DE 60/0.8N) (see Figs. 3 and 4).

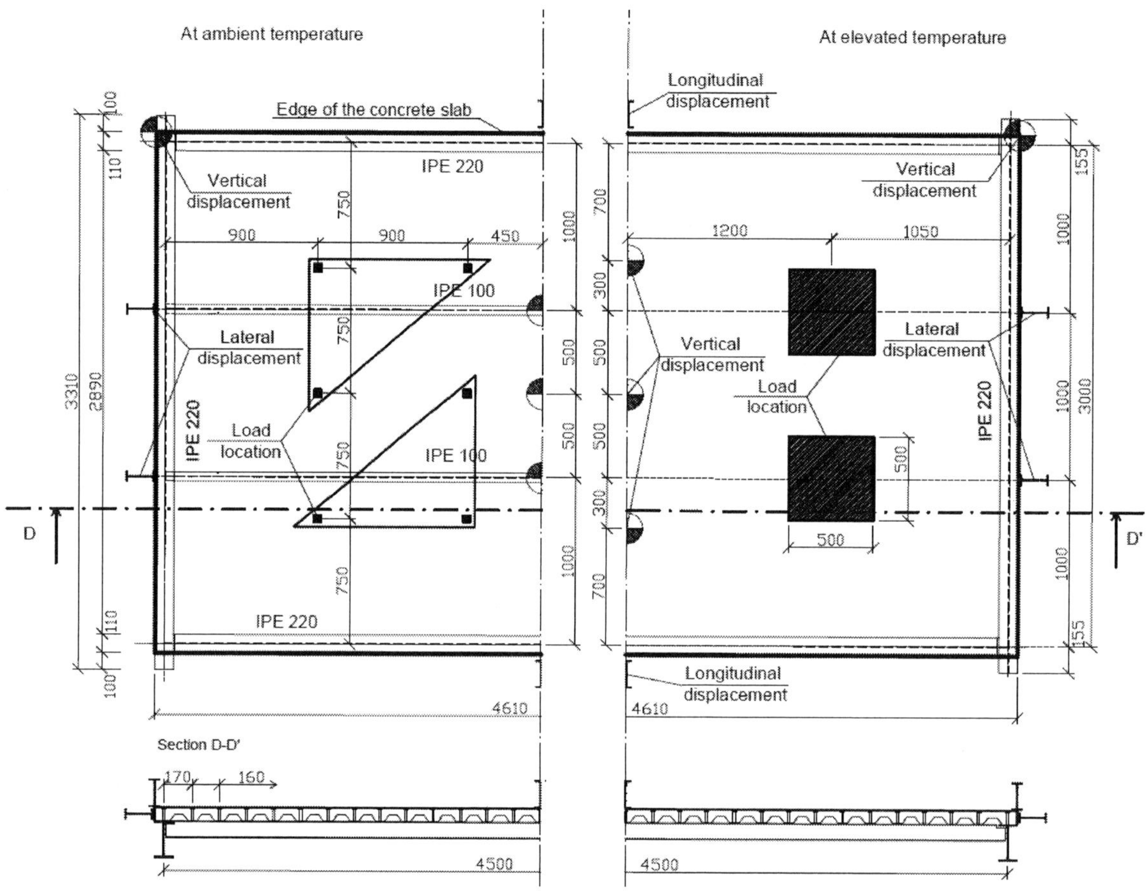

Figure 2. Sizes of the specimens, loading points, and location of the deflectometers on the floor slabs tests

Figure 3. View of slab AMB-PL-2 during a test at ambient temperature

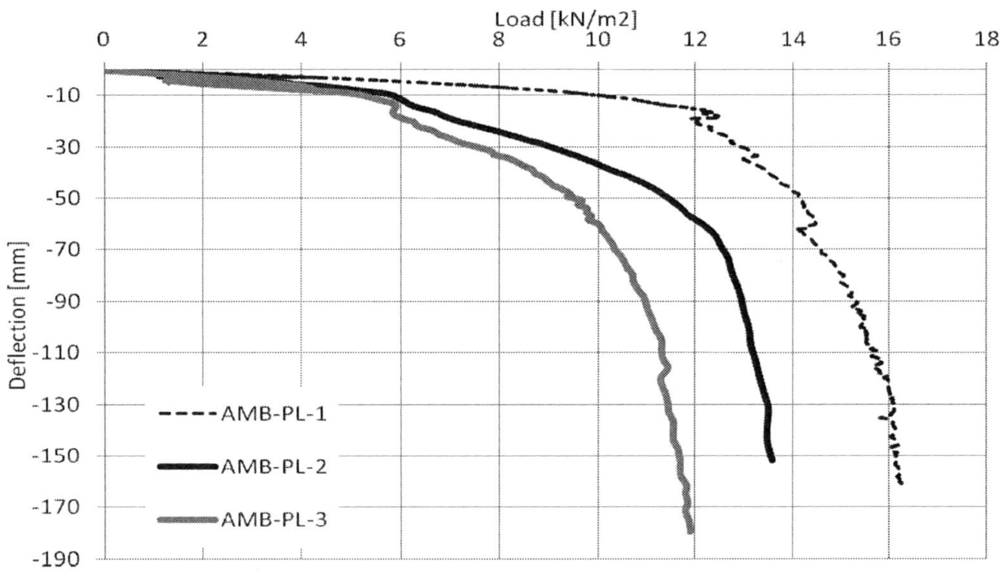

Figure 4. Comparison of deflections of the slabs during a tests at ambient temperature

3. TEST AT ELEVATED TEMPERATURE

The tests were performed at PAVUS, a testing laboratory in Prague. The behaviour of a composite slab on the furnace was recorded by 13 deflectometers and 40 thermocouples. Twelve thermocouples were cemented in the composite slab, four were located on steel beams and six recorded the gas temperature in the furnace. Seven deflectometers measured the vertical deflections and six measured the horizontal deflections. A simple design method for temperatures was calculated by a differential method.

At the end of the test at elevated temperature ELE-BE-1 created a final fracture of the fibre-concrete slab across a longer span in 198th min of the test. Displacement of the slab reached 300 mm, at span/10. Connections of the composite internal beams to edge beams remained intact. Shear connectors of the internal beams were suspended. The headed studs of the internal beams were fractured at the ends. The headed studs in the center of the beams were taken out of the slab with a cone of concrete. This damage corresponded to the development of compression and tension regions in the slab acting as a two-way spanning membrane.

Test ELE-BE-1 was stopped at 125th min without reaching collapse. The deflection was 196 mm, (see Fig.5 and 6).

Figure 5. Slab ELE-BE-1 during the fire test at 150th min

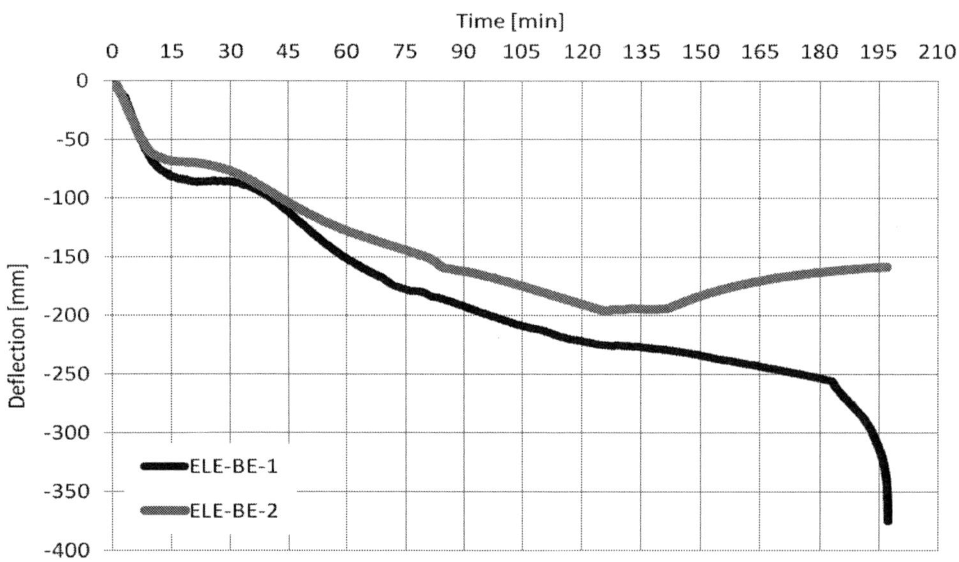

Figure 6. Comparison of deflections of the slabs during the tests at elevated temperature

4. MATERIAL TESTS

In concrete-producing plants, testing specimens were created for each mixture and cubes and scantlings were made. These tests produced a tensile strength at 20 °C, 500 °C and a temperature of 600 °C (see Fig. 7). Reduction coefficients were similar to coefficients for cold formed steel. The compressive strength of the fibre-concrete at ambient was obtained from tests. A reduction coefficient was used in compressive strength at elevated temperature in accordance EN 1992-1-2 (see [6]).

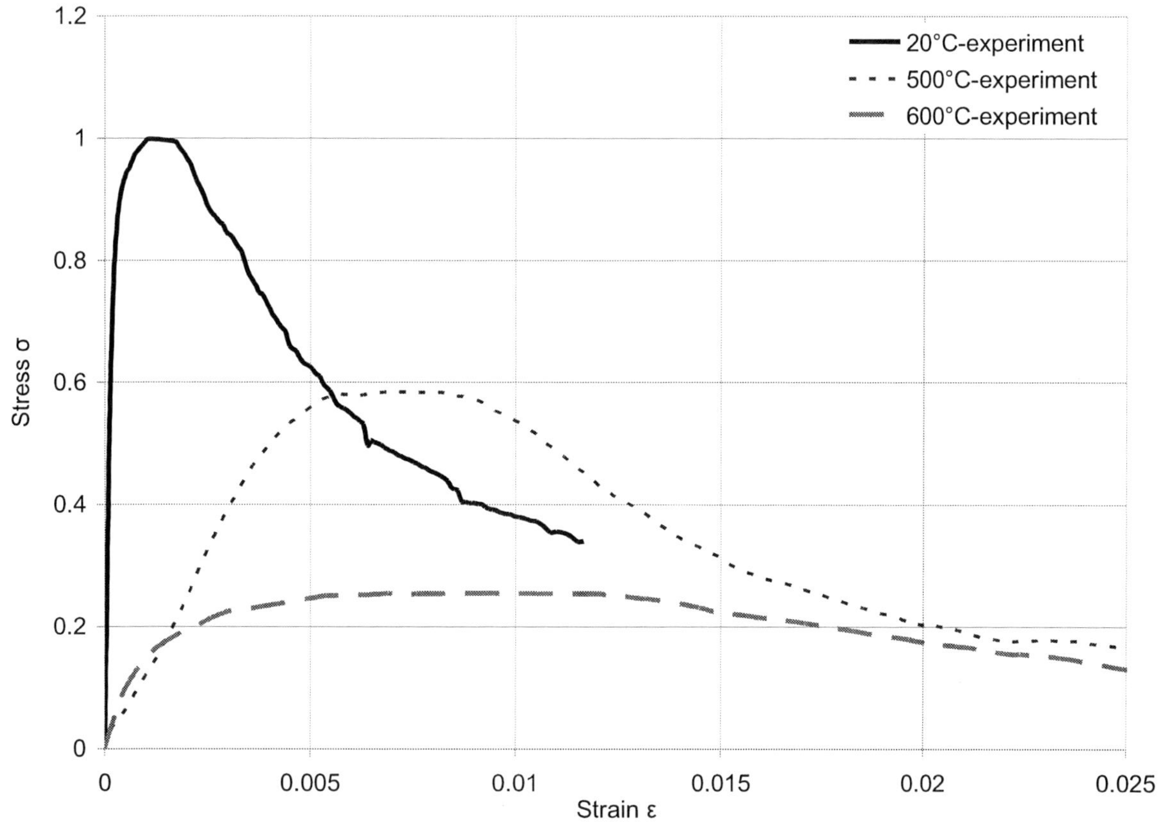

Figure 7. Stress-strain diagram of the fibre-concrete at 20 °C, 500 °C and 600 °C

5. FIRE RESISTANCE

Fire resistance of floor slabs is traditionally computed from the resistance of single elements. Properly designed structures make it possible to reach improved fire resistance. This occurs because the floor slab achieves large deflection and a new load-bearing system is started.

The capacity of this system was calculated by the simple prediction model. This model supposes two cases of the collapse. In the first case a crack developed through the full depth of the slab at its center. The second type of infraction crushed the concrete slab at the corners. The type of collapse depended on the strength of the concrete and amount and shape of the fibre. The first case of collapse in the fire state is more frequent. In the prediction model, the slab is divided by the plastic lines into four elastic parts. The pattern of the yield lines is given as:

$$n = \frac{\sqrt{3(L/B)^2 \mu + 1} - 1}{2\sqrt{\mu}(L/B)^2} \quad (1)$$

where L is the longer size of the slab; B is the shorter size of the slab; μ of the ration of the bending resistance in orthogonal directions

The ain effect on the load-bearing capacity was deflection. The deflection of the slab was due to the stretched fibre-concrete, which is given as:

$$w_c = \sqrt{0.5\varepsilon_c \frac{3L^2}{8}} \quad (2)$$

where L is the longer size of the slab; ε_c is the strain of the concrete and w_c is the deflection of the slab from a mechanical load.

Also temperature has a significant effect. It changes the material properties, but on the other hand, it enhances fire resistance due to the thermal gradient which enlarges the deflection of the slab. The deflection of the slab from the thermal gradient is given by:

$$w_t = \frac{\alpha_c(\theta_2 - \theta_1)B^2}{8h\beta} \quad (3)$$

where B is the shorter size of the slab; α_c is the coefficient of thermal elongation; θ_1 is the temperature of upper face; θ_2 is the temperature of the lower face; h is the depth of the slab; and β is the coefficient of a uniform temperature distribution.

6. CONCLUSION

From the tests, it is clear that the load-bearing capacity depended on the kind and strength of the used fibres. The low amount of the fibre (50kg/m^2) already assured the needed ductility and strength in tension.

At elevated temperatures, fire resistant R120 reached the both slabs. By the test ELE-BE-1, the integrity was maintained for 75 min. By test ELE-BE-2, the integrity of the slab was maintained during entire test.

Increasing the ductility of the fibre-concrete in high temperatures had a positive effect on the behaviour composite floor slab. The ductility of the fibre-concrete confirmed that the slabs create tension membrane action. This action transferred the effects of the load to the periphery of the fire resistant slab.

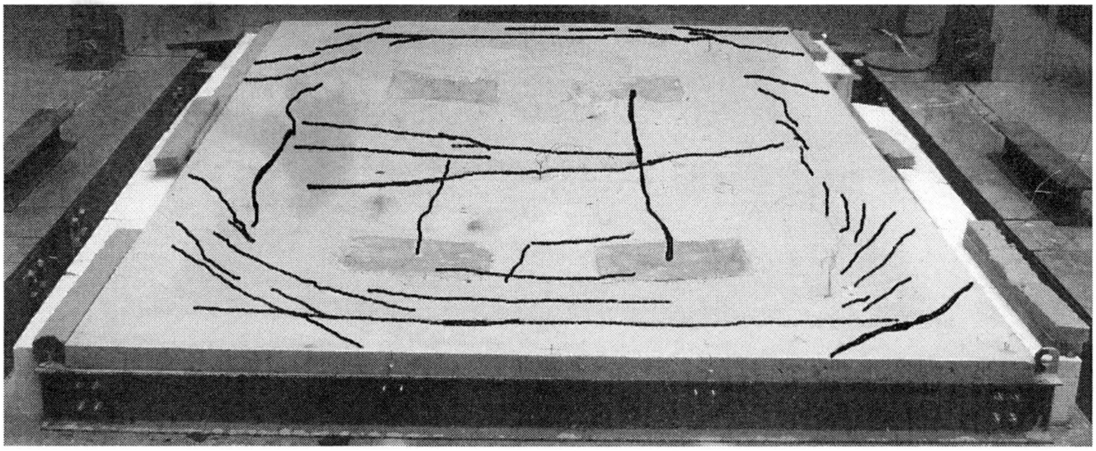

Figure 7. Slab ELE-BE-2 after the fire test with marked cracks

Acknowledgements

The authors express their gratitude to the Czech Science Foundation for the financial support of their research carried out within the project P105/10/2159, and the ITAM AS CR, v.v.i. for the support RVO: 68378297.

References

1. Bailey, C.G., Moore, D.B., "The structural behaviour of steel frames with composite floor slabs subjected to fire," Part1, Theory, *Struct. Eng.* 78, (11) (2000)19–27.
2. Bailey, C.G., Moore, D.B., "The structural behaviour of steel frames with composite floor slabs subjected to fire," Part2,: Design, *Struct. Eng.* 78, (11) (2000)28–35.
3. Bednář, J., Wald, F., Vodička, J., Kohoutková, A., "Experiments on membrane action of composite floors with steel fibre reinforced concrete slab exposed to fire," *Fire Safety Journal*, 59 (2013) 111–121.
4. Rustin, F., Kodur, V.R., "Enhancing the fire resistance of composite floor assemblies through the use of steel fibre reinforcedconcrete," *Struct. Eng.*, 33 (10)(2011)2870–2878.
5. Wasim, K., Khoury, V., "Thermal and mechanical properties of fibre reinforced high performance self-consolidating concrete at elevated temperatures," *Cem. Concr. Res.*, 41(11)(2011)1112–1122.
6. EN1992-1-2, "Eurocode 2: Design of concrete structures" – Part 1-2: *General rules – Structural fire design*, CEN, Brussels, 2006